Material-based Mechanobiology

Biomaterials Science Series

Titles in the Series:

1: Stimuli-responsive Drug Delivery Systems
2: Biodegradable Thermogels
3: Biofabrication and 3D Tissue Modeling
4: Biomaterial Control of Therapeutic Stem Cells
5: Antimicrobial Materials for Biomedical Applications
6: Decellularized Extracellular Matrix: Characterization, Fabrication and Applications
7: The Chemistry of Medical and Dental Materials, 2nd edition
8: Injectable Hydrogels for 3D Bioprinting
9: Extracellular Vesicles: Applications to Regenerative Medicine, Therapeutics and Diagnostics
10: Biomimetic Protein Based Elastomers: Emerging Materials for the Future
11: Phosphate and Borate Bioactive Glasses
12: Material-based Mechanobiology

Material-based Mechanobiology

Edited by

Jun Nakanishi
National Institute for Materials Science (NIMS), Japan
Email: nakanishi.jun@nims.go.jp

and

Koichiro Uto
National Institute for Materials Science (NIMS), Japan
Email: uto.koichiro@nims.go.jp

ROYAL SOCIETY
OF **CHEMISTRY**

Biomaterials Science Series No. 12

Print ISBN: 978-1-83916-185-8
PDF ISBN: 978-1-83916-537-5
EPUB ISBN: 978-1-83916-538-2
Print ISSN: 2397-1401
Electronic ISSN: 2397-141X

A catalogue record for this book is available from the British Library.

The Royal Society of Chemistry is a charity, registered in England and Wales, Number 207890, and a company incorporated in England by Royal Charter (Registered No. RC000524), registered office: Burlington House, Piccadilly, London W1J 0BA, UK, Telephone: +44 (0) 20 7437 8656.

For further information see our website at www.rsc.org

Printed in the United Kingdom by CPI Group (UK) Ltd, Croydon, CR0 4YY, UK

Preface

The advancement of molecular biology has introduced new concepts in medical treatment, such as gene therapy and molecular targeted drugs. Recently, interest has been growing in mechanobiology, which focuses on the role of physical force in biological processes. The sensory organs, including touch and hearing, provide familiar examples of this concept from our daily life. David Julius and Ardem Patapoutian were awarded a Nobel Prize in Physiology or Medicine in 2021 for their work on the identification of transducing molecules.[1] However, recent mechanobiological research has elucidated the involvement of physical forces more broadly in nonsensory cells and has identified various mechanosensory machineries in different scales. The understanding of mechanobiological mechanisms may provide another, new paradigm in medicine; therefore, it is crucial to develop new methodologies, distinct from those of conventional molecular biology, to manipulate and analyze this invisible entity.

Engineered materials are promising tools in mechanobiology research because of their interaction with forces and transmission of force within them. Considering the substantial advances in the field of material-based mechanobiology in the last two decades, we organized three international symposia on nanoarchitectonics for mechanobiology, in 2015, 2016, and 2019,[2–4] in order to further encourage collaboration between the fields of mechanobiology and materials research. Furthermore, we also received an invitation from the Royal Society of Chemistry to write a book on material-based mechanobiology, which provided the motivation for this edition.

This book is composed of 14 chapters written by promising researchers who are investigating various aspects of mechanobiology, such as the manipulation of mechanical properties or multiscaled structures of materials and the development of dynamic or smart materials. We express our deepest gratitude to all the authors and hope that this book will inspire

Biomaterials Science Series No. 12
Material-based Mechanobiology
Edited by Jun Nakanishi and Koichiro Uto

Published by the Royal Society of Chemistry, www.rsc.org

readers to enrich this new discipline through their imagination and creativity.

Last, but not the least, we would like to acknowledge Michelle, Connor, and Amina for their tremendous support during the production of this edition. We received the invitation to write the book in 2019, shortly before the COVID-19 outbreak that caused an unexpected delay in the publication of the book. We are grateful for their continuous support and patience.

Jun Nakanishi and Koichiro Uto

References

1. https://www.nobelprize.org/all-nobel-prizes-2021/.
2. https://www.nims.go.jp/mana/mechanobiology2015/index.html.
3. https://www.nims.go.jp/mana/mechanobiology2016/index.html.
4. https://www.nims.go.jp/mana/mechanobiology2019/index.html.

Contents

Chapter 1 An Introduction to Material-based Mechanobiology 1
Jun Nakanishi and Koichiro Uto

1.1 Overview of Material-based Mechanobiology 1
1.2 Historical Background of Mechanobiology 2
1.2.1 Before the Dawn of Molecular Mechanobiology 2
1.2.2 Dawn of Molecular Mechanobiology: Mechanosensory Molecules and Molecular Assemblies 3
1.3 Material-based Mechanobiology 6
1.3.1 Material-based Mechanobiology: Form 7
1.3.2 Material-based Mechanobiology: Matrix Mechanics 10
1.3.3 Material-based Mechanobiology: Force Detection 14
1.4 Future Directions of Material-based Mechanobiology 15
References 16

Chapter 2 On the Molecular Basis of Cellular Mechanobiology 21
Jorge Oliver-De La Cruz, Fabiana Martino, Stefania Pagliari and Giancarlo Forte

2.1 Mechanosensing at the Cell–Matrix Interface 21
2.1.1 Origin of the Mechanical Signals 21
2.1.2 Cellular Membrane as Prime Mechanosensing Structure 24

Biomaterials Science Series No. 12
Material-based Mechanobiology
Edited by Jun Nakanishi and Koichiro Uto

Published by the Royal Society of Chemistry, www.rsc.org

2.1.3 Mechanosensitive Ion Channels 24
2.1.4 Focal Adhesions (FAs) 26
2.2 Mechanotransduction at the Cytoskeleton 28
2.2.1 Cytoskeleton Network 28
2.2.2 Regulation of Cytoskeleton Intracellular Tension: Rho/ROCK Pathway 29
2.3 Transferring Mechanical Cues to the Nucleus 31
2.3.1 Nucleus as a Mechanosensor 31
2.3.2 Linker of Nucleoskeleton and Cytoskeleton (LINC) Complex 31
2.3.3 Nucleoskeleton and Transcriptional Activity Control 32
2.3.4 Nucleo-cytoskeleton Coupling in Cellular Adaptation to Mechanical Stress 33
2.3.5 Shuttling Mechanosensors and Mechanosensitive Transcription Factors 33
2.4 Conclusions and Perspectives 36
Acknowledgements 36
References 37

Chapter 3 Mechanotransduction at the Cell Surface and Methods to Study Receptor Forces 44
Rong Ma, Brendan R. Deal and Khalid Salaita

3.1 Introduction 44
3.2 Mechanotransduction at the Cell Surface 46
3.2.1 Mechanotransduction Through Integrins 46
3.2.2 Mechanotransduction Through T Cell Receptor 46
3.3 Methods to Study Receptor Forces 52
3.3.1 DNA Mechanics 52
3.3.2 DNA-based Molecular Force Sensing 55
3.4 Conclusion and Future Perspectives 60
Acknowledgements 61
References 61

Chapter 4 Measurement and Manipulation of Cellular Forces Using Silicone Elastomers 64
Hiroaki Hirata and Masahiro Sokabe

4.1 Introduction 64
4.2 Mechanical Measurement Using Silicone Elastomers 65
4.2.1 Measurement of Force at Cell–ECM Adhesion 65

4.2.2 Measurement of Force at Cell–Cell Adhesion 69
4.2.3 Cell Stiffness Measurement Using Microfluidics 70
4.3 Mechanical Manipulation by Stretching Silicone Elastomer Membranes 72
4.3.1 Distinct Strain Types 73
4.3.2 Sustained or Cyclic 75
4.4 Combination of Mechanical Manipulation and Measurement 77
4.5 Concluding Remarks 79
Acknowledgements 80
References 80

Chapter 5 Geometric Cues for Directing Cell Fate 85
Sara Romanazzo and Kristopher A. Kilian

5.1 Introduction 85
5.2 Microengineering Techniques to Control the Shape of Cells and Tissue 88
5.2.1 Photolithography 88
5.2.2 Soft Lithography 88
5.2.3 Inkjet Cell Printing 92
5.2.4 Optical Tweezers and Optoelectronic Tweezers 92
5.2.5 Laser-based Cell Patterning 92
5.2.6 Dielectrophoresis 93
5.2.7 Other Techniques 93
5.3 Controlling Tissue Geometry in 3D 94
5.4 Deconstructing Biological Behavior Through Micropatterning 96
5.4.1 Teasing out Mechanotransduction During Stem Cell Differentiation 96
5.4.2 Controlling Shape and Interfacial Topography in Cancer Cultures 100
5.5 Conclusion and Future Perspectives 103
Abbreviations 103
References 103

Chapter 6 Dynamic Interfaces for Mechanobiological Studies 110
Jun Nakanishi and Kazuhiro Tatematsu

6.1 Introduction 110

6.2 Switchable Interfaces 113
6.2.1 Switchable Interfaces for the Spatiotemporal Control of Cell Adhesion 113
6.2.2 Switchable Interfaces at Subcellular and Molecular Resolutions 117
6.2.3 Switchable Interfaces for Exploring Dynamic Mechanobiology 119
6.3 Compliant Interfaces 122
6.3.1 Dissipative Interfaces 122
6.3.2 Liquid–Liquid Interfaces 125
6.3.3 Force-actuating Materials 128
6.4 Conclusions and Future Perspectives 129
Acknowledgements 129
References 130

Chapter 7 Micro- and Nanopatterned Substrates for Studies on the Mechanobiology of Cell–Matrix Adhesions 135
Joel Christian and Elisabetta Ada Cavalcanti-Adam

7.1 Chemomechanical Coupling at the Cell–Extracellular Matrix Interface 135
7.2 Spatial Control of Cell–Extracellular Matrix Adhesion by Surface Patterning 137
7.2.1 Micropatterning of Matrix Proteins to Direct Cell Shape and Architecture 137
7.2.2 Nanopatterning of Matrix Ligands for the Control of Receptor Clustering 138
7.3 Surface Patterning for Cell-generated Force Measurements 141
7.3.1 Micro- and Nanopatterning of Soft Materials to Probe Cell Rigidity Sensing and Measure Cellular Traction Forces 142
7.3.2 Nanopatterning and Measurements of Molecular Forces at Cell–Matrix Adhesion Sites 145
7.4 Outlook 147
Abbreviations 147
Acknowledgements 148
References 148

Chapter 8 Role of Topographic Cues in Engineering the Muscle Niche 152
Jesse Macadangdang, Christian Mandrycky, Changho Chun, Nicholas A. Geisse, David L. Mack and Alec S. T. Smith

8.1 Introduction 152
8.1.1 Historical Background 153
8.1.2 Structure–Function Relationship in Muscle Biology 154
8.1.3 Importance of Substrate Patterning for Muscle Tissue Engineering and Its Applications 155
8.2 Fabricating Substrate Topographies for Use in Cell Culture 157
8.2.1 Template-assisted Fabrication 157
8.2.2 Template-free Fabrication 159
8.3 Use of Patterned Surfaces to Enhance Muscle Cells in Culture 163
8.3.1 Patterned Cardiomyocyte Case Studies 163
8.3.2 Replicating Skeletal Muscle Structure *In Vitro* Using Topographic Patterns for Disease Modelling and Drug Screening Applications 166
8.3.3 Other Applications: Patterned Smooth Muscle Cells for Vascular Modelling and Tissue Engineering 170
8.4 Future Directions 170
8.4.1 Topography Is Only a Piece of the Maturation Puzzle 170
8.4.2 High-throughput Assays and Assay Agnostic Applications 172
8.4.3 Fit for Purpose? 173
8.4.4 Three-dimensional Applications 173
8.5 Conclusions 175
Disclosures 176
References 176

Chapter 9 Engineered Substrates with Dynamically Tunable Topography 184
Koichiro Uto

9.1 Topography 184
9.2 Silicone Elastomer System for Micro- and Nanofabrication 188
9.2.1 Design of Lithography-free Reconfigurable Wrinkles Using PDMS 188

9.2.2 PDMS with Reconfigurable Topography to Guide Cell Functions 191
9.3 Surface Relief Grating System 194
9.3.1 Photoinduced Surface Relief as Reconfigurable Topography 195
9.3.2 Reconfigurable Topography Based on PSR for Guiding Cell Functions 196
9.4 Shape-memory Polymer System 199
9.4.1 Design of Programmed Topography Change Based on SMPs 200
9.4.2 SMP Platforms with Programmable Topography for Guiding Cell Functions 201
9.5 Conclusions and Future Perspective 207
Acknowledgements 208
References 208

Chapter 10 Curvature Mechanobiology **213**
Ryosuke Matsuzawa, Midori Takeuchi, Takuya Nishina and Tadahiro Yamashita

10.1 Introduction 213
10.1.1 Motivation to Investigate Cellular Behaviours on Curved Surfaces 213
10.1.2 Curvature as a Fundamental Geometrical Parameter 214
10.2 Engineering Curved Surface 216
10.2.1 Basic Moulding Techniques 216
10.2.2 Controlling Interfacial Geometries 219
10.2.3 Machining and Etching 220
10.2.4 Lithography 220
10.2.5 Technical Perspectives 221
10.3 Biomolecules Related to Cellular Curvature Sensing 221
10.3.1 What Senses Cellular-scale Curvature? 221
10.3.2 Curvature Recognition *via* Molecular Affinity 223
10.3.3 Role of Intracellular Components 225
10.3.4 Role of Intercellular Connections 227
10.3.5 Current Perspectives 227
10.4 Modelling Cellular Behaviours on Curved Surfaces 227
10.4.1 Continuum Modelling Approaches 228
10.4.2 Discrete Modelling Approaches 232
10.4.3 Future Subjects 232
10.5 Conclusions 233
References 233

Chapter 11 Dynamic Hydrogel **239**
Takeshi Ueki and Ryota Tamate

11.1 Hydrogel Classification and Switchable Hydrogel Design Principles 239
11.2 Hydrogels Exhibiting One-way Stiffness Changes 240
11.2.1 Design of Switchable Hydrogels with Decreasable Elastic Modulus 240
11.2.2 Design of Switchable Hydrogels with Increasable Elastic Modulus 245
11.3 Hydrogels Capable of Reversible Stiffness Changes 248
11.3.1 Reversible Stiffness Changes in Response to Chemical Stimuli 248
11.3.2 Reversible Stiffness Changes in Response to Physical Stimuli 253
11.4 Cell Scaffolds Capable of Stress Relaxation 257
11.5 Conclusions and Future Perspectives 262
References 263

Chapter 12 Manipulation of Durotaxis on a Matrix with Cell-scale Stiffness Heterogeneity **265**
Satoru Kidoaki

12.1 Introduction 265
12.2 Threshold Stiffness Gradient Required to Induce Durotaxis 266
12.2.1 Fabrication of a Stiffness Gradient Substrate 266
12.2.2 Determination of the Threshold Stiffness Gradient Required to Induce Durotaxis 268
12.3 Manipulation of Durotaxis 269
12.3.1 Rectified Durotaxis 269
12.3.2 Reverse Durotaxis 272
12.4 Combined Durotaxis-induced Nomadic Cell Migration 276
12.5 Conclusions and Perspectives 279
Abbreviations 279
References 280

Chapter 13 Engineered 3D Matrices with Spatiotemporally Tunable Properties **282**
Koichiro Uto, Riho Tanimoto and Cole A. DeForest

13.1 Introduction 282

13.2 The Importance of 3D Cell Culture 283
13.3 Design Strategy of Engineered Matrices for 3D Cell Culture 285
13.3.1 Azide–Alkyne Cycloaddition for Engineered Hydrogel Crosslinking 286
13.3.2 Diels–Alder Reaction for Engineered Hydrogel Crosslinking 288
13.3.3 Thiol–Ene Reaction for Engineered Hydrogel Crosslinking 289
13.3.4 Other Reactions for Engineered Hydrogel Crosslinking 290
13.4 Engineered Matrices for 4D Cell Culture 291
13.4.1 Programmable Hydrogels Based on Degradability 291
13.4.2 Engineered 3D Matrices with Spatiotemporally Programmable Biochemical Cues 298
13.4.3 Adaptable Engineered 3D Matrices for Mechanobiology 302
13.4.4 Conclusions and Future Perspective 304
Acknowledgements 305
References 305

Chapter 14 Combining Genetic and Mechanical Factors to Model Disease **309**
Jaimie Mayner, Elena Demeester and Adam J. Engler

14.1 Introduction 309
14.2 Modeling Disease with a Single Variable 310
14.2.1 Genetic Models 310
14.2.2 Environmental Effects on Cells 315
14.3 Combining Genetic and Environmental Factors to Model Disease 325
14.3.1 Heart Disease 325
14.3.2 Cancer 329
14.4 Conclusions and Perspectives 330
Acknowledgements 332
References 332

Subject Index **338**

CHAPTER 1

An Introduction to Material-based Mechanobiology

JUN NAKANISHI*[a,b,c] AND KOICHIRO UTO[a]

[a] Research Centre for Functional Materials, National Institute for Materials Science (NIMS), 1-1 Namiki, Tsukuba 305-0044, Japan; [b] Graduate School of Advanced Science and Engineering, Waseda University, 3-4-1 Okubo, Shinjuku-ku, Tokyo 169-8555, Japan; [c] Graduate School of Advanced Engineering, Tokyo University of Science, 6-3-1 Niijuku, Katsushika-ku, Tokyo 125-8585, Japan
*Email: NAKANISHI.Jun@nims.go.jp

1.1 Overview of Material-based Mechanobiology

"The force will be with you, always," a famous quote by master Yoda in the popular fiction film *Star Wars*, applies to our real lives. Forces are involved not only in responses within sensory tissues but also in any part of our body. The movement of skeletal muscles and organs, blood flow, and gravity generate mechanical stimuli, and cells use these mechanical cues to regulate their morphology, proliferation, migration, and functions in order to eventually control various physiological and pathological processes, such as development and cancer progression.[1–3] Conventional molecular biological studies have not paid attention to, or rather had forgotten (see Section 1.2), the role of these mechanical forces. However, the discovery of various mechanosensory molecules responsible for cellular force perception has led to increased research on the roles of biological forces over time. After the initial identification of mechanosensitive ion channels existing at the interface between the cell and the outside world, many new mechanosensing

Biomaterials Science Series No. 12
Material-based Mechanobiology
Edited by Jun Nakanishi and Koichiro Uto

Published by the Royal Society of Chemistry, www.rsc.org

machineries have been found, such as molecules involved in focal adhesion, cytoskeleton structures, and even nuclei as intracellular organelles. All such mechanosensing machineries perceive external forces and translate them into "molecular words" to regulate signal transduction and gene expression. At the same time, cells exert forces on substrates and adjacent cells *via* the functions of actin cytoskeleton and adhesion machineries and utilise them to sense the mechanical properties of their surroundings and dictate their activities and fates. Moreover, the forces exerted among neighbouring cells or those between cells and connective tissues or body fluids do not stay constant; rather, they are constantly fluctuating and play essential roles in time-evolving biological processes, including tissue formation, wound healing, and the development and metastasis of cancer. Another important feature is that forces can directly react with the multilayered architectures of living systems, ranging from nanoscale proteins, subcellular organelles, cells, tissues, and organs to whole bodies; we could coin the hierarchical nature as "mechanoarchitectonics". Therefore, it is necessary to develop new methodologies for quantitatively sensing the forces in real time or, the other way around, for loading time-evolving "dynamic" forces in various spatiotemporal scales. Such technologies will elucidate a complete picture of the relationships of forces in our body and eventually create a new paradigm in biology and medicine. However, the measurement and manipulation of these invisible forces are challenging for most molecular biologists, whose main targets have been the "molecular words". On the other hand, material scientists are more familiar with the interactions and transmission of forces with and through materials, which can be used as translators of forces in multiple spatiotemporal scales. This compatibility has led to a new discipline in mechanobiology studies, based on engineered materials, whose aim is to measure and manipulate molecular and cellular forces, wherein materials play an essential role. This direction of research can be called "material-based mechanobiology", which this book addresses. Mechanobiology research is heading toward the peak of its success due to the availability of new biological tools. The last two decades have envisioned the exponential growth of material-based methodologies, which has motivated us to collect a wide variety of work from the front-runners of this field. In this chapter, we will introduce the origin and a brief history of mechanobiology research in general, followed by depicting the developmental processes of material-based mechanobiology, whose details will be introduced in the following chapters.

1.2 Historical Background of Mechanobiology

1.2.1 Before the Dawn of Molecular Mechanobiology

The origin of mechanobiology goes back to the late 19th century when the German surgeon Julius Wolff hypothesised that bones had evolved to maintain mechanically optimal geometries with maximal strength and minimal weight. This concept later came to be known as Wolff's law of bone remodelling.[4] In 1917, D'Arcy Thompson, a biologist, published a book titled *On Growth and Form*, which describes the logic of physics and mathematics that can be applied to explain structure formation and the morphogenesis of plants and animals. Furthermore, there is a clear link between the morphology of living organisms and the mechanical forces they encounter. The age before DNA discovery and the emergence of molecular biology had no molecular insights and hence was discussed only from the mathematical and physical viewpoints.

In the 1950s, researchers started investigating the domain of cellular scale due to the establishment of cell culture technologies around that time. It was reported that cancer cells, unlike normal cells, maintain proliferative potential under nonadherent conditions such as on soft agar.[5,6] A separate study also showed that adding ionic drugs to erythrocytes caused morphological changes through changes in membrane tension.[7] These facts suggest that the cells can mechanically sense extracellular matrix (ECM) cues and the membrane tension. Further research to uncover these mechanisms led to the discovery of focal adhesion, where the protein complexes at the cell–ECM interface act as bridges between the ECM and the actomyosin machinery to regulate cell adhesion and migration.

Meanwhile, muscle biophysicists have elucidated the essential role of actomyosin in muscle contraction.[8,9] The findings from these studies indicate that the actomyosin machinery-driven mechanical force strongly contributes not only to muscle contraction but also to cell division and migration. In particular, the actomyosin machinery in nonmuscle cells is coupled to cell–matrix adhesion sites. Therefore, it is reasonable to infer that the cells exert forces on the substrate using intrinsic contractility. However, this cellular force is too weak and could not be detected by the conventional tools available at that time. To tackle these technical challenges, Harris *et al.* fabricated a thin elastic silicone fluid substrate made of polydimethylsiloxane (PDMS), cross-linked only at the top (~1 μm) by brief flaming, and they succeeded in visualising the traction forces generated by cells as a function of surface wrinkles on the cross-linked elastic substrate.[10] This study showed for the first time that even nonmuscle fibroblasts generate forces on substrates, which led to the development of the next generation of tools for measuring cell traction (discussed in Chapters 3 and 4). Moreover, this study probably marked the first report of material science encounters in the field of mechanobiology.

1.2.2 Dawn of Molecular Mechanobiology: Mechanosensory Molecules and Molecular Assemblies

Harris *et al.*'s report showed clear evidence that the cells are mechanically coupled to the ECM. However, the mechanisms with which these cells sense the mechanical properties of their surroundings still needed to be investigated. The first molecular evidence was obtained from analysing ion channels by patch-clamp recording (Figure 1.1A).[11] In this technique, the formation of tight sealing, called gigaseal, between the pipette and plasma membranes was

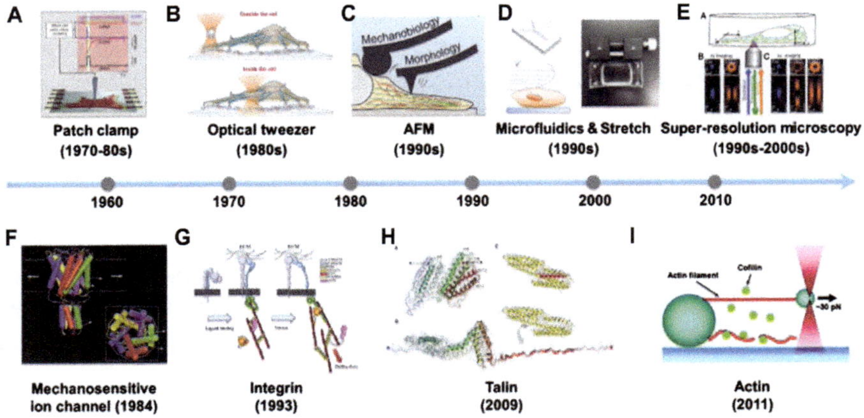

Figure 1.1 Timeline of (A–E) technologies and (F–I) discovered mechanosensory molecules in mechanobiology research: (A) Patch clamp method, developed in the late 1970s and early 1980s, allowed for the first time to record currents of single ion channel. Reproduced from ref. 11 with permission from Elsevier, Copyright 2021. (B) Optical tweezers, developed in the late 1980s, allowed for the first time to manipulate a single cell. Adapted from ref. 17 with permission from Springer Nature, Copyright 2019. (C) Atomic force microscopy (AFM), developed in 1990s, allowed for the first time the probe of living cells. Adapted from ref. 18 with permission from Elsevier, Copyright 2021. (D) Microfluidic and stretching devices, developed in the middle 1990s, applied for the first time in cell biology and biochemistry. Adapted from ref. 40, https://doi.org/10.3389/fphys.2018.01417, under the terms of the CC BY 4.0 license https://creativecommons.org/licenses/by/4.0/ and ref. 41 with permission from Bio-Protocol LLC, Copyright 2020. (E) Super-resolution microscopy, developed in the middle 1990s and early 2000s, and a variety of techniques, such as STED, STORM, and PALM, have been proposed during that period. Adapted from ref. 88 with permission from National Academy of Sciences, Copyright 2008. (F) Mechanosensitive ion channel: An ion channel that responds to membrane stretch in skeletal muscle using a patch-clamp technique was discovered in 1984. Adapted from ref. 14 with permission from AAAS, Copyright 1998. (G) Integrin: It also functions as a mechanosensory that mediates force transmission to the actin cytoskeleton, as reported in 1993. Reproduced from ref. 23 with permission from Elsevier, Copyright 2014. (H) Talin: Actomyosin contraction causes force-dependent unfolding of talin, leading to exposure of vinculin binding sites, which were found to function as mechanosensors in 2009. Adapted from ref. 27 with permission from AAAS, Copyright 2009. (I) Actin filaments were identified, in 2011, as negative tension sensors that increase their affinity for cofilin when relaxed, leading to their own disassembly. Reproduced from ref. 33 with permission from Taylor & Francis Group, Copyright 2009.

critical to reduce the background noise.[12] To achieve gigasealing, gentle suction was applied to the membrane before channel recording, which inevitably applied membrane tension to the patched membrane region. In 1984, Guharay *et al.* noticed single-channel conductance generation during this membrane sucking process and reported the presence of a mechanosensitive channel (MS channel), called a stretch-activated channel (Figure 1.1F);[13,14] ten years later, Sukharev *et al.* isolated the gene of an MS channel (MscL, MS channel of large conductance) from *E. coli.*[15] This was the first study to identify molecular mechanosensors from nonsensory cells. At the same time, the discovery of the molecular identity of MS channels made molecular biological studies possible. A later study identified other MS channels, including TREK and Piezo channels, from eukaryotic cells. In particular, Piezo channels have been associated with human diseases related to mechanopathologies, hereditary xerocytosis, and generalised lymphatic dysplasia, thereby becoming important targets in mechanobiological studies.[16]

Meanwhile, the force transmission mechanisms at the cell–ECM interface have also been extensively investigated. In contrast to the research on membrane tension mechanosensing, where the patch-clamp technique provides a relatively simple way to investigate the forces involved, various experimental tools such as atomic force microscopy and optical/magnetic tweezers are needed for mechanobiology studies (Figure 1.1B,C).[17,18] In the 1970s, the existence of transmembrane proteins connecting the extracellular matrix (ECM) to the actin-based cytoskeleton began to be debated. Hynes and other groups had reported that extracellular fibronectin fibres were physically attached to the intracellular actin filaments, but their identity was not confirmed until the 1980s.[19–21] In 1986, this membrane protein complex was identified to be integrins.[22] Since then, at least 24 human integrin subtypes have been discovered. As most of the forces, including contractional forces, shear stresses, and hydrostatic pressures, result in stress concentration at the anchorage points at the cell–ECM interface, it is reasonable to hypothesise that integrins are involved in the mechanosensing process (Figure 1.1G).[23] Integrins exhibit several unique mechanoresponsive characteristics at the molecular level, including catch-bond and conformational changes or clustering.[24]

In addition to integrins, another important intracellular mechanosensing molecule has been discovered in the focal adhesion regions. Talin is a protein associated with cytoskeleton (Figure 1.1H).[25–27] The N- and C-terminals of talin bind to integrins and actin cytoskeleton, respectively; thus the force from the actin cytoskeleton is also transmitted to talins. The talin rod domain has many binding sites for vinculin, most of which are hidden in the folded structure. Both *in vitro* and *in vivo* experiments have shown that the tensile force derived from the activity of myosin II in the cytoskeleton can cause them to unfold and elongate and expose the rod domain of the vinculin binding.[27,28] Therefore, such observations collectively indicate a mechanosensory role for talins.

Interestingly, actin cytoskeleton is not only the source of force generation through the activity of myosin, but it can also respond to forces by itself. More specifically, the degradation of actin filament can be observed upon relaxation of its tension,[29] which indicates its role as a mechanosensor against cellular tension (negative tension sensor).[30] The question then arises as to how does actin respond to such mechanical tension changes? In various actin-based intracellular architectures, such as bundles, lamellipodia, and cortex, actin filaments withstand mechanical loading against their global structures. This has been revealed by using diverse microscopy techniques, including electron microscopy, fluorescence resonance energy transfer (FRET), atomic force microscopy, and optical trapping (Figure 1.1B, C). The force application to the actin filament changes its three-dimensional structure[31] thereby changing the affinity with various actin-binding proteins (ABPs).[30] Particularly, the tensile force increases the length of the actin filaments parallel to the direction of stretching, reducing the binding affinity to cofilin, an actin cleavage factor, increasing that to myosin II (Figure 1.1I),[29,32,33] and eventually stabilising the tensile fibres as the basis of various mechanical signalling processes.[34] On the contrary, the reduction of tension increases cofilin binding and decreases myosin binding; thereby the decrease in cellular tension promotes the degradation of actin filaments. Tension-induced changes in actin structural dynamics also affect the binding of actin nucleation proteins, such as Arp2/3.[35] Mechanically induced changes in actin-based structures also affect gene expression in at least some cell types, such as mesenchymal stem cell and breast cancer cell. As more tension fibres are formed during mechanical stimulation, the transcriptional conjugator YAP is translocated to the nucleus and activated. YAP is essential for Hippo signalling and mediates the increased expression of genes involved in cell proliferation and differentiation. Thus it is thought to result in physiological and pathophysiological processes due to the response of the actin cytoskeleton to extracellular mechanical forces.[36,37] In addition, the existence of the transcription conjugators in the downstream of actin-based signalling, including the Rho/Rho-kinase pathway, suggests triggering of gene expression regulation by mechanical forces.

In addition to cellular mechanosensing at the focal adhesion sites and cytoskeletal structures as previously mentioned, it has recently been shown that forces and deformations applied to the cellular surface can alter the shape of the nucleus *via* the force transmission through the cellular cytoskeleton. This nuclear deformation has a significant impact on the accessibility of the nuclear factors with the genetic codes in the nucleus. Therefore, the possibility of using cell nuclei as mechanosensors has been considered. Similar to the plasma membrane and cortex at the cell–ECM boundary, the nuclear envelope is a dynamic and force-sensitive interface between the cytoplasm and chromatin. The main structural cortex of the nuclear envelope is the lamina, composed of helix-rich fibrillar lamin proteins that assemble just below the inner nuclear membrane.[38] Two types of lamins, A-type (lamin A and lamin C) and B-type (lamin B1 and lamin B2), are tethered to the nucleus *via* the LINC complex, a linker of the nucleoskeleton and cytoskeleton.

Similarly, many mechanosensors have been identified and are still being sought, in different sizes and molecular assemblies, *e.g.* the cytoskeleton and nucleus, and their details are discussed in Chapter 2.

1.3 Material-based Mechanobiology

Two studies laid a strong foundation for material-based mechanobiology from an engineering perspective. One proposed the control of cell shape and the other modulation of the elastic modulus of the substrate. In parallel, materials for measuring forces applied at various scales and locations have been developed, and these have become indispensable technologies for mechanobiological research. In this section, we outline the development of material-based mechanobiology research in three primary directions: shape control, stiffness control, and force measurement, with a focus on the material aspect. In particular, cellular mechanoarchitectonics and the dynamic nature of mechanobiological responses will be discussed in wide variety of scale and time. Here, we first introduce these technologies for the control of cell form (geometry) and substrate mechanics, from which a wide variety of innovative materials and sciences have emerged.

1.3.1 Material-based Mechanobiology: Form

As discussed in the previous section, the relationship between morphology and growth has been observed since the early stages of the history of mechanobiology. However, at that time, the central focus was on larger biological units, such as tissue and bones. In 1997, Chen *et al.* proved that a similar rule could be applied to single-cell growth.[39] They took advantage of microcontact printing based on PDMS elastomers to fabricate cell-adhesive islands in a nonadhesive background and demonstrated the relationship between the geometry of single cells and their growth. The cells switched from proliferative to apoptotic upon decreasing the spreading area, indicating that the shape of cells determines whether they live or die due to their local mechanical deformation. This success exemplified the usefulness of this fabrication technology for controlling cellular morphologies for mechanobiological studies. In microcontact printing technology, clean room environments are not needed after preparing the replica moulds, from which PDMS-based stamps can be produced repeatedly. Such user-friendly features are one of the biggest reasons for the widespread use of microcontact printing in mechanobiology studies.

Particularly, the development of soft lithography, including microcontact printing, after this period is attributed mainly to the characteristics of PDMS. PDMS is the most widely used silicone polymer due to the relatively lower-cost raw material. Furthermore, cross-linking is easy as it is a simple addition reaction in the presence of a catalyst. In addition, the glass transition temperature of PDMS is remarkably low (*ca.* -125 °C) among conventional polymers, and the cross-linked material is in a rubbery state at room temperature. The elastomeric feature of this material makes it easy to peel off from the master, and its high adhesiveness facilitates the fabrication of microfluidic devices that adhere to glass substrates. This easy processing technique enables its use in mechanobiology research, such as the development of micropillar devices for measuring cell traction force, cell response to shear stress from fluid flow and substrate stretching, cell manipulation using flow channels and stretch chamber, among others (Figure 1.1D).[40,41]

Moreover, it is possible to apply an elastic force to cells using its elastic properties directly. In addition, PDMS has high oxygen permeability owing to its large free volume, which enables oxygen supply to cells through the substrate when used as a scaffold material, thereby facilitating the culturing of iPS and ES cells.[42] This wide variety of mechanobiological research utilising the material characteristics of PDMS and the simplicity of soft lithography is described in more detail in Chapter 4.

On the other hand, a study focusing on cell shape showed that the control of geometrical cellular shape generates stress at the edge, where the active expression of lamellipodia is observed, which triggers cell migration and is associated with the regulation of biological activities by force.[43] A tensegrity model was proposed in which the actin skeleton was regarded as an elastic material and microtubules as rigid materials, and an explanation from the material mechanics point of view was attempted.[44,45] In addition to the effects of cell shape on proliferation, cell polarity, and migration, there have been gradual developments in the effects of cell differentiation,[46] cell collectives in 2D,[47] and 3D cell culture.[48] At the same time, new patterning techniques such as dip-pen nanolithography[49] and laser-based patterning[50] have been developed to improve accuracy and enable 3D modelling. Especially in 3D bioprinting of cells, inks are often required to have characteristics such as strain softening, so there was also a contribution to research on the development of such materials.[51,52] In addition, changes in intracellular stress distribution due to its shape, which are observed in single cells, can also be seen in cell clusters when cells are considered as a continuum model and in force-driven regioselective cell proliferation and phenotypic changes in tissues (such as epithelial–mesenchymal transition[53] and epigenetic factor reprogramming[54]). These studies played a significant role in promoting the development of stem cell and cancer mechanobiology by recapitulating physiological 3D environments. More details on functional regulation by cell geometry are provided in Chapter 5.

On the other hand, attempts to utilise dynamic patterning are major progress in the field. As detailed in Chapter 6, materials that respond to heat, light, electric potential, *etc.*, have been developed to control cell adhesion at the interface, the activity of adhesion ligands, and the highly specific click reaction. Such materials can be used to expand the cell adhesion area on demand, freely arrange multiple cells, and induce cell migration.[55] In particular, cell migration is a major issue in mechanobiology because the forces between cells and substrates are closely coupled to this phenomenon.[56] For example, after defining the stress distribution of cells and cell collectives, this method enabled mechanobiological investigation of the impact of initial geometry on the appearance of leader cells in subsequent migration behaviour, as well as the effect of the chemical and mechanical cues of the ECM on this behaviour, by defining the interfacial properties of functional materials.[57] Meanwhile, a new development of such dynamic materials is to utilise these materials to dynamically control the activity of ligands, rather than simply changing the geometry.[58] The reason behind such a development can be attributed to the dynamic change of force seen in the expansion and contraction of blood vessels *in vivo.* Moreover, ultrasound-based therapy has attracted much attention,[59] and it has recently been revealed that cells are more sensitive to the loading rate than to the equilibrium force.[60] The switchable interface is beneficial for dynamic mechanobiology research and leads to the development of materials with faster response times.

In addition to this approach of enlarging the size and dimension (*e.g.*, from 2D to 3D) to resemble biological tissues and focus on temporal dynamics, research has also been conducted to downsize the controlling form to subcellular and molecular levels. This strategy leads to constraints on the arrangement and dynamics of molecules and molecular assemblies, and it allows us to explore mechanonanoarchitectonics, in which interactions at the molecular level are translated into cell size and cell populations. In particular, the block copolymer micellar nanolithography described in Chapter 7 is a technique for arranging gold nanoparticles in a hexagonal lattice based on the principle of colloidal lithography.[61] The clustering behaviour of integrins can be controlled because these nanoparticles possess the size of a pair of integrin heterodimers, and the particle spacing can be precisely controlled according to the fabrication conditions. Their study demonstrated that integrins serve as a molecular ruler that can recognise the nanoscopic ligand spacings. Moreover, technological innovations such as applying substrates of various elastic moduli,[62] immobilisation of multiple ligands,[63] and the measurement of forces applied to molecules during endocytosis in combination with molecular force probes (Chapter 3) are also being investigated.

In the study of mechanobiology through geometric control using materials, it is clearly shown that controlling the cell shape and intracellular mechanical field is essential for the manifestation of cell functions. On the other hand, some evidence suggesting that the topography of the cell scaffold may influence cell polarisation and movement was reported in the early 20th century.[64] These results are not surprising considering that ECMs provide various topographies, such as fibre and groove structures, *in vivo*. However, those studies employed natural materials such as spider webs and fibrin, making it difficult to investigate the effects of topography on cell behaviour systematically. With the development of nano- and microfabrication technologies and material fabrication techniques, it has become possible to fabricate topographic substrates with various size and scale characteristics that mimic mechanoarchitecture *in vivo*, enabling us to study ECM topography's role *in vitro*. Synthetic topographic substrates can now be used to successfully replicate the contact guidance—the tendency of cells to shape and migrate along with topographic features and to create highly orientation-controlled cellular morphologies and tissue structures and functions, such as muscle and cardiac tissue found *in vivo*. Topography is currently being studied for its understanding as a major physical factor covered by mechanobiology research, similar to stiffness and geometry, and is discussed in detail in Chapter 8.

On the other hand, recent remarkable developments in electron microscopy and imaging techniques have contributed significantly to the multiscale structural analysis of ECM and ultrastructural analysis of tissues. Such developments have made it possible to design topographies based on biological principles. It is well known that changes in tissue stiffness accompany the development and progression of diseases such as cancer and fibrosis and that similar dynamic changes have been observed in ECM topography. To realise such dynamic topographic changes, it is difficult to use solely fabrication and material manufacturing technologies, as previously described; the development of new materials is required. If dynamic topography plays a role in pathological processes, as suggested by comparing ultrastructural analysis of normal and pathological tissues, this type of material could be a valuable tool for mechanobiology research and understanding physiological and pathological phenomena, including the development and progression of disease. Although this is still an emerging field, and the focus is on fundamental research on materials, such as materials development and the operating mechanism of variable topography, the mechanobiology research using materials with reconfigurable topography is described in detail in Chapter 9.

It is also known that cells sense the curvature of substrates larger than a single cell, leading to the alteration of their activities such as proliferation, migration, and differentiation, in contrast to the nanoscopic curvature and topography of surfaces smaller than that of a cell. Although this force perception should be closely interplayed with tissue formation, there has been no clear evidence of its effect until recently because the displacement is very small from the viewpoint of individual cells, combined with the technical difficulty of forming highly precise curved surfaces over a wide-range area. In particular, in mechanosensing for the nanoscopic curvature studied previously, it is reasonable to assume that molecules such as curvature sensing proteins and mechanosensitive (MS) ion channels, which are also involved in membrane trafficking systems, can sense the curvature. In contrast, a curvature larger than cell size was predicted to be below the detection limit of these sensors. Therefore, it is thought that the cytoskeleton, the nucleus, and cell adhesion cooperate to sense the substrate's curvature, but the mechanism is still not fully understood. Chapter 10 discusses the latest findings and approaches based on theoretical calculations and simulations.

1.3.2 Material-based Mechanobiology: Matrix Mechanics

In addition to the study of mechanobiology through the control of cell shape, another important aspect was discovering that mesenchymal stem cell differentiation depends on the mechanical properties of the extracellular matrix, as reported by Engler *et al.* in 2006.[65] Since the middle of the 20th century, there had been studies suggesting that the substrate's stiffness affects cell survival. However, the authors successfully demonstrated that the elastic modulus of the substrate directed differentiation lineage, including the neurogenesis, myogenesis, and osteogenesis of mesenchymal stem cells (which corresponds to the elastic modulus of actual biological tissues) by using acrylamide gel scaffolds with varying degrees of cross-linking. In addition, nonmuscle myosin II was shown to be involved in the mechanosensing of the cell. Cells sense the mechanical properties of the substrate *via* intrinsic cellular forces (active touch), indicating that there is an essential cellular force perception in conjunction with the mechanosensing of external forces. The range of cellular stress perception was at the same level as the elastic modulus of biological soft tissues, ranging from 0.1 kPa to 100 kPa, and thus the modulation of various cell and tissue activities by varying substrate elastic moduli was explored. In particular, the acrylamide gel used in this study was familiar to biologists since it is used for SDS-PAGE (sodium dodecylsulfate polyacrylamide gel electrophoresis). Furthermore, the degree of cross-linking can be easily adjusted, and the material can be treated as a linear elastic material for practical purposes.

The molecular clutch model is currently considered a promising model for describing the mechanism of cellular force perception concerning substrate stiffness. This model is based on the concept that the efficiency of the cellular traction force on the substrate is balanced by a molecular clutch (like that in a car), assuming that the contractile force of actomyosin is the driving force (Figure 1.2).[66,67] Mitchison and Kirschner proposed the original model in 1988 as an example of neurite outgrowth.[68] In 2008, Chan and Odde established a mathematical model for this phenomenon.[66] The adaptor protein connecting actin and integrin repeatedly binds and dissociates with the substrate as an elastic body, and the dissociation rate increases with force. The model was validated experimentally and was widely recognised. Later, to resolve the discrepancy between the intrinsic biphasic response of traction force to elastic modulus and the phenotypic diversity depending on the cell type, the existence (concept) of retrograde flow, and the elastic modulus (optimal stiffness) at which the traction force reaches an extreme value was introduced into the model.[69] Roca-Cusachs *et al.* demonstrated that this could be explained by the axis of actin–talin–integrin–ECM.[70]

On the other hand, the diversity of mechanical properties exists between different tissues and organs and fluctuates along the time axis of biological activities. For example, the mechanical microenvironment surrounding cells dynamically fluctuates during morphogenesis and pathological processes, such as substrate fluidisation during epithelial lumen formation and stiffening of the mesenchyme due to tumorigenesis and fibrosis.[71,72] To mimic such temporal changes in mechanical properties, materials whose elastic moduli change in response to external stimuli have been developed. In particular, hydrogels in which the degree of cross-linking can be controlled enable the design of dynamic substrates that can be softened[73] or stiffened[74] on demand. An interesting concept of mechanical memory is the differentiation behaviour of stem cells depending on the history of the mechanical environment in which the cells are placed was discovered by utilising these properties of the materials.[75] The design principles and applications of

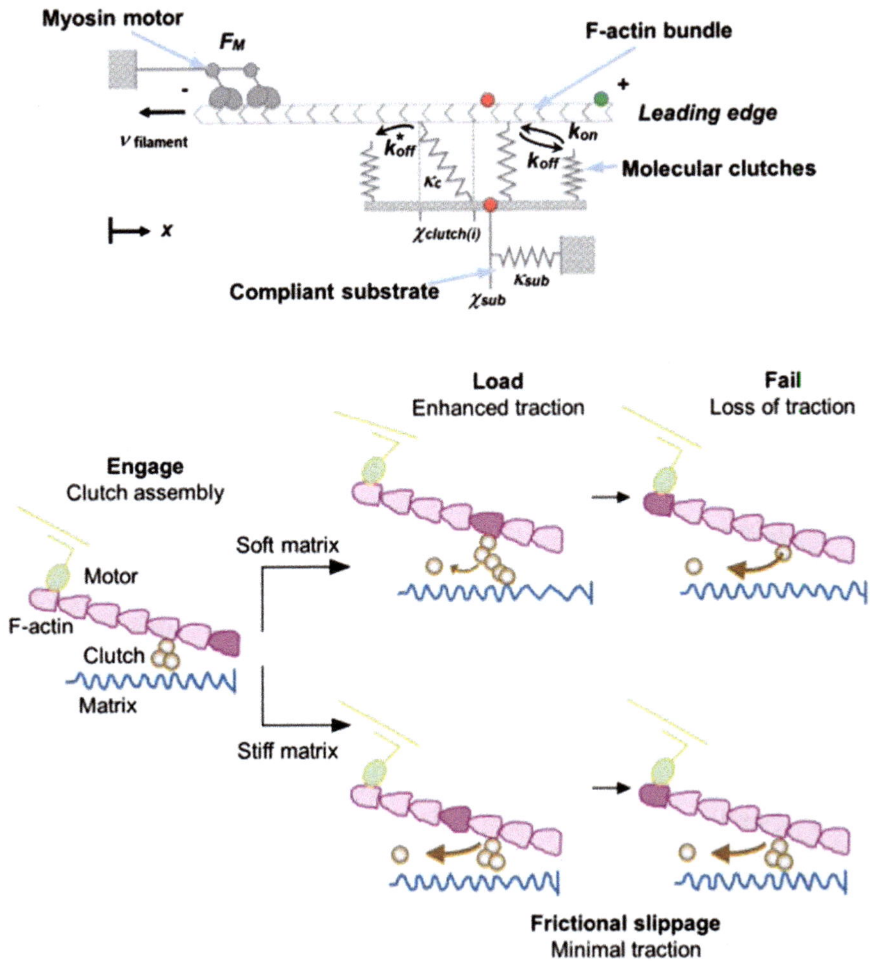

Figure 1.2 Schematic representation of molecular clutch model: (Top) The molecular clutches reversibly engage the actin filament with rates k_{on} and k_{off} to resist retrograde flow. Under loading, the clutches are stretched to strains $x_{clutch(i)}$ and tend to break at a force-dependent off-rate k^*_{off}. The transmission forces induce a local substrate strain x_{sub}. The mechanical stiffness of the clutch, κ_c, and the mechanical stiffness of the substrate, κ_{sub}, determine the mechanical resistance to load. Adapted from ref. 66 with permission from AAAS, Copyright 2008. (Bottom) On a soft matrix, the slow rate of tension build-up within clutch bonds maintains a low rate of bond dissociation (thin, brown arrow) and enhanced traction force on the extracellular matrix (stretching of the wavy line). Beyond a critical tension, the clutch bonds can no longer support the growth in tension and disengage from the extracellular matrix, resulting in loss of traction force and flux of F-actin (indicated by progression of the dark-coloured chevron to the left). On a stiff matrix, rapid tension build-up in clutch bonds results in enhanced dissociation of adhesive bonds (thick, brown arrow), minimal traction force, and movement (slippage) of the F-actin at all times. Adapted from ref. 67 with permission from AAAS, Copyright 2008.

various switchable materials that are responsive to various external stimuli or that can undergo reversible changes are introduced in Chapter 11.

In addition to temporal changes in the elastic modulus of the substrate, cells can sense small gradients in the spatial elastic modulus and appropriately change the direction of cell migration. This phenomenon is known as the durotaxis.[76] The phenomenon of cell migration towards or against the concentration gradient of molecules in the environment is also known as chemotaxis.[77] Research has been actively conducted to elucidate the mechanisms of directional sensing and polarity formation. The phenomenon of durotaxis itself is similar to that of chemotaxis; however, it is essential to develop materials that can precisely produce the elastic modulus gradient on a scale smaller than that of a cell to investigate the mechanism of durotaxis in detail. The mechanism by which gradients and absolute values of the elastic modulus of substrates modulate cell migration behaviour is discussed in Chapter 12. This chapter also deals with an emergent phenotype of stem cells, far from the tensional-equilibrated state in normal ones, which has been identified by their repetitive mechanical stimulation from the substrates.

On the other hand, mechanobiology studies on substrate mechanics, which originated from Engler's discovery, have mainly used elastic substrates, but biological tissues exhibit viscoelastic properties. This is because the extracellular matrix composed of collagen is formed by physical cross-linking and is different from elastic hydrogels based on chemical cross-linking, such as polyacrylamide gels. Therefore, studies using viscoelastic materials that mimic living tissues as cell culture substrates have attracted attention. It was found that imparting the viscous properties of the substrate increased cell elongation.[78] Here, the tuning of the relaxation time of the substrate requires the techniques and insights of material science, and this research has taken a leap forward by the merit of material scientists. Various methodologies have been developed, such as adding dissipative features to the cell culture interface described in Chapter 6 and incorporating dynamic noncovalent bonds into the hydrogel network, as described in Chapter 11. The previously mentioned molecular clutch concept has been extended to understand ECM density and cellular traction forces, and it applies to responses of the viscoelasticity of substrates as previously described.[79,80]

Mechanobiology studies focusing on substrate mechanics have mainly been conducted using two-dimensional (2D) culture systems. In contrast, three-dimensional (3D) culture systems are closer to physiological environments than conventional 2D culture systems and are a promising alternative to bridge the gap between existing cell culture models and biological tissue. Although extensive research has been conducted in tissue engineering, the focus was on properties such as biocompatibility and degradability rather than on the mechanical properties of these 3D scaffold materials. This subject is crucial for addressing and understanding the role of mechanoarchitectonics *in vivo* from a mechanics perspective. Recent developments in bio-orthogonal chemistry and materials science have led to the successful development of technologies to efficiently encapsulate target cells in hydrogels with tissue-relevant stiffness and high viability. The fact that cells encapsulated in stable hydrogels prepared by chemical cross-linking become mechanically insensitive and undergo a different cell morphology and fate than those in 2D culture systems emphasises the uniqueness of mechanobiology in 3D culture systems. In addition, it has been revealed that cells encapsulated in hydrogels with remodelling and degradation functions similar to those in living tissues can dynamically adapt to changes in the surrounding microenvironment and regulate their functions and fate. Therefore, there is a need to develop materials that can induce dynamic environmental changes in a programmable or adaptive manner in the presence of cells. Such 3D culture systems that enable temporal changes in the mechanical and biochemical environment mimicking *in vivo* can be regarded as a four-dimensional (4D) culture system and is the new trend in mechanobiology research. Chapter 13 discusses the effects of culture dimension on cultured cells, the design principle of 4D culture systems, and the novel cellular behaviours found in 4D culture systems using hydrogel-based approaches.

1.3.3 Material-based Mechanobiology: Force Detection

Harris *et al.* observed that wrinkles were formed at the periphery of cell adhesion when cells were cultured on compliant silicone rubber sheets, revealing the presence of cellular traction forces.[10] The detection of these forces applied to molecules and cells, as well as forces and stresses applied to specific sites of cells, is a central issue in mechanobiology, and tools and materials for this purpose have been developed. In detecting molecular dynamics in molecular biology, the approach is to bind to the target molecule and readout its complex formation by fluorescence or luminescence. However, it is necessary to develop various methodologies to detect forces and stresses depending on the application's location, extent, and scale, to resolve cellular mechanoarchitectonics. Many methods have been developed based on the work of Harris *et al.* to quantify the traction forces applied by a cell to its substrate. Traction force microscopy (TFM) and micropillars are versatile because they can detect force and stress based on finite element analysis from deformation and displacement information using elastic hydrogels and PDMS, as described in Chapter 4. In addition, the development of materials to detect interfacial stress between the cell and the substrate is currently being undertaken, and methodologies that combine the essence of materials science have been proposed, such as membrane-type surface stress sensors (MSS),[81] relying on a membrane-type cantilever, and materials that utilise the piezoelectric emission effect.[82]

A fluorescence resonance energy transfer– (FRET-) based fluorescent probe that links a spider-derived peptide sequence with known elasticity to GFP has a significant impact as an approach to detecting the force applied at the molecular level.[83] Since this probe is genetically encoded, it can be inserted into the target protein as a chimaera, making it possible to detect the force applied to vinculin, a group of proteins that accumulate in focal adhesions. Because of the convenience of being encoded in a gene, research has been conducted to introduce this module into other proteins, and it has become possible to detect the force applied to α-catenin,[84] E-cadherin,[85] *etc.* Salaita *et al.* developed a molecular force probe based on DNA nanotechnology, which led to a major revolution, and more details are discussed in Chapter 3.

In addition, it is now possible to detect cell membrane tension using fluorescence probes whose conjugate system expands and contracts in response to tension and measures forces in tissues using perfluorocarbon (PFC) developed by Campàs[86] and hydrogel-based cell-like microsensors developed by Cappello *et al.*[87]

As previously described, various materials and methodologies for detecting forces and stresses have been developed depending on the difference in the hierarchy of target molecules, cells, and tissues, a trend that will continue in the future.

1.4 Future Directions of Material-based Mechanobiology

In concluding this chapter, we would like to highlight the prospects of materials science research contributing to mechanobiology (Figure 1.3). While molecular biology studies unravel the molecular basis of life, mechanobiology considers the effects of "forces" on these phenomena. Mechanobiology complements molecular studies and should actively integrate technological innovations such as super-resolution microscopy (Figures 1.1E and 1.3A)[88,89] and next-generation sequencing. Technologies such as genome editing and iPSCs have brought about significant changes in biology. Chapter 14 describes good examples of using these state-of-the-art technologies to construct and understand pathological models of mechanorelated diseases.

In addition, most of the research in material-based mechanobiology has been conducted analytically *in vitro*, but future developments are expected to include technologies for controlling and measuring these forces directly *in vivo*. In such a case, it is essential to design bioinert biomaterials, and the chemical and mechanical design of the interface and bulk needs to be carefully thought out. Moreover, for organoid generation, it is essential to design materials that synchronise the time axis, such as the time evolution of biological phenomena and the mechanical properties of materials. We hypothesise that such "life-adaptive materials" will be the future of material-based mechanobiology (Figure 1.3B). This book focuses on the topic of material-based mechanobiology. Therefore, it mainly discusses materials that explore intracellular forces caused by the contractile force of actomyosin. The topics such as cellular force perception to external shear and stretching stimuli, which are essential and significant themes in mechanobiology, are not covered in detail. However, more active integration of the knowledge of materials science will lead to the development of therapeutic mechanobiological materials and technologies (Figure 1.3C).

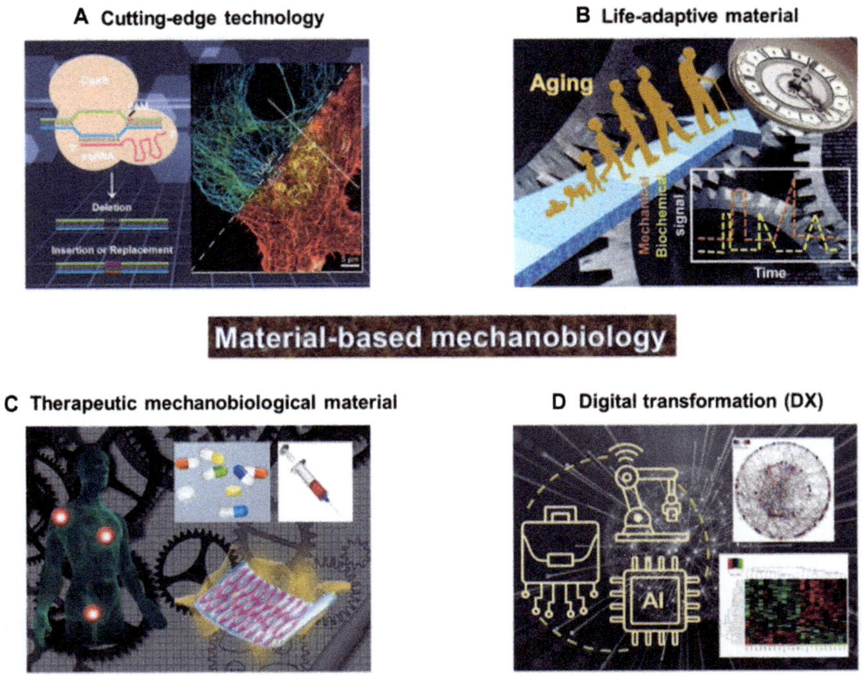

Figure 1.3 Future prospects of material mechanobiology: (A) Integration with cutting-edge technologies for understanding cellular phenomena at the molecular level and for more precise force control and measurement. Adapted from ref. 89, https://doi.org/10.1038/s41467-021-24901-3, under the terms of the CC BY 4.0 license https://creativecommons.org/licenses/by/4.0/. (B) Development of life-adaptive materials that can change their biochemical or physical properties in synchronisation with the time axis or that can adapt to developmental processes and aging to provide optimal fields for cells and tissues. (C) Development of new drugs and therapies based on the elucidation of the mechanism of cellular mechanics, and materials directly or indirectly involved in these drugs and therapies. (D) Development of omics approaches to accelerate mechanobiology research and transforming mechanobiology research with digital technology.

Moreover, while information science has been attracting the attention of all fields of science and technology, life science has greatly benefited from bioinformatics advancements in data analysis. The technology of acquiring mechanical information combined with its digital transformation (DX) (Figure 1.3D) and feeding it back to personalised medical technology and preventive medicine and further processing with artificial intelligence (AI) is expected to develop significantly in the next few years.

References

1. D. Ingber, *Ann. Med.*, 2003, **35**(8), 564–577.
2. T. Mammoto, A. Mammoto and D. E. Ingber, *Annu. Rev. Cell Dev. Biol.*, 2013, **29**(1), 27–61.

3. J. M. Phillip, I. Aifuwa, J. Walston and D. Wirtz, *Annu. Rev. Biomed. Eng.*, 2015, **17**(1), 113–141.
4. J. Wolff, P. Maquet and R. Furlong, *The Law of Bone Remodelling*, Springer Berlin Heidelberg, 1986.
5. K. K. Sanford, G. D. Likely and W. R. Earle, *J. Natl. Cancer Inst.*, 1954, **15**(2), 215–237.
6. H. M. Temin and H. Rubin, *Virology*, 1958, **6**(3), 669–688.
7. M. P. Sheetz and S. J. Singer, *Proc. Natl. Acad. Sci. U. S. A.*, 1974, **71**(11), 4457–4461.
8. H. Huxley and J. Hanson, *Nature*, 1954, **173**(4412), 973–976.
9. H. E. Huxley, *Science*, 1969, **164**(3886), 1356–1365.
10. A. Harris, P. Wild and D. Stopak, *Science*, 1980, **208**(4440), 177–179.
11. E. de Coulon, C. Dellenbach and S. Rohr, *iScience*, 2021, **24**(2), 102041.
12. A. Marty and E. Neher, in *Single-Channel Recording*, ed. B. Sakmann and E. Neher, Springer US, Boston, MA, 1995, pp. 31–52.
13. F. Guharay and F. Sachs, *J. Physiol.*, 1984, **352**(1), 685–701.
14. G. Chang, R. H. Spencer, A. T. Lee, M. T. Barclay and D. C. Rees, *Science*, 1998, **282**(5397), 2220–2226.
15. S. I. Sukharev, P. Blount, B. Martinac, F. R. Blattner and C. Kung, *Nature*, 1994, **368**(6468), 265–268.
16. P. Ridone, M. Vassalli and B. Martinac, *Biophys. Rev.*, 2019, **11**(5), 795–805.
17. C. Arbore, L. Perego, M. Sergides and M. Capitanio, *Biophys. Rev.*, 2019, **11**(5), 765–782.
18. L. Van der Meeren, J. Verduijn, D. V. Krysko and A. G. Skirtach, *iScience*, 2020, **23**(12), 101816.
19. R. O. Hynes, *Matrix Biol.*, 2004, **23**(6), 333–340.
20. R. O. Hynes, in *Cell Biology of Extracellular Matrix*, ed. E. D. Hay, Springer US, Boston, MA, 1981, pp. 295–334.
21. R. O. Hynes and K. M. Yamada, *J. Cell Biol.*, 1982, **95**(2), 369–377.
22. J. W. Tamkun, D. W. DeSimone, D. Fonda, R. S. Patel, C. Buck and A. F. Horwitz, *et al.*, *Cell*, 1986, **46**(2), 271–282.
23. R. Janoštiak, A. C. Pataki, J. Brábek and D. Rösel, *Eur. J. Cell Biol.*, 2014, **93**(10), 445–454.
24. J. Z. Kechagia, J. Ivaska and P. Roca-Cusachs, *Nat. Rev. Mol. Cell Biol.*, 2019, **20**(8), 457–473.
25. D. J. G. Rees, S. E. Ades, S. J. Singer and R. O. Hynes, *Nature*, 1990, **347**(6294), 685–689.
26. R. E. Gough and B. T. Goult, *FEBS Lett.*, 2018, **592**(12), 2108–2125.
27. A. del Rio, R. Perez-Jimenez, R. Liu, P. Roca-Cusachs, J. M. Fernandez and M. P. Sheetz, *Science*, 2009, **323**(5914), 638–641.
28. P. Kanchanawong, G. Shtengel, A. M. Pasapera, E. B. Ramko, M. W. Davidson and H. F. Hess, *et al.*, *Nature*, 2010, **468**(7323), 580–584.
29. K. Hayakawa, H. Tatsumi and M. Sokabe, *J. Cell Biol.*, 2011, **195**(5), 721–727.
30. A. R. Harris, P. Jreij and D. A. Fletcher, *Annu. Rev. Biophys.*, 2018, **47**(1), 617–631.

31. T. Shimozawa and S. Ishiwata, *Biophys. J.*, 2009, **96**(3), 1036–1044.
32. K. Hayakawa, S. Sakakibara, M. Sokabe and H. Tatsumi, *Proc. Natl. Acad. Sci.*, 2014, **111**(27), 9810–9815.
33. K. Hayakawa, H. Tatsumi and M. Sokabe, *Commun. Integr. Biol.*, 2012, **5**(6), 572–577.
34. K. Ohashi, S. Fujiwara and K. Mizuno, *J. Biochem.*, 2017, **161**(3), 245–254.
35. V. I. Risca, E. B. Wang, O. Chaudhuri, J. J. Chia, P. L. Geissler and D. A. Fletcher, *Proc. Natl. Acad. Sci.*, 2012, **109**(8), 2913–2918.
36. S. Dupont, L. Morsut, M. Aragona, E. Enzo, S. Giulitti and M. Cordenonsi, *et al.*, *Nature*, 2011, **474**(7350), 179–183.
37. G. Halder, S. Dupont and S. Piccolo, *Nat. Rev. Mol. Cell Biol.*, 2012, **13**(9), 591–600.
38. C. Y. Ho and J. Lammerding, *J. Cell Sci.*, 2012, **125**(9), 2087–2093.
39. C. S. Chen, M. Mrksich, S. Huang, G. M. Whitesides and D. E. Ingber, *Science*, 1997, **276**(5317), 1425–1428.
40. E. Ergir, B. Bachmann, H. Redl, G. Forte and P. Ertl, *Front. Physiol.*, 2018, **9**, 1417.
41. T. Hirashima, *Bio-Protoc.*, 2020, **10**(15), e3694.
42. S. Musah, N. Dimitrakakis, D. M. Camacho, G. M. Church and D. E. Ingber, *Nat. Protoc.*, 2018, **13**(7), 1662–1685.
43. K. K. Parker, A. L. Brock, C. Brangwynne, R. J. Mannix, N. Wang and E. Ostuni, *et al.*, *FASEB J.*, 2002, **16**(10), 1195–1204.
44. D. E. Ingber, *J. Cell Sci.*, 2003, **116**(7), 1157–1173.
45. D. E. Ingber, *J. Cell Sci.*, 2003, **116**(8), 1397–1408.
46. R. McBeath, D. M. Pirone, M. N. Celeste, K. Bhadriraju and C. S. Chen, *Dev. Cell*, 2004, **6**(4), 483–495.
47. C. M. Nelson, R. P. Jean, J. L. Tan, W. F. Liu, N. J. Sniadecki and A. A. Spector, *et al.*, *Proc. Natl. Acad. Sci. U. S. A.*, 2005, **102**(33), 11594–11599.
48. C. M. Nelson, M. M. Vanduijn, J. L. Inman, D. A. Fletcher and M. J. Bissell, *Science*, 2006, **314**(5797), 298–300.
49. R. D. Piner, J. Zhu, F. Xu, S. Hong and C. A. Mirkin, *Science*, 1999, **283**(5402), 661–663.
50. H. Yang, K. H. Yang, R. J. Narayan and S. Ma, *Essays Biochem.*, 2021, **65**(3), 409–416.
51. M. Shahbazi and H. Jäger, *ACS Appl. Bio Mater.*, 2021, **4**(1), 325–369.
52. S. V. Murphy and A. Atala, *Nat. Biotechnol.*, 2014, **32**(8), 773–785.
53. N. Gjorevski, E. Boghaert and C. M. Nelson, *Cancer Microenviron.*, 2012, **5**(1), 29–38.
54. J. Lee, T. G. Molley, C. H. Seward, A. A. Abdeen, H. Zhang and X. Wang, *et al.*, *Commun. Biol.*, 2020, **3**(1), 341.
55. J. Nakanishi, *Chem. – Asian J.*, 2014, **9**(2), 406–417.
56. S. van Helvert, C. Storm and P. Friedl, *Nat. Cell Biol.*, 2018, **20**(1), 8–20.
57. J. Nakanishi, *Chem. Rec.*, 2017, **17**(6), 611–621.
58. L. F. Kadem, K. G. Suana, M. Holz, W. Wang, H. Westerhaus and R. Herges, *et al.*, *Angew. Chem., Int. Ed.*, 2017, **56**(1), 225–229.

59. S. M. Z. Uddin, D. E. Komatsu, T. Motyka and S. Petterson, *J. Clin. Med.*, 2021, **10**(12), 2698.
60. A. Elosegui-Artola, X. Trepat and P. Roca-Cusachs, *Trends Cell Biol.*, 2018, **28**(5), 356–367.
61. M. Arnold, E. A. Cavalcanti-Adam, R. Glass, J. Blummel, W. Eck and M. Kantlehner, *et al.*, *ChemPhysChem*, 2004, **5**(3), 383–388.
62. R. Oria, T. Wiegand, J. Escribano, A. Elosegui-Artola, J. J. Uriarte and C. Moreno-Pulido, *et al.*, *Nature*, 2017, **552**, 219.
63. F. Posa, E. H. Baha-Schwab, Q. Wei, A. Di Benedetto, S. Neubauer and F. Reichart, *et al.*, *Biomaterials*, 2021, **267**, 120484.
64. P. Weiss, *J. Exp. Zool.*, 1934, **68**(3), 393–448.
65. A. J. Engler, S. Sen, H. L. Sweeney and D. E. Discher, *Cell*, 2006, **126**(4), 677–689.
66. C. E. Chan and D. J. Odde, *Science*, 2008, **322**(5908), 1687–1691.
67. Y. Aratyn-Schaus and M. L. Gardel, *Science*, 2008, **322**(5908), 1646–1647.
68. T. Mitchison and M. Kirschner, *Neuron*, 1988, **1**(9), 761–772.
69. L. Bangasser Benjamin, S. Rosenfeld Steven, Odde and J. David, *Biophys. J.*, 2013, **105**(3), 581–592.
70. A. Elosegui-Artola, R. Oria, Y. Chen, A. Kosmalska, C. Pérez-González and N. Castro, *et al.*, *Nat. Cell Biol.*, 2016, **18**, 540.
71. Y. Zhou, J. C. Horowitz, A. Naba, N. Ambalavanan, K. Atabai and J. Balestrini, *et al.*, *Matrix Biol.*, 2018, **73**, 77–104.
72. A. D. Theocharis, D. Manou and N. K. Karamanos, *FEBS J.*, 2019, **286**(15), 2830–2869.
73. A. M. Kloxin, A. M. Kasko, C. N. Salinas and K. S. Anseth, *Science*, 2009, **324**(5923), 59–63.
74. M. Guvendiren and J. A. Burdick, *Nat. Commun.*, 2012, **3**, 792.
75. C. Yang, M. W. Tibbitt, L. Basta and K. S. Anseth, *Nat. Mater.*, 2014, **13**(6), 645–652.
76. C.-M. Lo, H.-B. Wang, M. Dembo and Y.-L. Wang, *Biophys. J.*, 2000, **79**(1), 144–152.
77. P. J. Van Haastert and P. N. Devreotes, *Nat. Rev. Mol. Cell Biol.*, 2004, **5**(8), 626–634.
78. O. Chaudhuri, J. Cooper-White, P. A. Janmey, D. J. Mooney and V. B. Shenoy, *Nature*, 2020, **584**(7822), 535–546.
79. O. Chaudhuri, L. Gu, M. Darnell, D. Klumpers, S. A. Bencherif and J. C. Weaver, *et al.*, *Nat. Commun.*, 2015, **6**, 7365.
80. Z. Gong, S. E. Szczesny, S. R. Caliari, E. E. Charrier, O. Chaudhuri and X. Cao, *et al.*, *Proc. Natl. Acad. Sci.*, 2018, **115**(12), E2686–E2695.
81. G. Yoshikawa, T. Akiyama, S. Gautsch, P. Vettiger and H. Rohrer, *Nano Lett.*, 2011, **11**(3), 1044–1048.
82. Q. Zheng, M. Peng, Z. Liu, S. Li, R. Han and H. Ouyang, *et al.*, *Sci. Adv.*, 2021, **7**(22), eabe7738.
83. C. Grashoff, B. D. Hoffman, M. D. Brenner, R. Zhou, M. Parsons and M. T. Yang, *et al.*, *Nature*, 2010, **466**(7303), 263–266.

84. T.-J. Kim, S. Zheng, J. Sun, I. Muhamed, J. Wu and L. Lei, *et al.*, *Curr. Biol.*, 2015, **25**(2), 218–224.
85. N. Borghi, M. Sorokina, O. G. Shcherbakova, W. I. Weis, B. L. Pruitt and W. J. Nelson, *et al.*, *Proc. Natl. Acad. Sci.*, 2012, **109**(31), 12568–12573.
86. O. Campas, T. Mammoto, S. Hasso, R. A. Sperling, D. O'Connell and A. G. Bischof, *et al.*, *Nat. Methods*, 2014, **11**(2), 183–189.
87. M. E. Dolega, M. Delarue, F. Ingremeau, J. Prost, A. Delon and G. Cappello, *Nat. Commun.*, 2017, **8**(1), 14056.
88. B. Hein, K. I. Willig and S. W. Hell, *Proc. Natl. Acad. Sci.*, 2008, **105**(38), 14271–14276.
89. V. Navikas, S. M. Leitao, K. S. Grussmayer, A. Descloux, B. Drake and K. Yserentant, *et al.*, *Nat. Commun.*, 2021, **12**(1), 4565.

CHAPTER 2

On the Molecular Basis of Cellular Mechanobiology

JORGE OLIVER-DE LA CRUZ,[a,b] FABIANA MARTINO,[a,b,c] STEFANIA PAGLIARI*[a,b] AND GIANCARLO FORTE*[a,b]

[a] Center for Translational Medicine (CTM), International Clinical Research Center (ICRC), St. Anne's University Hospital, Brno, Czechia; [b] Competence Center for Mechanobiology in Regenerative Medicine, INTERREG ATCZ133, Brno, Czechia; [c] Faculty of Medicine, Department of Biology, Masaryk University, Brno, Czechia
*Emails: stefania.pagliari@fnusa.cz; giancarlo.forte@fnusa.cz

2.1 Mechanosensing at the Cell–Matrix Interface

2.1.1 Origin of the Mechanical Signals

Mechanical signals in biological tissues are generated within the extracellular matrix (ECM). ECM is defined as the noncellular component of tissues and organs providing the cells with physical support and functioning as a reservoir for biochemical cues. Although being evolutionarily conserved and constantly composed of structural adhesion proteins and polysaccharides, the ECM chemical structure is dynamically defined in a tissue and stage-specific fashion, thanks to the reciprocal crosstalk among its acellular components and the embedded cells. Together with elastins, osteonectin, osteopontin, periostin, and thrombospondin, collagens are by far the most represented ECM structural proteins.[1] The proteins ensuring cell adhesion are represented mainly by fibronectin, laminins, vitronectin, fibrillins, and tenascin, while hyaluronan, versican, and aggrecan are the most abundant

Biomaterials Science Series No. 12
Material-based Mechanobiology
Edited by Jun Nakanishi and Koichiro Uto

Published by the Royal Society of Chemistry, www.rsc.org

glycosaminoglycans and proteoglycans among the polysaccharides.[2] The chemical composition and three-dimensional organization of the ECM network determine its bulk mechanical properties (used hereafter as a synonym of tissue stiffness or elasticity), which are commonly quantified by *Young's modulus* (*E*). *E* is a tissue-specific parameter defined as a measure of the ability of a material to withstand changes in length when exposed to lengthwise tension or compression (www.britannica.com).

Cells perceive local stiffness, presumably with nanoscale resolution.[3] Local stiffness confers heterogeneity to the ECM and can be also defined as *microelasticity*. Given the prevalence of collagen among the structural ECM proteins and its unique structure and mechanical properties,[4] ECM microelasticity is a direct function of collagen concentration and correlates inversely with the water content (*i.e.* hydration) of the given tissue.[5] Collagen post-translational modifications (*i.e.* hydroxylation, glycosylation) are largely responsible for its crosslinking state and determine the bulk tissue mechanical properties. However, their contribution to microelasticity is considered modest, if any.[6] Even though the absolute and accurate quantification of biological tissue mechanical properties is problematic due to the plethora of distinct techniques, strategies, and mathematical models employed by different investigators, a common agreement exists that bulk elasticity is characteristic of the given biological tissues[7] (Figure 2.1).

The cellular response to modifications in ECM mechanics can be considered as a feedback loop operated by the cells in concert with ECM microelasticity. It can be regarded as a continuous trial-and-error process, in which the cell probes the ECM elasticity and adjusts its shape and size by tuning the intracellular forces generated.

The molecular mechanisms through which living cells perceive mechanical cues are still far from completely understood, but our understanding can be roughly broken down as follows: (1) ECM microelasticity is perceived by

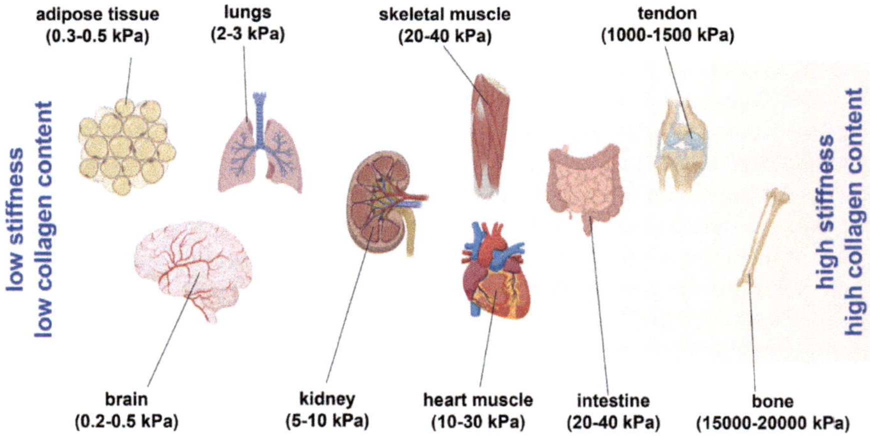

Figure 2.1 Biological tissues have a distinct bulk stiffness, which correlates with their collagen content. Image generated by using www.biorender.com.

the cell at the membrane, this step being referred to as *mechanosensing*; (2) the mechanical information is promptly transmitted inward through the modulation of cytoskeleton- and nucleoskeleton-propagated intracellular tension toward the nucleus. This part of the mechanism is usually defined as *mechanotransduction* and is accompanied by the activation of numerous *mechanosensitive signaling pathways*; (3) the long-term cell response is guaranteed by *mechanically actuated transcription complexes*, which eventually trigger the activation of specific genetic programs (also defined as *mechanosensitive genes*)[8,9] (Figure 2.2). Recent lines of evidence suggested that the application of force on the nuclear membrane can induce a deformation of the nucleus, which in turn determines the exposure of cryptic sites in the otherwise tightly packed heterochromatin.[10]

The mechanosensing–mechanotransduction apparatus, also defined as *mechanosome*, is composed by groups of proteins and cellular structures which preside over the cellular adaptation to the continuous dynamic modifications of the surrounding microenvironment. Mechanical stimuli determine distinct responses in mechanosensitive structures, which include protein conformational changes,[11,12] the translocation of transcription factors and cofactors,[13,14] translational and post-transcriptional modifications,[15,16] and alterations in chromatin organization.[17] In the present chapter, the term

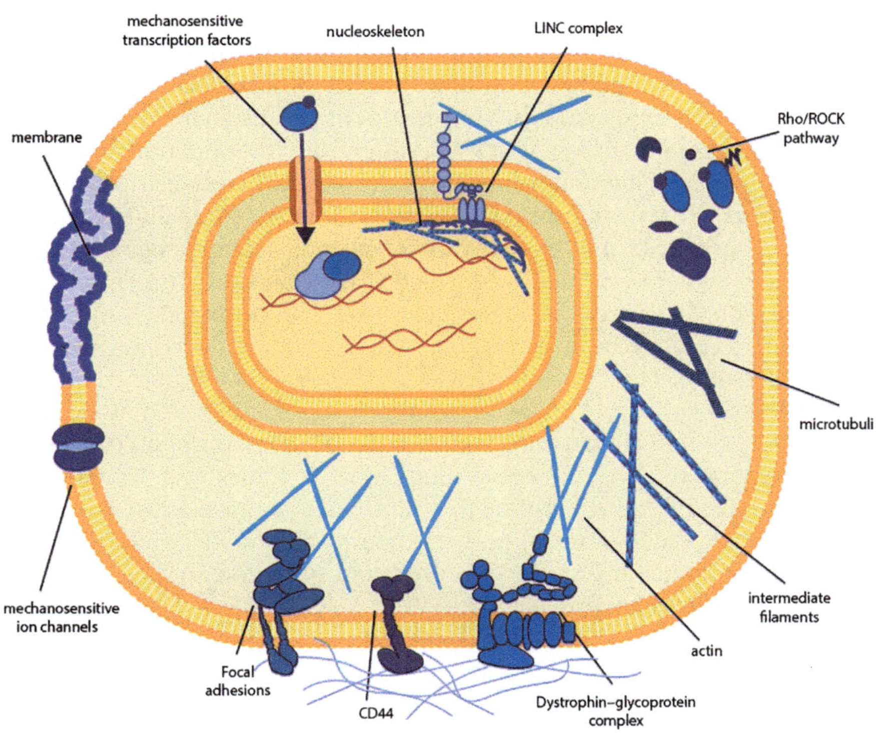

Figure 2.2 Schematic representation of the cellular mechanosome and its individual components.

"mechanosensor" will be used with a broad connotation, so as to define any intracellular macromolecule whose structure, localization, or function is perturbed in response to changes in tissue mechanics and/or geometry.

2.1.2 Cellular Membrane as Prime Mechanosensing Structure

The plasma membrane is a complex, heterogeneous structure composed of a fluid lipid bilayer, with interspersed, diffusing proteins, representing a semipermeable barrier physically separating the inner space of the cell from the outer environment. The membrane plays a role in several biological functions, including adhesion, migration, endocytosis, and secretion.[18,19]

Given its localization at the interface between the cell and the ECM and its highly dynamic nature, the plasma membrane has been recently proposed as a mechanosensitive structure. Membrane tension, defined as the force per unit length acting on a cross section of membrane,[20] is a measure of the resistance of the plasma membrane to deformations. It depends on lipid composition and on the interaction with the underlying actin cortex[20,21] and is promptly adjusted when an osmotic stress is applied or when cell shape needs restructuring. Cells were recently shown to compensate changes in lateral tension by unfolding or bending, loosening membrane interaction with the underlying actomyosin cortex and by sacrificing protrusions and invaginations such as microvilli and caveolae.[22] As a consequence, changes in membrane tension determine the opening of channels, the regulation of endoexocytosis processes and the re-arrangements in discrete domains[23] (Figure 2.3A).

Centripetal and centrifugal forces arising from the internal cytosolic pressure, the cytoskeleton dynamics, or the external cell–cell and cell–substrate interactions, induce stretching and/or bending of the lipid bilayer, causing substantial changes in membrane lateral tension. These events are thought to produce conformational changes in the embedded mechanosensors[24] (Figure 2.3B). Recently, the exit of murine ESCs from the pluripotency state has been directly associated with reduced membrane tension, as determined by the release of the cell membrane from the underlying actin cortex.[25] Similar responses to shear stress and stretching were proposed to induce the rapid modification of the physical properties of plasma membrane (*i.e.* lipid order, fluidity, and cholesterol content) in vascular endothelial cells.[26] Although the aspects of mechanosensing involving the plasma membrane appear still poorly understood, modifications in plasma membrane tension are now predicted to be able to directly affect several intracellular signaling pathways.[27]

2.1.3 Mechanosensitive Ion Channels

Mechanosensitive ion channels provide one of the fastest mechanisms to transform mechanical stimuli into electrochemical intracellular signals, and therefore most of them are associated with sensing structures in

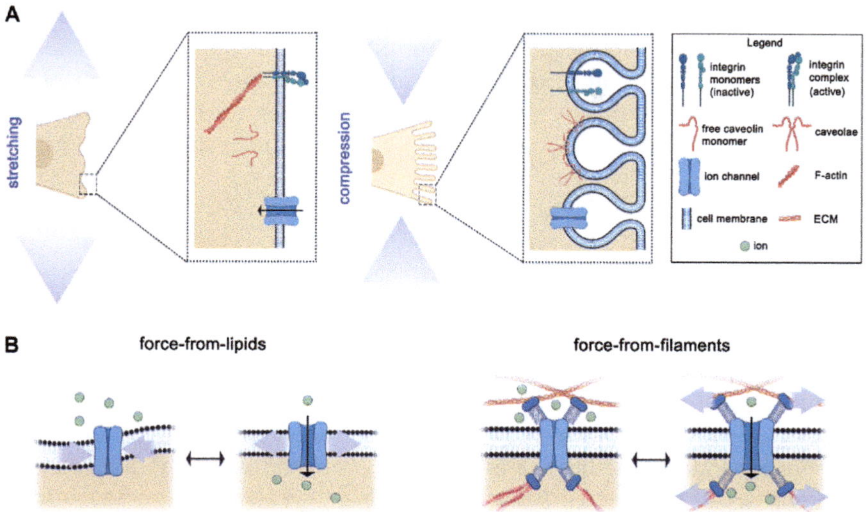

Figure 2.3 (A) Schematic representation of the alterations induced by the application of force (here exemplified by pale blue arrows) to domains (integrin heterodimer formation interacting with actin cytoskeleton (up), caveolae (middle), and ion channels (down) as a result of changes in plasma membrane tension. (B). Schematic representation of the force from lipid (left) and force from filament (right) mechanisms proposed to explain the biophysical principles of the mechanosensitive ionic channels. Schemes generated by www.biorender.com.

multicellular organisms.[28] While being very well characterized in bacteria,[29] the list of *mechanopores* is only partially resolved in mammals. As mechanoreceptors, they are specialized in sensing a specific type of mechanical force (*e.g.* shear stress, membrane stretching) according to the physiological context, although they can be activated by different stimuli when isolated. Nevertheless, most of the channels have a multimodal behavior, having their activity regulated by several stimuli, which include voltage, temperature, pH, osmolarity, chemicals, and mechanical stress.[30] Heterotetramers constituting the epithelial Na^+ channel (ENaC)[31] (SCNN1A, SCNN1B, SCNN1G, SCNN1), the potassium channels TREK1, TREK2, and TRAAK,[32–34] and the nonselective ion channels Piezo1/Piezo 2[35] have all been shown to act as mechanosensitive channels. Finally, several members of the transient receptor potential (TRP) family[36] were found to be sensitive to mechanical cues[37–39] but, in contrast to their *Drosophila* homologous nompC,[35] failed the test to be considered as true primary mechanoreceptors.

Different mechanisms have been proposed to explain the biophysical principles of the ionic channels mechanosensing (Figure 2.3B). The force from lipids paradigm implies that the mechanosensitive channels undergo conformational changes that modify the gating in direct response to lipid membrane deformations. This theory has been proven in experiments performed with bacterial and eukaryotic channels, isolated and reconstituted into proteoliposomes.[40] The protein domains required for the tight coupling between the lipid bilayer and the protein have been described for some of the mechanopores (Piezo1, TREK1).[41] Another possible system by which ion transport can be controlled is through the exclusion of lipid fatty acids from the pocket that constitute the pore channel (*e.g.* TRAAK1[41]). Dynamic accumulation/exclusion within lipid rafts could provide an additional way to exploit the asymmetrical properties of the membrane.[40]

The tethered model (or force from filaments),[42] instead, proposes that changes in gating occur as a result of the direct connections between channels and cytoskeletal or ECM proteins, being the ion channels responsible for bearing the paradigm of this archetype in mammals. Intermediate conditions that do not contradict any of the models have been also described: while changes in cell membrane tension could potentially provide the energy required for the activation of the mechanosensitive ion channels, the presence of more rigid structures such as cytoskeleton or surrounding ECM would hinder the deformation required to reach that threshold in physiological conditions. Nevertheless, the interaction with those structures could reduce the energy threshold needed. In fact, although retaining mechanosensing potential in the form of isolated structures, some of the ion channels were shown to be more sensitive to weaker mechanical stimuli when associated with ECM (*e.g.* Piezo1 to collagen IV[43]) or cytoskeleton (*e.g.* ENAC and actin[44]).

2.1.4 Focal Adhesions (FAs)

The discrete macromolecular complexes connecting the ECM with the cell cytoskeleton are defined as *focal adhesions* (FAs).[45] Their structure is described in detail elsewhere.[8] For simplicity, the hundreds of proteins participating in FA formation can be divided into layers: at the outer layer lie the ECM receptors, integrins. The integrins are the proteins providing direct linkage between the ECM and cell actin cytoskeleton. They are physically connected to the ECM through their extracellular domain (α-subunit), while their cytoplasmic tail (β-subunits) contacts F-actin *via* docking proteins (talin, vinculin, zyxin, tensin), which belong to the mechanosensing layer. The continuous rearrangement of actin cytoskeleton in response to external stimuli is ensured by the layers of mechanosignaling proteins (paxillin, focal adhesion kinase, Src, p130Cas) and actin regulators (ena-VASP, alpha-actinin, *etc.*).[46] A schematic representation of the different layers of focal adhesions is given in Figure 2.4.

Regardless of the proposed layer schematics, most FA components have been described to change either their localization or their conformation following the application of physical and biochemical stimuli generated at the ECM. The cooperative nature of these complexes, in fact, makes it difficult to determine the specific mechanosensitivity of FA members. Additionally, FA proteins like zyxin, talin, vinculin, tensin, paxillin, FAK,

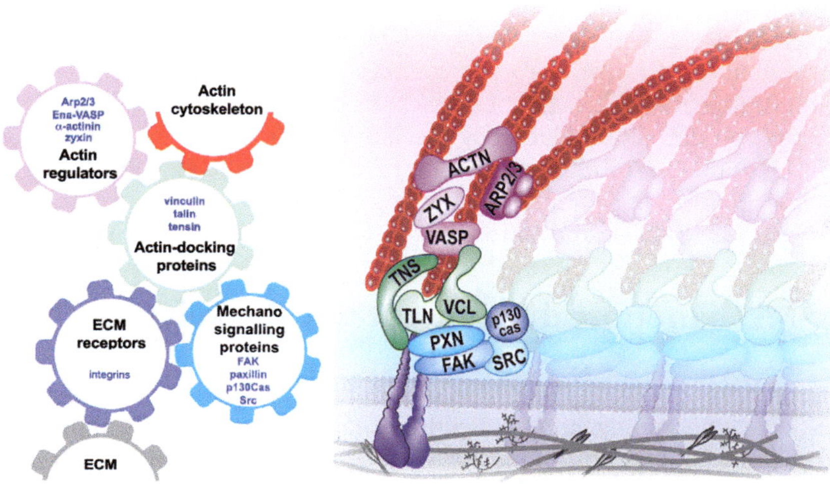

Figure 2.4 Schematic representation of the layers of focal adhesion proteins that bridge the extracellular matrix (ECM) and the actin cytoskeleton. Few components of the layers are indicated in blue text.

vasodilator-stimulated phosphoprotein (VASP), and alpha-actinin have been described as playing slightly different roles depending on the conditions and on the cell type investigated.[47]

Under mechanical load, integrins are induced to diffuse through the lipid bilayer and cluster.[11] The composition of FA core is dynamically remodeled following integrin clustering: They assemble and disassemble in a time scale of seconds to minutes by responding quickly to changes in the structure and mechanics of the extracellular milieu.[48,49] Among the proteins recruited to the FAs in response to integrin clustering is vinculin. The molecular basis of vinculin mechanosensitivity is still heavily debated. Cells lacking vinculin, in fact, show reduced contractility, that could be restored by transfecting the vinculin tail domain but not the head domain.[50] Upon reaching the FA site, vinculin binds to talin, which is connected to the integrin intracellular tail, *via* its head domain and undergoes quick conformational changes.[51,52] Vinculin recruitment to the FAs depends on the balance between the forces generated at the integrin–ECM and those propagated through the cytoskeleton. Thus its integration into the adhesion complex correlates directly with substrate stiffness.[53] Talin is a 270 kDa protein whose mechanosensitive properties were demonstrated unequivocally by experiments in isolated proteins. It is composed of an N-terminal globular head, a number of flexible rod domains, and C-terminal helices. The rod domains undergo force-dependent stochastic unfolding and refolding cycles with a piconewton resolution. This specific conformation and force sensitivity make talin a very effective force buffer.[54] Talin interacts physically with both β-integrin cytoplasmic domain and F-actin through its globular head, and recruits linker proteins 4.1, ezrin, radixin, and moesin at the FA–cytoskeleton interface.

Talin dimers have been described that are formed through C-terminal helices and that adopt different orientations, elongating or collapsing in response to ECM mechanical stimulation.[55] The complex nanomechanical properties of the talin flexible rod have been recently characterized:[56] The unfolding of the numerous rod domains induced by mechanical load was shown to expose cryptic hydrophobic binding sites able to interact with vinculin.[51,57] The successful binding of vinculin to talin is considered essential to stabilize talin–F-actin interaction and thus transfer the mechanical signal inward.[58]

2.2 Mechanotransduction at the Cytoskeleton

2.2.1 Cytoskeleton Network

The cytoskeleton is a dynamic network of filamentous polymers and crosslinkers acting as an intracellular scaffold. It ensures the propagation of extracellular and cell-generated forces and controls cell mobility, shape, and integrity through its rearrangement and dynamics.[59] Altered cytoskeleton organization can affect several signaling pathways, thus resulting in the aberrant activation of given genetic programs.[60] The cytoskeleton is mainly composed of actin fibers, microtubules (MTs), and intermediate filaments (IFs), which rearrange upon mechanical stress and contribute to the maintenance of its mechanical properties.[61] Cytoskeleton integrity and homeostasis are kept by actin-binding proteins such as nucleation-promoting factors (Arp2/3, profilin), capping proteins, depolymerizing factors (ADF/cofilin), stabilizing proteins, and crosslinkers (actinins).[62] The actin cytoskeleton is composed of actin filaments (F-actin), globular actin molecules (G-actin), and actin-binding proteins. F-actin is physically connected to FAs at the cell–ECM interface and to adherens junctions at cell–cell contact sites.[63] The generation of contractile forces is largely determined by the interaction of F-actin with the motor protein myosin II.[64] Crosslinkers such as filamin, alpha-actinin, and fascin strengthen the interaction between F-actin and myosin and together form complex structures called *stress fibers* (SFs). SFs generate and align in the same direction of the applied force and work as sensors of tension. SF–FA interplay is highly dynamic and ensures the continuous bidirectional transduction of force between the ECM and the cell.[65] During the mechanotransduction process, SFs contribute to FAs' maturation by regulating the recruitment of specific proteins to the FA complexes. For example, vinculin recruitment to the FA sites is favored by SFs' contractility,[66] which also prompts zyxin shuttling from cell periphery to the SFs. Both events are needed to build cytoskeletal tension.[67] SFs have been categorized into different subtypes according to their structure, assembly, and FA linkage.[68] Dorsal SFs, which are anchored to FAs only at one end, act as stabilizers and have no contractile properties, since they do not interact with myosin II.[69] Only indirectly connected to FAs, *transverse arcs* are instead contractile SFs displaying a periodic pattern of alpha-actinin and myosin II. *Ventral SFs* originate from the dynamic interaction between dorsal

SFs and transverse arcs. They are defined as contractile actomyosin bundles, being anchored at both ends to FAs.[68] The most recently described type of SFs is the perinuclear actin cap: it consists of actomyosin bundles surrounding the nucleus and ensuring the propagation of mechanical forces from FAs, at the cell periphery, to the nuclear envelope.[70] SFs directly interact with *microtubules.*[71] The stiffest components of the cytoskeleton network, MTs are involved in several biological processes such as intracellular trafficking, cell polarity, cilium formation, and cell division.[59,72] MTs are sensitive to mechanical stress, as demonstrated by the force-directed orientation and positioning of the mitotic spindle observed in dividing cells exposed to mechanical loading.[73] MTs promote FA turnover during 2D migration, most likely *via* endocytosis of the FA component for posterior rearrangement.[41] *Intermediate filaments* are the more flexible and stable component of the cytoskeleton. Their function and dynamics are regulated by post-translational modifications.[74] When subjected to mechanical forces, IFs can perceive the strength and the direction of the force and undergo intense deformation.[75] The most studied IF is vimentin. Vimentin is defined as a mechanoshield, increasing cell stiffness and protecting the cell from compressive forces.[76] Vimentin polymerization is affected by substrate stiffness,[77] probably contributing to the homeostatic balance between cell and substrate stiffness.

2.2.2 Regulation of Cytoskeleton Intracellular Tension: Rho/ROCK Pathway

The contractile properties of the actin cytoskeleton are mainly determined by Rho-associated coiled-coil kinases (ROCK) pathway,[78] which plays a central role in cell response to mechanical load. Rho GTPases act as molecular switches, alternating their active GTP-bound and inactive GDP-bound forms. The active/inactive switch is regulated by GTPase-activating protein (GAP) and guanine nucleotide exchange factor (GEF). Several Rho-GEFs and Rho-GAPs have been identified as regulators of actin cytoskeletal organization upon mechanical stimulation controlling the activity of diverse Rho family members.[79] Rho proteins are involved in several biological functions such as cell polarity, proliferation, differentiation, and cytoskeleton remodeling.[80] Mechanical load triggers RhoA activation, which regulates cytoskeleton contractility through its downstream effectors: Rho-associated protein kinase (ROCK) and formin diaphanous (mDia). RhoA-targeting GEFs such as Vav2, GEF-H1, and RhoGEF12 (Larg) have been identified as essential factors for RhoA activation in cells exposed to mechanical loading.[81] The RhoA/ROCK pathway controls intracellular contractile force generation through myosin II activation. In fact, RhoA induces ROCK, which regulates myosin II activity increasing myosin regulatory light chain (MLC) phosphorylation directly or by inhibition of MLC phosphatase (MLCP).[82] MLC phosphorylation promotes myosin II ATPase activity, thus favoring the generation of contractile force at the stress fibers.[83]

Additionally, the RhoA/ROCK pathway is involved in cytoskeletal stabilization by regulating the activity of actin-severing protein cofilin.[16] In its dephosphorylated active form, cofilin severs F-actin fibers, promoting actin depolymerization and thus reducing cytoskeletal tension. Upon mechanical stimulation, RhoA/ROCK activation induces cofilin phosphorylation by LIM kinase, thus inhibiting cofilin-mediated actin disassembly.[16] Moreover, mechanically induced RhoA activity directly controls cytoskeleton stabilization, promoting actin filament assembly *via* activation of the actin nucleating protein mDia (Figure 2.5).[84] MT network contributes to the regulation of RhoA/ROCK signaling. Disruption of MTs can indeed increase GEF-H1 availability, thus favoring RhoA activation, SFs generation, and cell contractility.[85] In a similar way, IFs including keratins and vimentin affect RhoA-mediated actin organization interacting with RhoA-targeting GEFs.[86] Perturbations of Rho/ROCK/myosin II pathway have been associated with

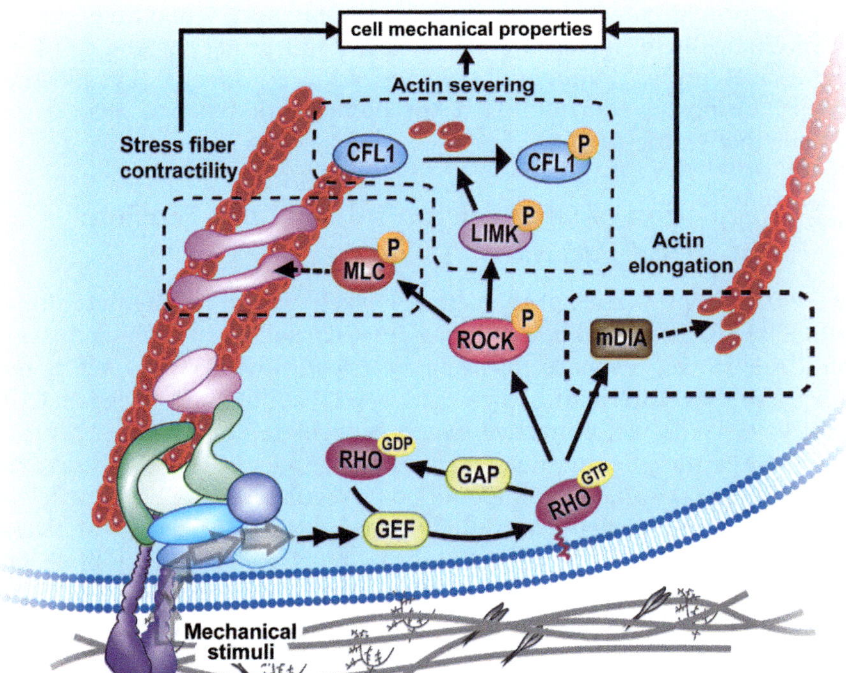

Figure 2.5 GTP-bound RhoA regulates F-actin elongation and integrity, as well as stress fiber contractility through its downstream effectors, diaphanous-related formin-1 (mDia) and Rho-associated protein kinase (ROCK). The latter activates LIM kinase (LIMK) to halt the severing action of cofilin proteins, while promoting the phosphorylation of myosin light chain (MLC), in turn increasing stress fibers' contractility. The transition between the active (RhoA-GTP) and the inactive (RhoA-GDP) forms is favored by GAP (GTPase-activating protein) and GEF (guanine nucleotide exchanging factor) proteins.

defective cytoskeletal mechanics and altered intracellular tension but can also affect many of its downstream signaling pathways leading to altered cell responsiveness to mechanical cues.[87] For example, inhibition of Rho/ROCK signaling pathway components affects SFs positioning and orientation in response to tensile forces.[64] Moreover, myosin II mutants display a reduced number of FAs, SFs disorganization, and contractile defects.[82,88]

2.3 Transferring Mechanical Cues to the Nucleus

2.3.1 Nucleus as a Mechanosensor

A growing body of evidence indicates that mechanical load at the cell periphery affects gene expression. The molecular mechanisms ensuring the transmission of mechanical forces from cell periphery into the nucleus are still far from being completely understood.[89] The nucleus, the largest and stiffest organelle in the cell, has been recently proposed as a mechanosensitive unit itself.[90] Few studies have suggested a role for the nuclear envelope in mediating the force propagation at the nucleus–cytoskeleton interface. The nuclear envelope is composed of outer and inner nuclear membranes, a perinuclear space, and the nuclear lamina. It contains several membrane-bound proteins and nuclear pore complexes (NPCs), which regulate the shuttling of molecules to the nucleus and access to the DNA.[91] Nuclear membrane is directly connected to the cellular cytoskeleton and perceives the mechanical force through its own mechanosensitive apparatus.[92,93] The application of a mechanical force on integrin receptors at the cell membrane results in nuclear envelope deformation, nuclear shape definition, and chromatin perturbation, which might modify gene expression.[94] In fact, evidence supporting the mechanoresponsiveness of the nucleus came from a recent study showing how mechanical stimulation is followed by chromatin stretching and by the activation of cryptic transcription sites.[95]

2.3.2 Linker of Nucleoskeleton and Cytoskeleton (LINC) Complex

The complex responsible for the nuclear-cytoskeletal coupling and for the transmission of cytoskeletal tension into the nucleus is the linker of nucleoskeleton and cytoskeleton (LINC) complex (Figure 2.6). LINC perturbation has been associated with several defects in nuclear responses, which include nuclear shaping and movement, signal transduction, and chromatin remodeling.[96,97] The core components of the LINC complex include the giant proteins nesprins at the outer membrane and Sad1 and UNC-84 domain–containing proteins (SUNs), which are anchored to the nuclear lamina.[97] Nesprins are spectrin repeated proteins containing a unique evolutionary conserved C-terminal Klarsicht ANC-1 and Syne Homology (KASH) transmembrane domain spanning the outer nuclear membrane (ONM).[92,94] When nesprins experience actomyosin-generated cytoskeletal tension, they

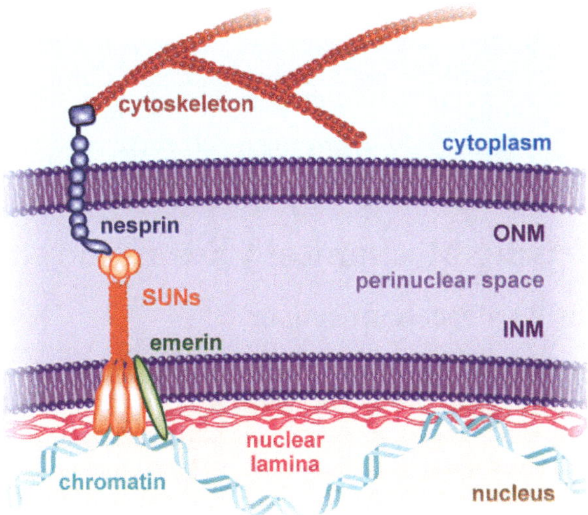

Figure 2.6 Schematic representation of the connection between the cytoskeleton, the nucleoskeleton, and the nuclear chromatin. Nesprin and SUN proteins span over the nuclear membrane and connect with the nuclear lamina, which is largely responsible for keeping nucleus shape and integrity.

diffuse through the nuclear envelope toward the cytoplasmic side.[98] Nesprins connect the nuclear envelope to the cytoskeleton either directly or through crosslinker proteins such as dynein, plectin, or kinesin-1.[99] In the perinuclear space, nesprins bind to the C-terminal domain of SUN proteins through their KASH domain. SUN proteins organize in trimers, span through the inner nuclear membrane (INM), and interact with the nuclear lamina and chromatin *via* their N-terminal nucleoplasmic domain.[100] SUNs can also interact with nuclear pore complexes (NPC) at the nuclear envelope, controlling their organization and distribution.[101] LINC complex components, nesprins, and SUNs cluster at the apical region of the nucleus and constitute the transmembrane actin-associated nuclear (TAN) line, which displays a discrete distribution similar to that found for FAs at the cell membrane: Upon mechanical stimulation, components of the LINC complex align on TAN lines, where they contact actin stress fibers SFs.[102]

2.3.3 Nucleoskeleton and Transcriptional Activity Control

On the nucleoplasmic side of the INM, SUN proteins physically bind to the intermediate filament lamin A to connect LINC complex to the nuclear lamina.[103] The direct connection cytoskeleton–LINC–nuclear lamina ensures the propagation of the external and intracellular mechanical forces to the nucleus.[103] Nuclear lamina is key to orchestrate nuclear shape, transcription factors activity and translocation, and chromatin organization and to provide structural and mechanical stability to the nucleus.[94] Both SUNs and lamin A bind to chromatin and participate in the regulation of gene expression upon mechanical stimulation.[103] Lamin A binds to chromatin either directly or through the regulatory protein emerin, an integral membrane actin-capping protein.[104] The activation of transcription has been associated with the chromatin rearrangements occurring at the nuclear periphery; in particular, NPCs are recognized sites of active transcription and interact with both DNA and cytoskeletal components.[105] Chromatin stretching in response to mechanical stimuli applied on cell surface was recently demonstrated and associated with a rapid induction of gene expression.[95] On the contrary, nuclear periphery has been described as a resting site for transcription factors, cloistering them and impeding their activation and interaction with target genes.[106] Examples of these events are given by transcription factors, including c-FOS and retinoblastoma protein (pRB), being sequestered by lamin A at the nuclear lamina.

2.3.4 Nucleo-cytoskeleton Coupling in Cellular Adaptation to Mechanical Stress

Given the crucial role of LINC and nucleoskeleton components in force transmission from cell periphery to the nucleus, it is not surprising that mutations in genes coding for proteins involved in this system may cause a failure in cell adaptation to mechanical load. As demonstrated by knockout experiments, the absence of LINC or nucleoskeleton components is associated with altered signal transduction and chromatin dynamics, nuclear distortion, and cytoskeleton derangement.[96,97] Upon mechanical stimulation, nuclear envelope components can undergo conformational changes and post-translational modifications. Modifications in ECM stiffness are in fact associated with altered lamin A levels: soft surfaces induce the phosphorylation of lamin A and its detachment from the nucleoskeleton.[107] Lamin A phosphorylation and its increased degradation, induced by reduced cytoskeletal tension, reduces cell contractility.[108] Lamin A dissociation from the nucleoskeleton leads to nuclear frailty and defective nuclear mechanics.[109] Lamin A mutants display defects in nucleocytoskeletal force transmission, cytoskeletal dynamics, and nuclear stability, thus failing to dictate a proper biological response when mechanical load is applied.[94,109] The INM protein emerin has also been recognized as a tension sensor, which undergoes phosphorylation upon mechanical stimulation and determines lamin A accumulation at the nuclear envelope.[110]

2.3.5 Shuttling Mechanosensors and Mechanosensitive Transcription Factors

The presence of cytoplasmic proteins in the nucleus has been known for decades, even though the biological relevance of this phenomenon is still poorly understood. Some of these proteins are transported to the nucleus in the absence of a nuclear localization signal and lack a DNA binding domain.[111] Recently, some of these shuttling proteins have been recognized as mechanosensitive elements, being released either from cell–cell and cell–ECM junctions or from the cytoskeleton following mechanical stress and accumulating in the nucleus. Hence they are identified as part of the *mechanotransduction* apparatus providing a direct connection between the surrounding milieu and gene expression. β-catenin was among the first proteins found to detach from the cadherin cell junction complex to act as a transcriptional coactivator within the Wnt signaling pathway.[112] Focal adhesion-associated kinase (FAK) can also shuttle from cell surface to the nucleus in response to changes in cell adhesion in order to interact with p53[113] and TFIID[114] transcription factors. Several actin-associated proteins, such as filamin A,[115] gelsolin,[116] and alpha-actinin 4,[112] translocate to the nucleus when the cell is exposed to mechanical stress, while CD44 can undergo a mechanically induced proteolytic cleavage, which releases an intracellular domain able to regulate the expression of stemness factors[117] and CD44 itself.[118] Arguably, the most paradigmatic and well studied example of mechanoregulated signaling pathway entailing the shuttling of its downstream effectors is Hippo. Originally identified among the pathways controlling organ shape and size in *Drosophila*, its core architecture is conserved in mammals. Ste20-like kinases (MST1/2) and large tumor suppressor kinases (LATS1/2) regulate the phosphorylation of the transcriptional coactivators, Yes-associated protein (YAP) and transcriptional coactivator with a PDZ-binding motif (WWTR1, also known as TAZ).[119] When phosphorylated, YAP/TAZ are retained in the cytoplasm or degraded; otherwise, they can shuttle to the nucleus to exert their co-transcriptional activity. Apart from phosphorylation and nuclear import/export, the fate and function of YAP/TAZ can be regulated by the abundance of transcriptional binding partners, controlled in a stage- and cell-specific fashion, and by alternative splicing.[120] YAP and TAZ are paralog proteins sharing common structural features, such as WW and PDZ domains, but executing rather different genetic programs.[121] Literature revision returns the image of a complex and concerted regulation of YAP/TAZ activity, which appear to be a hub for numerous signaling pathways. Given their role in organogenesis, it is not surprising that they are responsive to several mitogens such as epidermal growth factor (EGF), insulin, thrombin and lysophosphatidic acid (LPA).[122,123] YAP/TAZ are best known for their sensitivity to mechanical cues, being accumulated in the nuclei of cells cultured in conditions that allow cell spreading[124] such as 2D stiff substrates[60] and permissive adhesion areas[125] or 3D-compliant hydrogels.[124] In 2D, their shuttling is also determined by dynamic changes in substrate nanostructures,[126] an indication of their prompt responsiveness to ECM pathological remodeling.[127] YAP nuclear shuttling can be dynamically activated by fluid shear stress[112] and mechanical stretching.[112] YAP/TAZ relocation to the nucleus requires cytoskeleton stability, as controlled by Rho/ROCK-activity.[128] Lately, actomyosin-generated contractile force in connection with LINC was shown to flatten the nucleus and open the nuclear pores, hence favoring YAP nuclear import[129] (Figure 2.7).

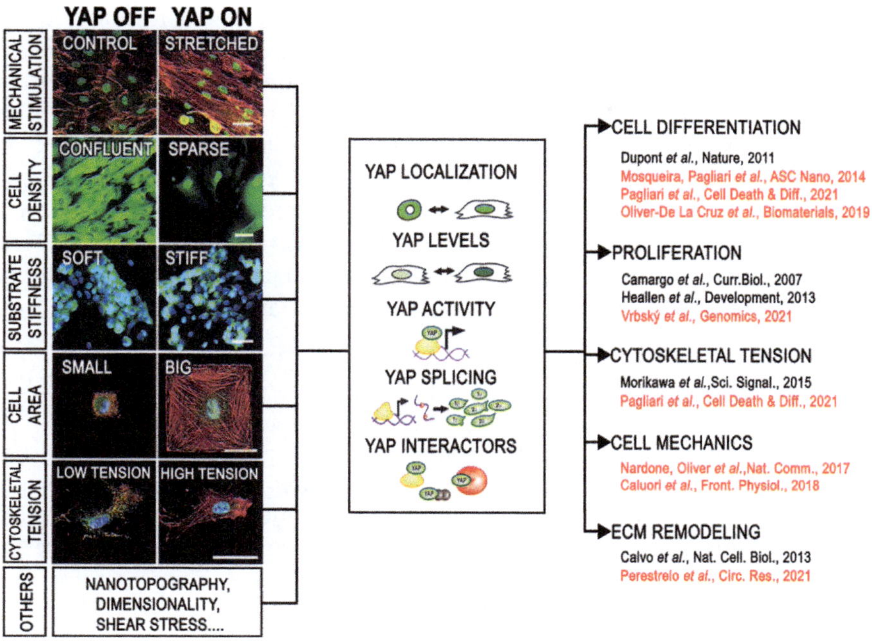

Figure 2.7 Schematic representation of the different mechanical cues (left) controlling YAP signaling (middle) and the acknowledged effects on cell function (right). Seminal references are indicated for each function. Immunostaining images represent DAPI (blue), YAP (green), and F-actin (red). CAL51 cells were used for substrate stiffness experiments; the other experiments were performed with immortalized mesenchymal stem cells. Scale bar represents 40 μm.

Nesprin-1 depletion has been demonstrated to prevent the mechanically induced shuttling of YAP to the nucleus,[130] while alterations in YAP signaling have been also found in lamin A/C gene (LMNA) mutants.[41] Finally, while the mechanosensitivity of YAP mostly relies on the interaction with other molecules, additional levels of regulation have been recently added: (1) YAP mRNA undergoes alternative splicing during development and in disease to generate proteins with different affinity for binding partners, distinct mechanosensitivity, and a rather unique transcriptional fingerprint.[120] (2) Moreover, YAP is a mechanically labile protein, whose unfolding is easily achieved to promote its translocation through the nuclear pores.[129]

Apart from the physical properties of the ECM, its composition has also an impact on YAP activity. Among others, YAP is regulated by the presence of fibronectin (through/FAK/Src signaling),[131] agrin (through the dystroglycan/dystrophin axis[132]), and hyaluronan (*via* CD44/NF2[133] or CD44/PAR1b[134]). Distinct reports have indicated that YAP/TAZ can be regulated by cell density, being sequestered in the cytoplasm of cells grown at high density through the interaction with components of the adherens junctions (cadherin-catenin system,[132] AMOT[132]) and/or tight junctions (ZO-1[135]), while accumulating in the nuclei of sparse cells. Interestingly, a recent body of evidence was provided that ECM–cell mechanics and YAP establish a feed-forward system by which YAP can be controlled by mechanical forces and, in turn, promote the transcription of genes coding for FA components,[125] cytoskeletal associated proteins,[136,137] or Rho-associated guanine exchange factors.[138] These data are likely to explain how YAP participates in the regulation of cytoskeletal tension[136,137] and cell mechanics.[125,139] In turn, ECM composition[140,125] and mechanics[141] are also regulated by YAP/TAZ. The effects on cell mechanics have a profound impact on important cellular events such as proliferation[120,142,143] and differentiation.[60,124,137,144]

2.4 Conclusions and Perspectives

Experimental evidence accumulated over the last decades demonstrating that the role of *mechanosensing* and *mechanotransduction* is not restricted to specialized circumstances (*e.g.* hearing or muscle contraction) but can affect several biological functions in virtually every cellular context. Forces are active in all biological scenarios, as they orchestrate cellular contractility, motility, cell–cell and cell–ECM adhesion. They also participate in ECM remodeling. Due to their connections with classical biochemical pathways, the mechanosensing complexes described in this chapter provide cells with the ability to integrate signals and define a concentric model in which the mechanical cues arising from the extracellular environment affect specific genetic programs. In this view, *mechanosensors* allow the cell to adapt to the external environment in a broad time scale, ranging from a quick response (in the millisecond range in the case of mechanosensitive ion channels opening[28] or conformational changes of talin and vinculin[138]) to long-lasting effects on gene expression or even epigenetic modifications.[138] Changes in the biophysical microenvironment are often associated with pathological conditions. The function of cartilage and bone tissues, which are intensively exposed to mechanical stress, is therefore strictly dependent on their mechanical integrity.[145,146] Additionally, the change in mechanical properties associated with the formation of a fibrotic tissue is known to underlie and worsen aging pathologies, such as cardiovascular diseases[140] and cancer.[147] In the latter, the mechanical properties of the microenvironment orchestrated by cancer-associated fibroblasts offer a protective shield against the immune system and chemotherapy. Deciphering the contribution of *mechanobiology* to cell fate and function is therefore crucial to offer new perspectives for the design of novel therapeutic solutions.

Acknowledgements

The study was supported by the European Social Fund and European Regional Development Fund Project MAGNET (Number CZ.02.1.01/0.0/0.0/15_003/0000492).

References

1. T. J. McKee, G. Perlman, M. Morris and S. V. Komarova, *Sci. Rep.*, 2019, **9**, 10542.
2. C. Frantz, K. M. Stewart and V. M. Weaver, *J. Cell Sci.*, 2010, **123**, 4195–4200.
3. S. Sen, A. J. Engler and D. E. Discher, *Cell. Mol. Bioeng.*, 2009, **2**, 39–48.
4. M. D. Shoulders and R. T. Raines, *Annu. Rev. Biochem.*, 2009, **78**, 929–958.
5. L. R. Smith, S. Cho and D. E. Discher, *Physiology*, 2018, **33**, 16–25.
6. M. A. Chapman, R. Pichika and R. L. Lieber, *J. Biomech.*, 2015, **48**, 375–378.
7. C. F. Guimarães, L. Gasperini, A. P. Marques and R. L. Reis, *Nat. Rev. Mater.*, 2020, **5**, 351–370.
8. F. Martino, A. R. Perestrelo, V. Vinarský, S. Pagliari and G. Forte, *Front. Physiol.*, 2018, **9**, 1–21.
9. J. L. Alonso and W. H. Goldmann, *AIMS Biophys.*, 2016, **3**, 50–62.
10. D. Camozzi, C. Capanni, V. Cenni, E. Mattioli, M. Columbaro, S. Squarzoni and G. Lattanzi, *Nucleus*, 2014, **5**, 427–440.
11. S. J. Shattil, C. Kim and M. H. Ginsberg, *Nat. Rev. Mol. Cell Biol.*, 2010, **11**, 288–300.
12. M. Yao, B. T. Goult, B. Klapholz, X. Hu, C. P. Toseland, Y. Guo, P. Cong, M. P. Sheetz and J. Yan, *Nat. Commun.*, 2016, **7**, 1–11.
13. O. Huber, R. Korn, J. McLaughlin, M. Ohsugi, B. G. Herrmann and R. Kemler, *Mech. Dev.*, 1996, **59**, 3–10.
14. M. A. Trembley, P. Quijada, E. Agullo-Pascual, K. M. Tylock, M. Colpan, R. A. Dirkx, J. R. Myers, D. M. Mickelsen, K. De Mesy Bentley, E. Rothenberg, C. S. Moravec, J. D. Alexis, C. C. Gregorio, R. T. Dirksen, M. Delmar and E. M. Small, *Circulation*, 2018, **138**, 1864–1878.
15. J. M. Dong, L. S. Lau, Y. W. Ng, L. Lim and E. Manser, *Biochem. J.*, 2009, **418**, 173–184.
16. K. Hayakawa, H. Tatsumi and M. Sokabe, *J. Cell Biol.*, 2011, **195**, 721–727.
17. Y. A. Miroshnikova, M. M. Nava and S. A. Wickström, *J. Cell Sci.*, 2017, **130**, 2243–2250.
18. A. Diz-Muñoz, D. A. Fletcher and O. D. Weiner, *Trends Cell Biol.*, 2013, **23**, 47–53.
19. K. Jacobson, P. Liu and B. C. Lagerholm, *Cell*, 2019, **177**, 806–819.
20. B. Pontes, P. Monzo and N. C. Gauthier, *Semin. Cell Dev. Biol.*, 2017, **71**, 30–41.
21. R. Nambiar, R. E. McConnell and M. J. Tyska, *Proc. Natl. Acad. Sci. U. S. A.*, 2009, **106**, 11972–11977.
22. A. Pietuch, B. R. Brückner and A. Janshoff, *Biochim. Biophys. Acta, Mol. Cell Res.*, 2013, **1833**, 712–722.
23. A. L. Le Roux, X. Quiroga, N. Walani, M. Arroyo and P. Roca-Cusachs, *Philos. Trans. R. Soc., B*, 2019, **1779**, 20180221.
24. W. Rawicz, B. A. Smith, T. J. McIntosh, S. A. Simon and E. Evans, *Biophys. J.*, 2008, **94**, 4725–4736.

25. M. Bergert, S. Lembo, S. Sharma, L. Russo, D. Milovanović, K. H. Gretarsson, M. Börmel, P. A. Neveu, J. A. Hackett, E. Petsalaki and A. Diz-Muñoz, *Cell Stem Cell*, 2020, 1–8.
26. K. Yamamoto and J. Ando, *Circ. J.*, 2018, **82**, 2691–2698.
27. G. Licari, K. Strakova, S. Matile and E. Tajkhorshid, *Chem. Sci.*, 2020, **11**, 5637–5649.
28. B. Martinac and C. D. Cox, in *Reference Module in Life Sciences*, Elsevier, 2017.
29. B. Martinac, J. Adler and C. Kung, *Nature*, 1990, **348**, 261–263.
30. B. Minke and B. Cook, *Physiol. Rev.*, 2002, **82**, 429–472.
31. C. M. Canessa, A. M. Merillat and B. C. Rossier, *Am. J. Physiol.*, 1994, **267**, 1682–1690.
32. J. T. Brennecke and B. L. de Groot, *Biophys. J.*, 2018, **114**, 1336–1343.
33. F. Maingret, M. Fosset, F. Lesage, M. Lazdunski and E. Honoré, *J. Biol. Chem.*, 1999, **274**, 1381–1387.
34. A. J. Patel, E. Honoré, F. Maingret, F. Lesage, M. Fink, F. Duprat and M. Lazdunski, *EMBO J.*, 1998, **17**, 4283–4290.
35. R. G. Walker, A. T. Willingham and C. S. Zuker, *Science (80-.)*, 2000, **287**, 2229–2234.
36. K. Venkatachalam and C. Montell, *Annu. Rev. Biochem.*, 2007, **76**, 387–417.
37. K. Katanosaka, S. Takatsu, K. Mizumura, K. Naruse and Y. Katanosaka, *Sci. Rep.*, 2018, **8**, 1–10.
38. T. Mochizuki, T. Sokabe, I. Araki, K. Fujishita, K. Shibasaki, K. Uchida, K. Naruse, S. Koizumi, M. Takeda and M. Tominaga, *J. Biol. Chem.*, 2009, **284**, 21257–21264.
39. M. A. Corrigan, G. P. Johnson, E. Stavenschi, M. Riffault, M.-N. Labour and D. A. Hoey, *Sci. Rep.*, 2018, **8**, 3824.
40. B. Martinac, P. R. Rohde, A. R. Battle, E. Petrov, P. Pal, A. F. Foo, V. Vásquez, T. Huynh and A. Kloda, *Methods mol. biol.*, 2010, **606**, 31–53.
41. D. E. Ingber, *FASEB J.*, 2006, **20**, 811–827.
42. O. P. Hamill and D. W. McBride, *Annu. Rev. Physiol.*, 1997, **59**, 621–631.
43. B. M. Gaub and D. J. Müller, *Nano Lett.*, 2017, **17**, 2064–2072.
44. C. Mazzochi, J. K. Bubien, P. R. Smith and D. J. Benos, *J. Biol. Chem.*, 2006, **281**, 6528–6538.
45. N. Strohmeyer, M. Bharadwaj, M. Costell, R. Fässler and D. J. Müller, *Nat. Mater.*, 2017, **16**, 1262–1270.
46. K. A. Jansen, P. Atherton and C. Ballestrem, *Semin. Cell Dev. Biol.*, 2017, **71**, 75–83.
47. R. S. Malik-Sheriff, S. Imtiaz, H. E. Grecco and E. Zamir, *Sci. Rep.*, 2018, **8**, 1–15.
48. H. Wolfenson, I. Lavelin and B. Geiger, *Dev. Cell*, 2013, **24**, 447–458.
49. H. B. Schiller and R. Fässler, *EMBO Rep.*, 2013, **14**, 509–519.
50. C. T. Mierke, P. Kollmannsberger, D. P. Zitterbart, J. Smith, B. Fabry and W. H. Goldmann, *Biophys. J.*, 2008, **94**, 661–670.
51. A. Rio, R. Perez-jimenez, R. Liu, P. Roca-cusachs, J. M. Fernandez and M. P. Sheetz, *Science (80-.)*, 2009, **323**, 638–641.

52. A. Carisey, R. Tsang, A. M. Greiner, N. Nijenhuis, N. Heath, A. Nazgiewicz, R. Kemkemer, B. Derby, J. Spatz and C. Ballestrem, *Curr. Biol.*, 2013, **23**, 271–281.
53. D. W. Zhou, T. T. Lee, S. Weng, J. Fu and A. J. García, *Mol. Biol. Cell*, 2017, **28**, 1901–1911.
54. M. Yao, B. T. Goult, B. Klapholz, X. Hu, C. P. Toseland, Y. Guo, P. Cong, M. P. Sheetz and J. Yan, *Nat. Commun.*, 2016, **7**, 1–11.
55. J. Golji and M. R. K. Mofrad, *Biophys. J.*, 2014, **107**, 1802–1809.
56. J. Neumann and K. E. Gottschalk, *Protein Eng., Des. Sel.*, 2016, **29**, 503–512.
57. K. Maki, N. Nakao and T. Adachi, *Biochem. Biophys. Res. Commun.*, 2017, **484**, 372–377.
58. J. D. Humphries, P. Wang, C. Streuli, B. Geiger, M. J. Humphries and C. Ballestrem, *J. Cell Biol.*, 2007, **179**, 1043–1057.
59. D. A. Fletcher and R. D. Mullins, *Nature*, 2010, **463**, 485–492.
60. S. Dupont, L. Morsut, M. Aragona, E. Enzo, S. Giulitti, M. Cordenonsi, F. Zanconato, J. Le Digabel, M. Forcato, S. Bicciato, N. Elvassore and S. Piccolo, *Nature*, 2011, **474**, 179–184.
61. B. Fabry, G. N. Maksym, J. P. Butler, M. Glogauer, D. Navajas and J. J. Fredberg, *Phys. Rev. Lett.*, 2001, **87**, 1–4.
62. B. Bugyi and M. F. Carlier, *Annu. Rev. Biophys.*, 2010, **39**, 449–470.
63. T. D. Pollard and G. G. Borisy, *Cell*, 2003, **112**, 453–465.
64. K. Ohashi, S. Fujiwara and K. Mizuno, *J. Biochem.*, 2017, **161**, 245–254.
65. S. Pellegrin and H. Mellor, *J. Cell Sci.*, 2007, **120**, 3491–3499.
66. A. M. Pasapera, I. C. Schneider, E. Rericha, D. D. Schlaepfer and C. M. Waterman, *J. Cell Biol.*, 2010, **188**, 877–890.
67. M. Yoshigi, L. M. Hoffman, C. C. Jensen, H. J. Yost and M. C. Beckerle, *J. Cell Biol.*, 2005, **171**, 209–215.
68. P. Hotulainen and P. Lappalainen, *J. Cell Biol.*, 2006, **173**, 383–394.
69. S. Tojkander, G. Gateva and P. Lappalainen, *J. Cell Sci.*, 2012, **125**, 1855–1864.
70. D. H. Kim, S. B. Khatau, Y. Feng, S. Walcott, S. X. Sun, G. D. Longmore and D. Wirtz, *Sci. Rep.*, 2012, **2**, 555.
71. Y. Jiu, J. Lehtimäki, S. Tojkander, F. Cheng, H. Jäälinoja, X. Liu, M. Varjosalo, J. E. Eriksson and P. Lappalainen, *Cell Rep.*, 2015, **11**, 1511–1518.
72. J. Zhang, W. H. Guo and Y. L. Wang, *Proc. Natl. Acad. Sci. U. S. A.*, 2014, **111**, 16383–16388.
73. J. Fink, N. Carpi, T. Betz, A. Bétard, M. Chebah, A. Azioune, M. Bornens, C. Sykes, L. Fetler, D. Cuvelier and M. Piel, *Nat. Cell Biol.*, 2011, **13**, 771–778.
74. N. T. Snider and M. B. Omary, *Nat. Rev. Mol. Cell Biol.*, 2014, **15**, 163–177.
75. L. Kreplak, H. Bär, J. F. Leterrier, H. Herrmann and U. Aebi, *J. Mol. Biol.*, 2005, **354**, 569–577.
76. M. G. Mendez, D. Restle and P. A. Janmey, *Biophys. J.*, 2014, **107**, 314–323.
77. M. E. Murray, M. G. Mendez and P. A. Janmey, *Mol. Biol. Cell*, 2014, **25**, 87–94.

78. Y. Fukata, K. Kaibuchi, M. Amano and K. Kaibuchi, *Trends Pharmacol. Sci.*, 2001, **22**, 32–39.
79. E. C. Lessey, C. Guilluy and K. Burridge, *Biochemistry*, 2012, **51**, 7420–7432.
80. S. Etienne-Manneville and A. Hall, *Nature*, 2002, **420**, 629–635.
81. F. Peng, B. Zhang, A. J. Ingram, B. Gao, Y. Zhang and J. C. Krepinsky, *Cell. Signal.*, 2010, **22**, 34–40.
82. K. Burridge and M. Chrzanowska-Wodnicka, *Annu. Rev. Cell Dev. Biol.*, 1996, **12**, 463–519.
83. K. Burridge, E. Monaghan-Benson and D. M. Graham, *Philos. Trans. R. Soc., B*, 2019, **374**, 20180229.
84. A. F. Palazzo, T. A. Cook, A. S. Alberts and G. G. Gundersen, *Nat. Cell Biol.*, 2001, **3**, 723–729.
85. M. Krendel, F. T. Zenke and G. M. Bokoch, *Nat. Cell Biol.*, 2002, **4**, 294–301.
86. Y. Jiu, J. Peränen, N. Schaible, F. Cheng, J. E. Eriksson, R. Krishnan and P. Lappalainen, *J. Cell Sci.*, 2017, **130**, 892–902.
87. H. Darenfed, B. Dayanandan, T. Zhang, S. H. K. Hsieh, A. E. Fournier and C. A. Mandato, *Cell Motil. Cytoskeleton*, 2007, **64**, 97–109.
88. S. Even-Ram, A. D. Doyle, M. A. Conti, K. Matsumoto, R. S. Adelstein and K. M. Yamada, *Nat. Cell Biol.*, 2007, **9**, 299–309.
89. K. N. Dahl and A. Kalinowski, *J. Cell Sci.*, 2011, **124**, 675–678.
90. S. Cho, J. Irianto and D. E. Discher, *J. Cell Biol.*, 2017, **216**, 305–315.
91. T. J. Kirby, *J. Appl. Physiol.*, 2019, **127**, 608–618.
92. M. Crisp, Q. Liu, K. Roux, J. B. Rattner, C. Shanahan, B. Burke, P. D. Stahl and D. Hodzic, *J. Cell Biol.*, 2006, **172**, 41–53.
93. G. R. Fedorchak, A. Kaminski and J. Lammerding, *Prog. Biophys. Mol. Biol.*, 2014, **115**, 76–92.
94. T. J. Kirby and J. Lammerding, *Nat. Cell Biol.*, 2018, **20**, 373–381.
95. A. Tajik, Y. Zhang, F. Wei, J. Sun, Q. Jia, W. Zhou, R. Singh, N. Khanna, A. S. Belmont and N. Wang, *Nat. Mater.*, 2016, **15**, 1287–1296.
96. B. Burke and K. J. Roux, *Dev. Cell*, 2009, **17**, 587–597.
97. H. N. Fridolfsson and D. A. Starr, *J. Cell Biol.*, 2010, **191**, 115–128.
98. P. T. Arsenovic, I. Ramachandran, K. Bathula, R. Zhu, J. D. Narang, N. A. Noll, C. A. Lemmon, G. G. Gundersen and D. E. Conway, *Biophys. J.*, 2016, **110**, 34–43.
99. A. Méjat and T. Misteli, *Nucleus*, 2010, **1**, 40–52.
100. B. A. Sosa, A. Rothballer, U. Kutay and T. U. Schwartz, *Cell*, 2012, **149**, 1035–1047.
101. Q. Liu, N. Pante, T. Misteli, M. Elsagga, M. Crisp, D. Hodzic, B. Burke and K. J. Roux, *J. Cell Biol.*, 2007, **178**, 785–798.
102. G. W. Gant Luxton, E. R. Gomes, E. S. Folker, H. J. Worman and G. G. Gundersen, *Nucleus*, 2011, **2**, 173–181.
103. F. Haque, D. J. Lloyd, D. T. Smallwood, C. L. Dent, C. M. Shanahan, A. M. Fry, R. C. Trembath and S. Shackleton, *Mol. Cell. Biol.*, 2006, **26**, 3738–3751.

104. M. Plessner, M. Melak, P. Chinchilla, C. Baarlink and R. Grosse, *J. Biol. Chem.*, 2015, **290**, 11209–11216.
105. A. Akhtar and S. M. Gasser, *Nat. Rev. Genet.*, 2007, **8**, 507–517.
106. S. Heessen and M. Fornerod, *EMBO Rep.*, 2007, **8**, 914–919.
107. J. Swift, I. L. Ivanovska, A. Buxboim, T. Harada, P. C. D. P. Dingal, J. Pinter, J. D. Pajerowski, K. R. Spinler, J.-W. Shin, M. Tewari, F. Rehfeldt, D. W. Speicher and D. E. Discher, *Science (80-.)*, 2013, **341**, 1240104.
108. A. Buxboim, J. Irianto, J. Swift, A. Athirasala, J. W. Shin, F. Rehfeldt and D. E. Discher, *Mol. Biol. Cell*, 2017, **28**, 3333–3348.
109. J. Lammerding, C. L. Stewart, R. T. Lee, J. Lammerding, P. C. Schulze, T. Takahashi, S. Kozlov, T. Sullivan, R. D. Kamm, C. L. Stewart and R. T. Lee, *J. Clin. Invest.*, 2004, **113**, 370–378.
110. C. Guilluy, L. D. Osborne, L. Van Landeghem, L. Sharek, R. Superfine, R. Garcia-Mata and K. Burridge, *Nat. Cell Biol.*, 2014, **16**, 376–381.
111. M. Lu, M. R. Muers and X. Lu, *Nat. Rev. Mol. Cell Biol.*, 2016, **17**, 523–532.
112. S. Orsulic, O. Huber, H. Aberle, S. Arnold and R. Kemler, *J. Cell Sci.*, 1999, **112**, 1237–1245.
113. V. M. Golubovskaya, R. Finch and W. G. Cance, *J. Biol. Chem.*, 2005, **280**, 25008–25021.
114. A. Serrels, T. Lund, B. Serrels, A. Byron, R. C. McPherson, A. Von Kriegsheim, L. Gómez-Cuadrado, M. Canel, M. Muir, J. E. Ring, E. Maniati, A. H. Sims, J. A. Pachter, V. G. Brunton, N. Gilbert, S. M. Anderton, R. J. B. Nibbs and M. C. Frame, *Cell*, 2015, **163**, 160–173.
115. F. B. Berry, M. A. O'Neill, M. Coca-Prados and M. A. Walter, *Mol. Cell. Biol.*, 2005, **25**, 1415–1424.
116. S. K. Archer, C. Claudianos and H. D. Campbell, *BioEssays*, 2005, **27**, 388–396.
117. Y. Cho, H. W. Lee, H. G. Kang, H. Y. Kim, S. J. Kim and K. H. Chun, *Oncotarget*, 2015, **6**, 8709–8721.
118. I. Okamoto, Y. Kawano, D. Murakami, T. Sasayama, N. Araki, T. Miki, A. J. Wong and H. Saya, *J. Cell Biol.*, 2001, **155**, 755–762.
119. T. Oka and M. Sudol, *Genes Cells*, 2009, **14**, 607–615.
120. J. Vrbský, V. Vinarský, A. R. Perestrelo, J. O. De La Cruz, F. Martino, A. Pompeiano, V. Izzi, O. Hlinomaz, V. Rotrekl, M. Sudol, S. Pagliari and G. Forte, *Genomics*, 2021, **113**, 1349–1365.
121. S. W. Plouffe, K. C. Lin, J. L. Moore, F. E. Tan, S. Ma, Z. Ye, Y. Qiu, B. Ren and K. L. Guan, *J. Biol. Chem.*, 2018, **293**, 11230–11240.
122. R. Fan, N. G. Kim and B. M. Gumbiner, *Proc. Natl. Acad. Sci. U. S. A.*, 2013, **110**, 2569–2574.
123. J. W. Haskins, D. X. Nguyen and D. F. Stern, *Sci. Signal.*, 2014, **7**, 1–11.
124. J. Oliver-De La Cruz, G. Nardone, J. Vrbsky, A. Pompeiano, A. R. Perestrelo, F. Capradossi, K. Melajová, P. Filipensky and G. Forte, *Biomaterials*, 2019, **205**, 64–80.

125. G. Nardone, J. Oliver-De La Cruz, J. Vrbsky, C. Martini, J. Pribyl, P. Skládal, M. Pešl, G. Caluori, S. Pagliari, F. Martino, Z. Maceckova, M. Hajduch, A. Sanz-Garcia, N. M. Pugno, G. B. Stokin and G. Forte, *Nat. Commun.*, 2017, **8**, 15321.
126. D. Mosqueira, S. Pagliari, K. Uto, M. Ebara, S. Romanazzo, C. Escobedo-Lucea, J. Nakanishi, A. Taniguchi, O. Franzese, P. Di Nardo, M. J. Goumans, E. Traversa, P. Pinto-do-Ó, T. Aoyagi and G. Forte, *ACS Nano*, 2014, **8**, 2033–2047.
127. A. R. Perestrelo, A. C. Silva, J. Oliver-De La Cruz, F. Martino, V. Horváth, G. Caluori, O. Polanský, V. Vinarský, G. Azzato, G. de Marco, V. Žampachová, P. Skládal, S. Pagliari, A. Rainer, P. Pinto-do-Ó, A. Caravella, K. Koci, D. S. Nascimento and G. Forte, *Circ. Res.*, 2021, **128**, 24–38.
128. J.-S. Mo, F.-X. Yu, R. Gong, J. H. Brown and K.-L. Guan, *Genes Dev.*, 2012, **26**, 2138–2143.
129. A. Elosegui-Artola, I. Andreu, A. E. M. Beedle, A. Lezamiz, M. Uroz, A. J. Kosmalska, R. Oria, J. Z. Kechagia, P. Rico-Lastres, A. L. Le Roux, C. M. Shanahan, X. Trepat, D. Navajas, S. Garcia-Manyes and P. Roca-Cusachs, *Cell*, 2017, **171**, 1397–1410, e14.
130. T. P. Driscoll, B. D. Cosgrove, S. J. Heo, Z. E. Shurden and R. L. Mauck, *Biophys. J.*, 2015, **8**, 15321.
131. D. H. Kim and D. Wirtz, *Biomaterials*, 2015, **48**, 161–172.
132. S. Chakraborty, K. Njah, A. V. Pobbati, Y. B. Lim, A. Raju, M. Lakshmanan, V. Tergaonkar, C. T. Lim and W. Hong, *Cell Rep.*, 2017, **18**, 2464–2479.
133. H. Morrison, L. S. Sherman, J. Legg, F. Banine, C. Isacke, C. A. Haipek, D. H. Gutmann, H. Ponta and P. Herrlich, *Genes Dev.*, 2001, **15**, 968–980.
134. T. Ooki, N. Murata-Kamiya, A. Takahashi-Kanemitsu, W. Wu and M. Hatakeyama, *Dev. Cell*, 2019, **49**, 590–604, e9.
135. N. G. Kim, E. Koh, X. Chen and B. M. Gumbiner, *Proc. Natl. Acad. Sci. U. S. A.*, 2011, **108**, 11930–11935.
136. Y. Morikawa, M. Zhang, T. Heallen, J. Leach, G. Tao, Y. Xiao, Y. Bai, W. Li, J. T. Willerson and J. F. Martin, *J. Sci. Signal*, 2015, **375**, ra41.
137. S. Pagliari, V. Vinarsky, F. Martino, A. R. A. Perestrelo, J. Oliver De La Cruz, G. Caluori, J. Vrbsky, P. Mozetic, A. Pompeiano, A. Zancla, S. G. Ranjani, P. Skladal, D. Kytyr, Z. Zdrahal, G. Grassi, G. Forte, P. Mozetic, A. Pompeiano, A. Zancla, S. G. Ranjani, P. Skladal, D. Kytyr, Z. Zdráhal, G. Grassi, M. Sampaolesi, A. Rainer and G. Forte, *Cell Death Differ.*, 2020, 1–15.
138. C. Lin, E. Yao, K. Zhang, X. Jiang, S. Croll, K. Thompson-Peer and P. T. Chuang, *eLife*, 2017, **6**, e21130.
139. G. Caluori, J. Pribyl, M. Pesl, J. Oliver-De La Cruz, G. Nardone, P. Skladal and G. Forte, *Front. Physiol.*, 2018, **9**, 1121.
140. A. R. Perestrelo, A. C. Silva, J. Oliver-De La Cruz, F. Martino, V. Horváth, G. Caluori, O. Polanský, V. Vinarský, G. Azzato, G. de Marco, V. Žampachová, P. Skládal, S. Pagliari, A. Rainer, P. Pinto-do-Ó, A. Caravella, K. Koci, D. S. Nascimento and G. Forte, *Circ. Res.*, 2021, **128**, 24–38.

141. F. Calvo, N. Ege, A. Grande-Garcia, S. Hooper, R. P. Jenkins, S. I. Chaudhry, K. Harrington, P. Williamson, E. Moeendarbary, G. Charras and E. Sahai, *Nat. Cell Biol.*, 2013, **15**, 637–646.
142. F. D. Camargo, S. Gokhale, J. B. Johnnidis, D. Fu, G. W. Bell, R. Jaenisch and T. R. Brummelkamp, *Curr. Biol.*, 2007, **17**, 2054–2060.
143. T. Heallen, Y. Morikawa, J. Leach, G. Tao, J. T. Willerson, R. L. Johnson and J. F. Martin, *Development*, 2013, **140**, 4683–4690.
144. D. Mosqueira, S. Pagliari, K. Uto, M. Ebara, S. Romanazzo, C. Escobedo-Lucea, J. Nakanishi, A. Taniguchi, O. Franzese, P. Di Nardo, M. J. Goumans, E. Traversa, P. Pinto-do-Ó, T. Aoyagi and G. Forte, *ACS Nano*, 2014, **8**, 2033–2047.
145. S. J. Gilbert and E. J. Blain, in *Mechanobiology in Health and Disease*, Elsevier, 2018, pp. 99–126.
146. A. Scheuren, E. Wehrle, F. Flohr and R. Müller, *Biomech. Model. Mechanobiol.*, 2017, **16**, 2017–2034.
147. S. Huang and D. E. Ingber, *Cancer Cell*, 2005, **8**, 175–176.

CHAPTER 3

Mechanotransduction at the Cell Surface and Methods to Study Receptor Forces

RONG MA,[a] BRENDAN R. DEAL[a] AND KHALID SALAITA*[a,b]

[a] Department of Chemistry, Emory University, Atlanta, GA, USA; [b] Wallace H. Coulter Department of Biomedical Engineering, Georgia Institute of Technology and Emory University, Atlanta, GA, USA
*Email: k.salaita@emory.edu

3.1 Introduction

Mechanical forces play a critical role in modulating many cellular processes.[1] Once cells experience mechanical stimuli, including substrate rigidity, external forces, and endogenous forces, they dynamically transduce the mechanical input into biochemical signals. With these signals, cells adapt and respond to their microenvironments, as well as make decisions involving activation, migration, proliferation, differentiation, and apoptosis (Figure 3.1A).[2] These processes, known as mechanosensing and mechanotransduction, have been observed in various cells, including platelets, cancer cells, stem cells, and immune cells.[3] Receptors on the cell surface, including integrins, notch, T cell receptor (TCR), and B cell receptor (BCR), participate intensively in mechanically sensing the environment and guide the cell through decision-making processes.[4–7] This chapter will start with an introduction of cell receptor force transduction, specifically force interplay between integrins and T cell receptor and their ligands, as well as its role in the cellular response. Then we will introduce tools that have been

Biomaterials Science Series No. 12
Material-based Mechanobiology
Edited by Jun Nakanishi and Koichiro Uto

Published by the Royal Society of Chemistry, www.rsc.org

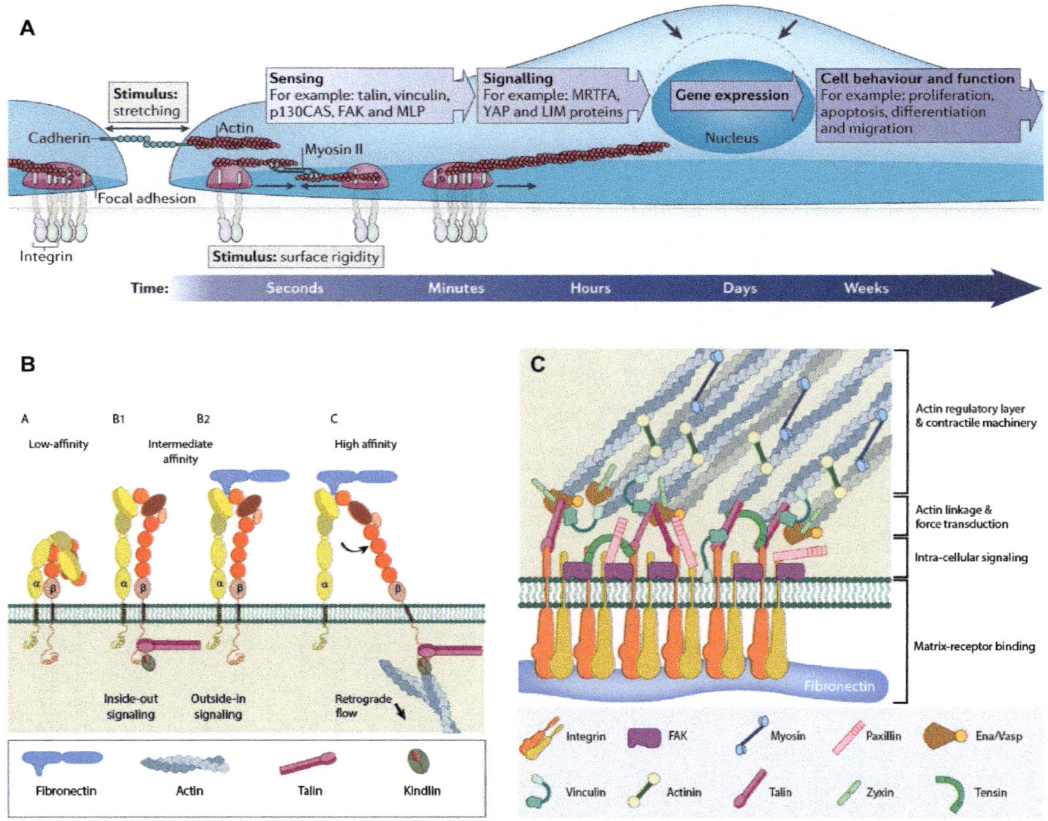

Figure 3.1 (A) Mechanotransduction modulates various cellular processes, integrin activation, and focal adhesion assembly. Reproduced from ref. 2 with permission from Springer Nature, Copyright 2014. (B) Integrin receptor in closed conformational A extending its extracellular domain to change into conformation B1. The cytoplasmic legs are further separated, resulting in another structural change from the extended closed conformation B1/2 to extended open conformation C, which has higher affinity binding to its ligand and facilitate mechanosensing and transduction. (C) Upon matrix-receptor binding, multiple proteins are recruited to assemble focal adhesions, which bridge the ECM with the cytoskeleton. Image provided courtesy of the Mechanobiology Institute, National University of Singapore [https://www.mechanobio.info/].

developed to study the receptor forces. Finally, this chapter will briefly discuss the future perspectives of using DNA-based tools to answer questions in the field of mechanobiology.

3.2 Mechanotransduction at the Cell Surface

3.2.1 Mechanotransduction Through Integrins

Integrins are adhesion proteins that bind to the extracellular matrix (ECM). An integrin molecule consists of an alpha and a beta subunit and forms a heterodimer. The integrin heterodimer has a large ectodomain, a transmembrane domain, and a short cytoplasmic tail. There are ~24 known integrin heterodimers mediating binding to diverse ECM molecules. One general ligand that these integrin ectodomains can recognize is a short amino acid sequence, arginine-glycine-aspartic acid (RGD motif), which is abundant in ECM proteins such as fibronectin and fibrinogen.[8] Studies suggest that integrins exist in different conformations that exhibit different ligand affinities. The integrins in the inactive bent form have lower affinity for their ligand, and the integrins in the active extended open form have high affinity (Figure 3.1B,C). The integrin constantly and rapidly changes between the different conformations to bind a ligand. Integrin signaling can be initiated by either proteins (talin and kindlin) binding on the cytoplasmic tail ("inside out") or ECM ligands binding to its ectodomain ("outside in"), and these two pathways can drive integrin activation cooperatively.[9] When activated and bound to its ligand, integrin molecules are fully extended, with the cytoplasmic tails separated. The activated integrins can cluster and recruit multiple proteins inside the cells, including talin, vinculin, and paxillin, to form focal adhesions (FA).[10] FAs are coupled to the cytoskeleton and can transmit forces around tens of piconewtons (pN) to sense the environment.[11,12] As focal adhesions grow and mature, force transduction activates pathways such as Rho kinase-mediated phosphorylation of myosin II, which can then stabilize large focal adhesions. These stable focal adhesions aid in actin bundle assembly and stress fiber formation.[13–16] The actin stress fibers coupled to the FAs can transduce higher forces through actomyosin contractions and further facilitate mechanical signaling.[17,18] Consequently, through formation of FAs, integrins act as a main hub for mechanosensing and mechanotransduction in various cell types and facilitate activities like cell adhesion and migration.

3.2.2 Mechanotransduction Through T Cell Receptor

3.2.2.1 General Overview of T Cell Activation

T cells are critical in the adaptive immune system, as they constantly search for antigens and provide surveillance against viral infections and cancer. The T cell receptor (TCR) is a surface receptor that scans and tests the antigens by

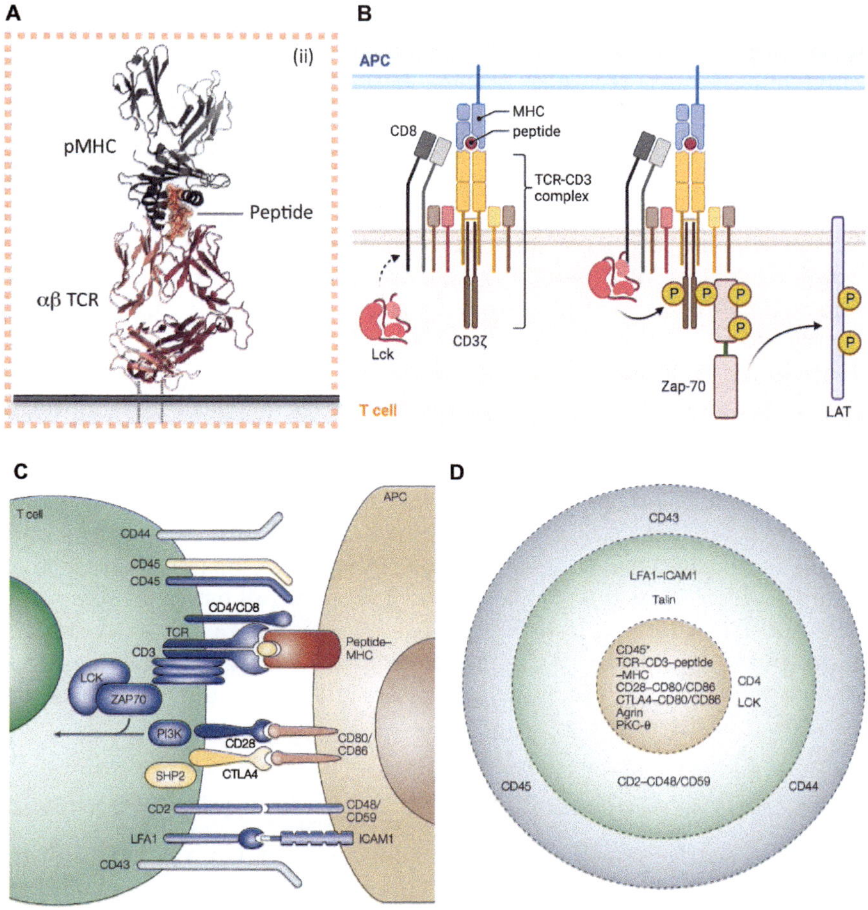

Figure 3.2 TCR–pMHC binding and its signaling pathway: (A) Structure of TCR binding to pMHC. Reproduced from ref. 19 with permission from Elsevier, Copyright 2018. (B) Signaling pathway after TCR triggering. (C) Side view showing the receptors and signaling proteins at the immune synapse formed between a T cell and an antigen presenting cell. (D) Top-down view showing the spatial organization of the proteins at the immune synapse. Reproduced from ref. 20 with permission from Springer Nature, Copyright 2003.

binding to them. TCR has high specificity for antigenic peptides that are presented by a major histocompatibility complex (pMHCs) on the surface of the antigen-presenting cell (Figure 3.2A), meaning it only recognizes antigenic peptides, exempting cells from autoimmune responses. TCR is a protein composed of a highly variable alpha chain and a beta chain, forming a heterodimer. TCR is usually expressed together with six invariant CD3 chain molecules that contain the signaling domains, forming a TCR–CD3 complex (Figure 3.2B). When it encounters a potent pMHC, TCR binds to the pMHC and initiates a discriminatory signaling cascade. The CD4 or CD8 coreceptors of TCR recruit Lck kinase, which can phosphorylate the signature tyrosine residues located at the signaling domain of the CD3 cytoplasmic tail, immunoreceptor tyrosine-based activation motifs (ITAMs). Phosphorylated ITAMs can bind to SH2 domain of the Zap70 kinase, and recruit Zap70 from the cytoplasm to the plasma membrane. Zap70 then phosphorylates linker for activation of T cells (LAT), which further recruits PLCγ1 and induces Ca^{2+} signaling. LAT also recruits adaptor proteins Grb2 and Gads, which bind to SOS and SLP-76 and lead to Ras, Rac, Rho GTPase activation.[19] This is an incredibly sensitive signaling cascade, which responds to an antigenic peptide within a few minutes after the first encounter with exceptional specificity.

Upon initial TCR triggering, the immune synapse (Figure 3.2C,D) is formed at the interface of the T cell and the antigen presenting cell (APC). The immune synapse is a highly organized dynamic structure, with a central region of the supramolecular activation complex (cSMAC), a peripheral region (pSMAC), and a distal ring (dSMAC). Together they form a "bull's-eye" pattern, with distinct spatiotemporal distribution of surface receptors and signaling molecules. The TCR–CD3 complex, as well as the CD28 coreceptor, is primarily located in the cSMAC upon activation. Adhesion ligands such as CD2 and LFA-1 (lymphocyte function-associated antigen 1) are relocated into the pSMAC, and the phosphatase CD45 is usually segregated to the dSMAC. Together, they facilitate T cells signaling and function, such as direct killing of the target cell.[20]

3.2.2.2 *Proposed Mechanism for Initial TCR–pMHC Recognition*

TCR binds to antigens while patrolling, and, upon antigen recognition, it can trigger a rapid whole cell response. This process can be initiated with as few as 1–2 cognate pMHC and within seconds to minutes. However, the affinity between TCRs and pMHCs in solution is usually between 1 μM and 10 μM, sometimes up to 100 μM, which does not provide much opportunity for antigen discrimination.[19–22] Therefore, a question that puzzles the field is how the TCR recognizes an antigen from the massive number of endogenous ligands that are displayed on antigen-presenting cell surfaces with explicit specificity and sensitivity.

Several models have been proposed to explain the remarkable ability of T cells to recognize specific antigens, including (1) kinetic segregation, (2) kinetic proofreading, (3) serial triggering, and (4) conformational change triggering model.[19,23] The kinetic segregation model proposed that the triggering is dependent on the segregation of large surface proteins like phosphatase CD45 from proteins that have small extracellular domains, such as TCR. The kinetic proofreading model suggests that the triggering is based on the differential binding durations, which provides the discrimination between strong and weak antigens. The serial engagement hypothesis suggested that successive rapid binding and unbinding events between clustered TCRs and a few pMHC is the mechanism of amplifying the discrimination signal that drives TCR triggering. The conformational change model argues that the binding to pMHC induces a structural change in the TCR ectodomain, which exposed the signaling ITAMs inside the cells.

3.2.2.3 Mechanical Forces in TCR Triggering

Originally, these models were proposed without the involvement of mechanical forces. However, over the past decade, growing evidence suggests that molecular mechanical forces play an important role in T cell activation. Single-molecule force spectroscopy studies found that when a defined force is applied to immunoreceptors, the dissociation kinetics between the receptors and ligands is regulated by force (Figure 3.3). Moreover, the T cell triggering correlates with the duration of the receptor–ligand binding under a certain force (~10–20 pN). Specifically, the potent TCR–pMHC bindings exhibit a "catch-bond" behavior, *i.e.* the application of force prolongs the lifetime of the binding. Conversely, with weak agonist or antagonist peptides, the TCR exhibits a "slip-bond" behavior, *i.e.* the lifetime decreases upon force application (Figure 3.3E).[24] The catch-bond model can explain the differential potency of pMHCs when their affinities to TCR are similar.

These findings suggest that immunoreceptors are mechanosensors. Evident from the forces revealed by traction force microscopy, micropillars, and molecular tension probes, the T cells generate traction forces and transmit defined pN forces through TCR to pMHC, and the force is correlated with functional output like cytotoxic cell killing.[24,25] These complementary methods have provided new insight into the TCR triggering mechanism and further connected the role of cytoskeleton coordination in the T cell activation. Accordingly, the revised hypotheses of triggering mechanisms now including the role of force are discussed in the field and are shown in Figure 3.4.[23] Briefly, for the kinetic segregation model, the forces generated by cell cytoskeleton and transmitted through TCR can drive the segregation of CD45. And for the kinetic proofreading model, the catch-bond observation with single-molecule methods can explain the amplified discrimination between potent and weak antigens in terms of its 2D binding kinetics. For the serial engagement model, the engagement of the cytoskeleton would drive the clustering and potentially the successive bindings through the protrusive microvilli on the T cell surface.[26] For the conformational change model, a study using optical tweezers showed that the force can extend the FG loop on the TCR beta chain, prolong the lifetime, and initiate signaling,[27] though whether force induces conformational change on the cytoplasmic tail of TCR–CD3 complex is still unclear.

Taken together, mechanical force is a crucial parameter that regulates TCR triggering. However, as one of the most sensitive, specific, and efficient recognition-activation processes in biology, TCR triggering is more likely to be explained by a collective effort rather than by a single mechanism. For example, the initial fast bindings could extend the FG loop, prolong the lifetime, press on the CD3ε, and induce a TCR–CD3 conformational change,

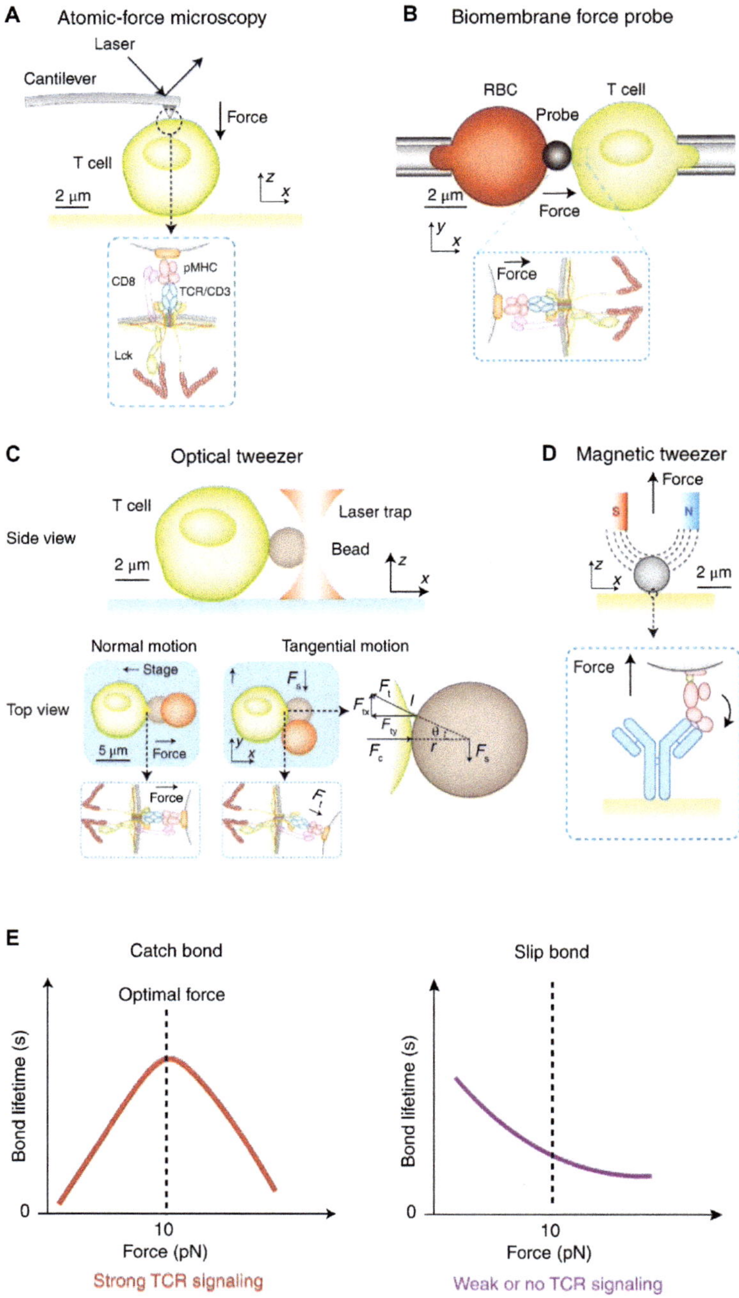

Figure 3.3 Single-molecule force spectroscopy methods to apply force to T cells: (A) Atomic force microscopy, (B) biomembrane force probe, (C) optical tweezer, and (D) magnetic tweezer are all used to apply an external force and interrogate the force contribution in TCR–pMHC binding. (E) Illustration of catch-bond and slip-bond. Reproduced from ref. 24 with permission from Springer Nature, Copyright 2019.

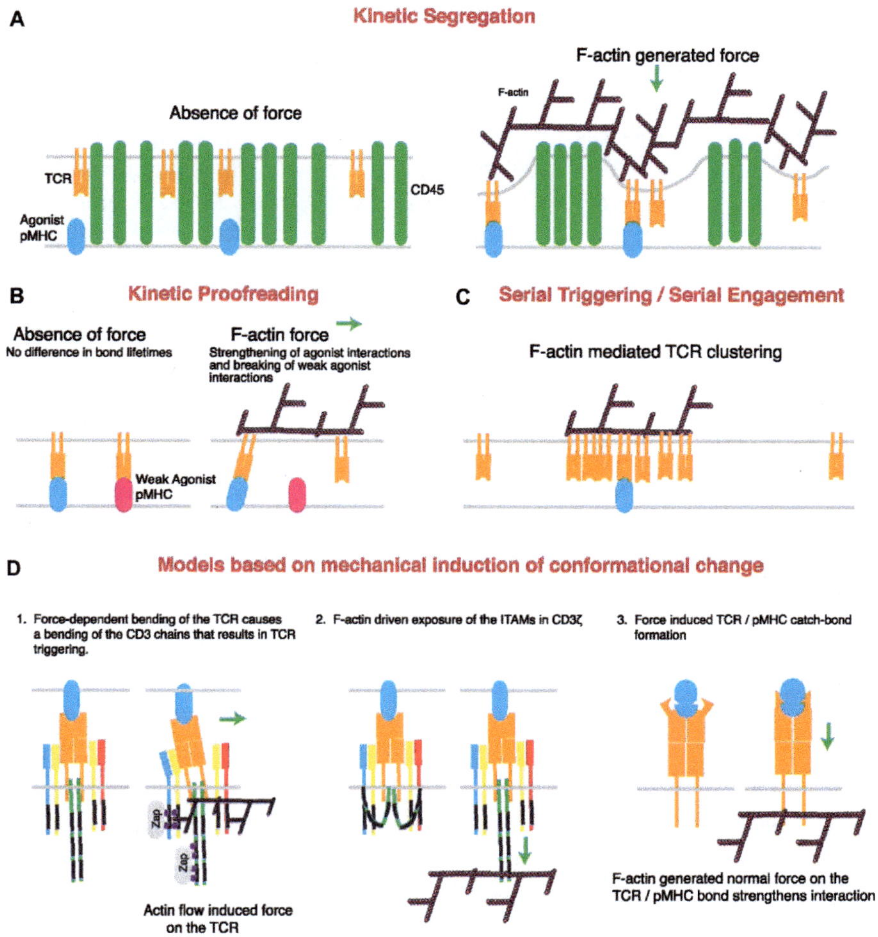

Figure 3.4 Revised models for TCR triggering with force and cytoskeleton engagement: (A) TCR force may contribute to the kinetic segregation of CD45. (B) Presence of force may contribute to the kinetic proofreading of the ligand by the TCR. (C) Cytoskeletal forces could drive TCR clustering and contribute to the serial engagement of TCR–antigen. (D) Forces transmitted through TCR could result in TCR conformational change, which may prolong the lifetime of TCR–antigen binding and facilitate phosphorylation. Reproduced from ref. 23, https://doi.org/10.3389/fimmu.2016.00068, under the terms of the CC BY 4.0 license https://creativecommons.org/licenses/by/4.0/.

which could further release the ITAMs and allow phosphorylation in the cytoplasmic tail. As the phosphorylation rapidly signals through Ca^{2+} to the cytoskeleton, the actin network could engage and mediate the rearrangement of the receptors, including the segregation of phosphatases and the clustering of TCRs, strengthen the mechanical test on the TCR–pMHCs, and contribute to a significantly amplified discrimination by exposing more phosphorylation sites.

3.3 Methods to Study Receptor Forces

Methods like traction force microscopy and micropillar arrays have been widely used in the field of mechanobiology to reveal these mechanical events during biological processes.[28,29] However, their relatively poor spatiotemporal resolution, as well as force sensitivity, has long been the bottleneck of uncovering more details of cell receptor mechanics (Figure 3.5).[30] The emergence of molecular tension probes in 2011 greatly enhanced the resolution of receptor force mapping. Our lab pioneered the use of entropic polyethylene glycol (PEG) as the force-sensing spring to report force in the range of 1–20 pN, but the "analog" force-responsive fluorescence signal (fluorescence intensity increases as force magnitude increases) could be confused with the percentage of opened probes in force mapping.[31,32] Conveniently, using nucleic acids for force sensing is advantageous, as it improves force mapping resolution and reports the "digital" force-responsive fluorescence signal (fluorescence switches on as duplex mechanically melts), simplifying data processing and interpretation.[33] Moreover, DNA offers unrivaled flexibility in multiple ways. First, the probe assembly is simple yet still allows for a tunable force threshold. Second, different chemical modifications can be incorporated easily for conjugation with a ligand, fluorophore, quencher, or surface. Third, the simple rules of Watson–Crick base pairing also allow the probes to be highly modular and programmable with other additional features.

3.3.1 DNA Mechanics

Because DNA is extensively used a mechanical probe, it is imperative to provide some background on the response of nucleic acids to external molecular forces. DNA is an elastic polymer, and its mechanical properties are often described

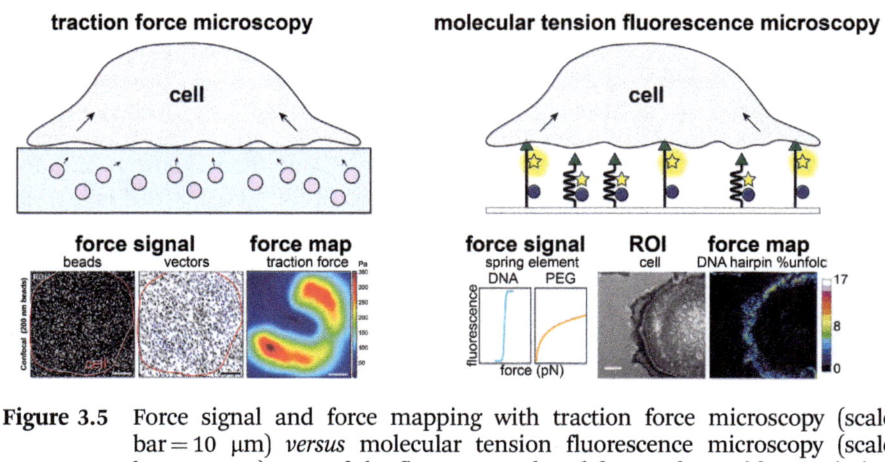

Figure 3.5 Force signal and force mapping with traction force microscopy (scale bar = 10 μm) *versus* molecular tension fluorescence microscopy (scale bar = 10 μm). Part of the figure, reproduced from ref. 29 with permission from American Chemical Society, Copyright 2020; part of the figure, reproduced from ref. 33 with permission from Springer Nature, Copyright 2014.

using the worm-like chain model (WLC) and specifically single-stranded DNA. The mechanical response of DNA under tensile force is usually characterized by single-molecule force spectroscopy methods, including optical tweezers, magnetic tweezers, and atomic force microscopy.[34,35] Various DNA structures have been examined under force, and among them, a few fundamental structural transitions have been intensively used for tool development in the field of DNA nanotechnology for mechanobiology studies.[36]

3.3.1.1 DNA Hairpin Unfolding under Force

For a simple DNA hairpin, its energy landscape can be simplified as two energy minima corresponding to the closed and open states that are separated by a single energy barrier. The application of stretching force "tilts" the energy landscape and shifts the free energy such that the closed state is destabilized, in contrast to the open state, which is stabilized. The external force that results in an equal probability of the open and closed states (free energy difference between the two states = zero) is described as the $F_{1/2}$ (Figure 3.6A). During the mechanical unfolding of the hairpin, the applied external force (work) must make up for the free energy of hybridization, as well as the free energy associated with stretching of the nucleotides.[35] Therefore, the $F_{1/2}$ can be defined as

$$F_{1/2} = (\Delta G_{\text{unfold}} + \Delta G_{\text{stretch}})/\Delta x \tag{3.1}$$

where ΔG_{unfold} is the free energy of hairpin unfolding at zero force, $\Delta G_{\text{stretch}}$ is the free energy of stretching the unfolded ssDNA at $F_{1/2}$, and Δx is the opening distance from folded to unfolded state. The $\Delta G_{\text{stretch}}$ can be calculated from a simple worm-like-chain (WLC) model,

$$\Delta G_{\text{stretch}} = \left(\frac{k_{\text{b}}T}{L_{\text{p}}}\right)\left[\frac{L_0}{4\left(\dfrac{1-x}{L_0}\right)}\right]\left[3\left(\frac{x}{L_0}\right)^2 - 2\left(\frac{x}{L_0}\right)^3\right], \tag{3.2}$$

where k_{b} is Boltzmann constant, L_{p} is the persistent length of DNA, T is temperature, L_0 is ssDNA contour length, and x is the hairpin extension. The Δx can be estimated from the contour length following 0.44 ± 0.02 nm per nt (with a 2.0 nm width of duplex DNA correction). The ΔG_{unfold} increases linearly with increasing stem GC% and stem length; however, increasing stem length also allows more energy to be stored, which affects $\Delta G_{\text{stretch}}$, and collectively defines the range of $F_{1/2}$ to within ~2–20 pN.[28]

3.3.1.2 DNA Duplex Rupture under Force

Unlike DNA hairpins, DNA duplex melting is irreversible. The same DNA duplex can have identical chemical and thermal stability but drastically

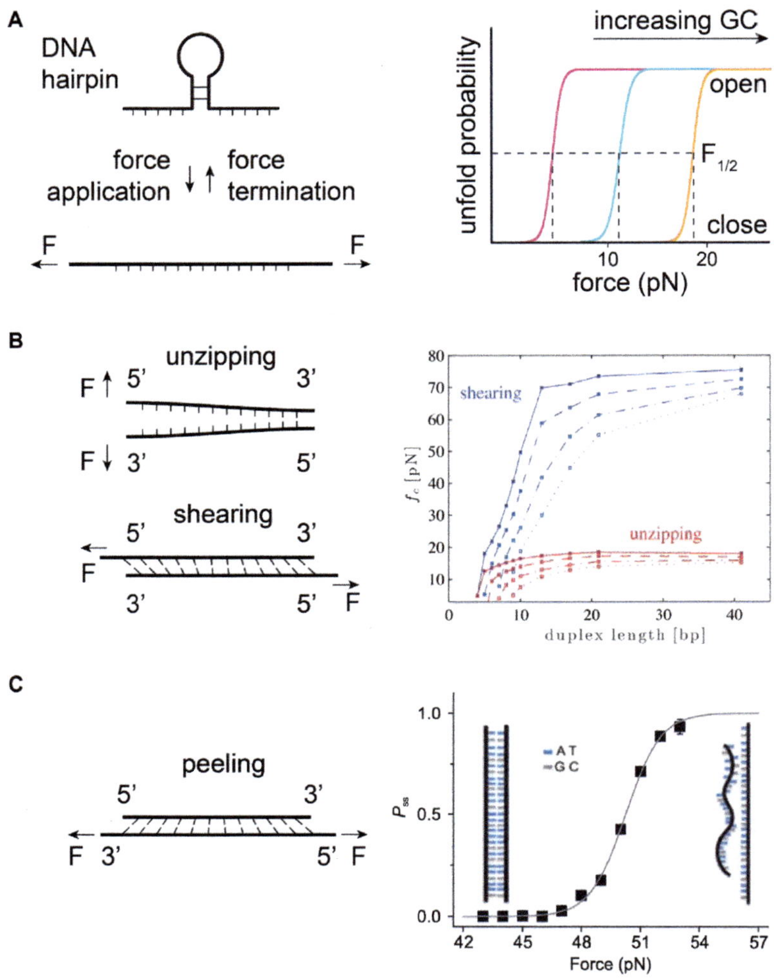

Figure 3.6 (A) Mechanical unfolding of DNA hairpin. (B) Mechanical rupture of DNA duplex. Part of the figure, reproduced from ref. 36 with permission from American Chemical Society, Copyright 2020. (C) Mechanical peeling of DNA duplex. Plot shows the transition and T_{tol} for a 28mer DNA duplex. Reproduced from ref. 41 with permission from Springer Nature, Copyright 2016.

different mechano-stability depending on the geometry of how force is applied (Figure 3.6B). When a short DNA duplex is stretched in an antiparallel manner (shearing, 5′–5′ or 3′–3′ pulling), the critical force at which 50% of the duplexes rupture (in a fixed time 2 s) increases with duplex length and plateaus at a certain length.[37] It can be described by the de Gennes model, which treats a DNA duplex as an elastic ladder and takes into account the hydrogen bond between base pairs. As the force is not evenly distributed on the base pairs, the base pairs that are bearing the load is a finite number, which results in a finite length x^{-1} that can be described with the spring constant of the stretched DNA backbone Q and the spring constant of the stretched hydrogen bonds between base pairs R, where

$$x^{-1} = \sqrt{Q/2R} \tag{3.3}$$

Therefore, in the de Gennes model, the rupture force for a DNA duplex in the shearing geometry can be described as

$$F = 2f_c\left[x^{-1}\tanh\left(x\frac{L}{2}\right) + 1\right] \tag{3.4}$$

where f_c is the critical force for separating a single base pair (~3.9 pN), $x^{-1} = 6.8$ bp by magnetic tweezer characterizations, and L is the number of DNA base pairs.[7,37]

Another geometry of short DNA duplex rupture, unzipping mode (5′–3′ pulling perpendicularly to the duplex), can be treated as a single bp shearing, while the remaining base pairs are only thermally stabilizing the duplex. According to eqn (3.4), the rupture force is ~12 pN, which is very close to what has been measured experimentally.[38]

3.3.1.3 DNA Duplex Peeling under Force

DNA duplexes can also melt under force that is applied on the same strand (5′–3′ pulling), through a force-driven strand separation process known as peeling. Magnetic tweezers characterization showed that for short DNA duplexes under 5′–3′ pulling on the same strand, the melting is a two-state system with a transition state consisting of several base pairs (Figure 3.6C). The application of force contributes to the free energy needed for the DNA to transition from dsDNA to ssDNA. The critical force that induces this transition, T_{tol}, is defined as the critical force at which there is 50% ssDNA at a given time and is dependent on the GC%. For AT-rich short duplexes, the T_{tol} is usually below 65 pN, and peeling occurs before the DNA B-form to S-form transition.[39,40] Specifically, for a 24mer or a 28mer duplex in a magnetic tweezer study, the T_{tol} was found to be ~41 pN and ~50 pN for the 24mer and 28mer duplexes, respectively.[41]

3.3.2 DNA-based Molecular Force Sensing

Taking advantage of the mechanical melting properties of DNA molecules, DNA-based force probes have been invented to map and study the forces exerted from a receptor on the cell surface to its ligand. Typically, the force probes are comprised of a mechanical melting region, a fluorophore-quencher pair that reports mechanical melting events, and the ligand that the receptor of interest can recognize. These DNA-based molecular probes are often anchored on 2D surfaces to present the ligands to cells. After cells are plated onto these force probe-functionalized surfaces, the receptors recognize, bind, and exert forces to the ligands. As the forces are transmitted through the DNA construct, if the force is greater than the $F_{1/2}$ or T_{tol} of the DNA construct, mechanical melting will occur, resulting in a change in the distance between the fluorophore and quencher, yielding a fluorescent signal to report this mechanical melting event. These probes can be reversible or irreversible depending on the structure of the DNA force-sensing region (Figure 3.7).

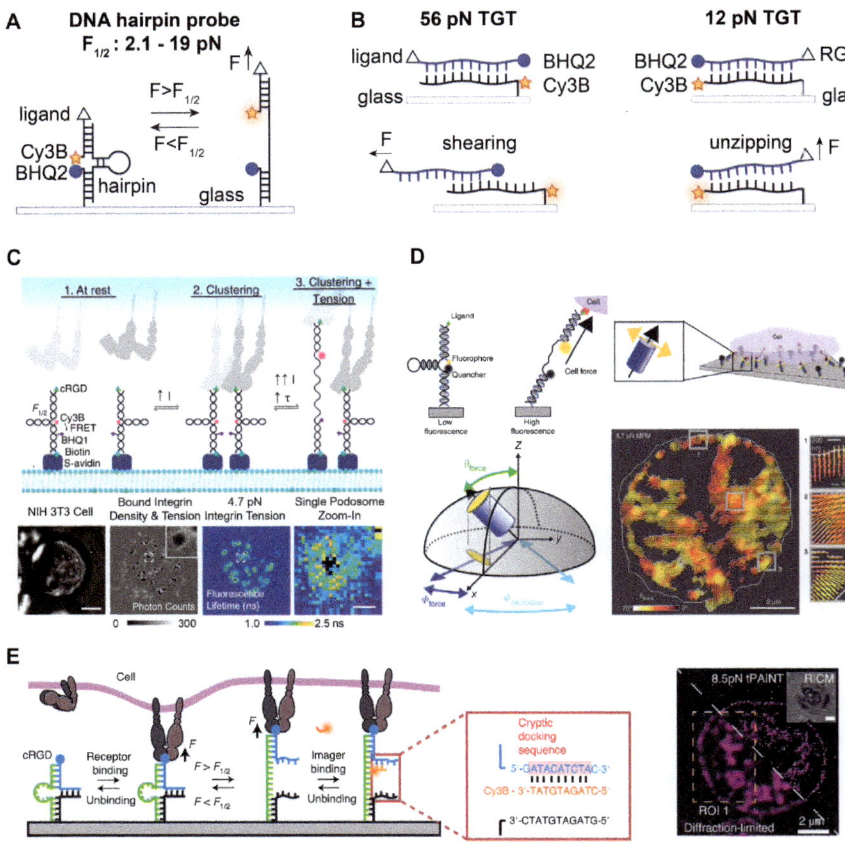

Figure 3.7 Classic reversible and irreversible DNA-based tension probes: (A) Reversible DNA-based hairpin tension probe with turn-on fluorescence upon hairpin unfolding. (B) Irreversible DNA-based tension gauge tether with turn-on fluorescence upon duplex rupture in different geometries. (C) Fluorescent lifetime imaging enables mapping of receptor clustering and force signal. Reproduced from ref. 49 with permission from Springer Nature, Copyright 2019. (D) Molecular force microscopy (MFM) with DNA hairpin probes reports the integrin force orientation. Reproduced from ref. 52 with permission from Springer Nature, Copyright 2018. (E) Live-cell super-resolved PAINT imaging of piconewton cellular traction forces. Reproduced from ref. 54 with permission from Springer Nature, Copyright 2020.

3.3.2.1 Reversible DNA Force Probes

DNA hairpin structures are incorporated as the force-sensing module due to their dynamic ability to unfold and refold with or without the presence of the force. DNA hairpin tension probes report the forces exceeding the threshold, $F_{1/2}$. As $F_{1/2}$ is dependent on the ΔG_{unfold} and $\Delta G_{\text{stretch}}$ as discussed in Section 3.3.1.1, one can easily tune the force threshold that the probes map by tuning the GC% and length of the hairpin stem. The first DNA hairpin tension probes were reported in 2014 by two groups independently. Chen and colleagues reported an all-covalent system by using a single-strand DNA with a stem-loop region, with modifications of a fluorophore, a quencher, and a peptide ligand RGD. With this molecular tension probe, they mapped the heterogeneous integrin forces in fibroblast focal adhesions.[42] Meanwhile, the Salaita lab pioneered a highly modular DNA hairpin tension probe system, comprised of three strands, a top strand with a ligand and a fluorophore, a bottom strand with a quencher and an anchoring moiety, and a hairpin strand that links them together. This multistrand design provided more flexibility for force mapping as opposed to the single-strand design, as one can easily change the force threshold of the probe by swapping the hairpin strand, which does not require any additional chemical modifications. One benefit of using a reversible force probe is that it provides low force threshold mapping. Our lab reported a small library of DNA hairpin tension probes ranging from 2.1 pN to 19 pN.[5,33] While providing great temporal resolution of the force dynamics, one disadvantage of the hairpin probes is that it only reports forces in real time. The force history information is difficult to capture because it refolds within microseconds once force is withdrawn. Another limitation is that the narrow force detection limit is not sufficient for receptors that can produce stronger forces. The latest generation of DNA tension probes now address these limitations and have expanded the capabilities of DNA tension probe technology.

3.3.2.2 Irreversible DNA Tension Probes

Irreversible DNA tension probes fail to provide temporal resolution of the force dynamics; however, they provide information on the receptor force history. Ha and coworkers first reported manipulation of the tension exerted through receptor–ligand binding with tension gauge tethers (TGT). The two geometries, unzipping and shearing, detect and tolerate peak tension of 12 pN and 56 pN, respectively.[7] Compared to hairpin probes, TGTs have a bigger range of force detection, which is tuned by positioning the ligand at different positions throughout the length of the duplex.[43] As a receptor mechanotransduction manipulating tool, the TGT is well suited, since the chemical cues it provides to cells are the same, while the mechanical cues are very different. However, as a force reporter, TGTs have two major drawbacks. First, the lowest T_{tol} for TGTs is 12 pN. Though there have been attempts to reach a lower force regime using a single-strand DNA wrapping around the protein SSB, which achieved ~ 4 pN of T_{tol};[44] the other drawback remains: Once TGTs are ruptured, the mechanical signaling is terminated as well, inevitably affecting the function of the cell.

3.3.2.3 Other Variations of the Probes to Answer Specific Mechanobiology Questions

Building on the basic structure of DNA hairpins and TGTs, more DNA structures have been designed to investigate different biophysical questions at the cell–substrate interface. Wang and coworkers further developed the TGT library by positioning the ligand on different bases on the top strand, incorporating fluorophore-quencher pairs, and assembling multiplexing substrates, which further calibrated the integrin forces inside and outside FAs to a narrower rage.[43,45] To amplify the receptor force readout, a mechanically triggered isothermal polymerization reaction based on the TGT structure was reported, which has potential in evaluating mechanomodulatory drugs.[46] Additionally, DNA origami structures with multivalent hairpin force probes were reported, capitalizing on the ability of this self-assembly technique to present multiple ligands with precisely controlled spacing, proving to be a great tool to study the relationship between the force and clustering.[47] This method would also allow for investigating the cross-talk between different receptors during mechanotransduction.

To study forces on membranes, DNA force probes were also redesigned to be compatible with supported lipid bilayers (SLB). SLB is a fluid system, which means an increase in intensity cannot be directly attributed to mechanical pulling by the receptors. To solve this issue, two strategies have been reported to distinguish receptor clustering from receptor forces. In 2016, a ratiometric probe was published, which can report TCR clustering and TCR forces simultaneously on a fluid membrane.[48] In 2019, a fluorescent lifetime imaging method was published that was able to distinguish the receptor clustering and force signals (Figure 3.7C), in addition to investigating the force during podosome formation on a fluid membrane.[49] While FLIM can be a useful solution for mechanobiology studies in fluid systems, it also has a relatively poor temporal resolution (~60 s), which limited its application to less dynamic receptor forces. Apart from using SLBs to mimic the cell–cell interface, there are also attempts of mapping the force at the cell–cell junction. In 2017, a membrane DNA tension probe was tested at the cell–cell junction. Instead of being anchored on a 2D substrate, the DNA tension probe was anchored to the membrane of a cell through two cholesterol modifications *via* hydrophobic interaction.[50] It is, however, important to be aware of the stability of cholesterol probe anchoring, which could potentially cause false negative signals. Cholesterol can also become involved in cellular uptake, also potentially causing false signals.

To improve the receptor force signal produced by immune cells, which is generally difficult to detect, a mechanical information-storing DNA tension probe based on mechanically selective hybridization was developed.[51]

The ability of this probe to toggle between measuring real-time tension and accumulative tension made it versatile, allowing for applications in both static and migratory cells. The application in T cells revealed that they transmit >4.7 pN TCR forces to altered peptide ligands (APL) and that the mechanical sampling of APLs is correlated with their potency. In addition, it revealed that in activated T cells, PD1 transmits >4.7 pN forces to its ligand. This tension probe is among the most sensitive molecular force detection methods, ideal for visualizing weak and transient forces, as well as forces that are transmitted through low-density receptors; however, that the forces accumulated and amplified with this method are limited to 4.7–19 pN.

In addition to mapping the magnitude of the receptor forces and percentage opening of the probes, measuring the force orientation provides direct proof to support the hypotheses of whether the force orientation affects receptor mechanotransduction. To measure the receptor force orientation at the molecular level, molecular force microscopy (MFM) was invented by coupling fluorescence polarization microscopy with DNA hairpin probes.[52] Taking advantage of the stacking between the dye Cy3B and the DNA base pairs, MFM successfully reported the axial integrin force alignment during platelets activation, supporting a hypothesis of the involvement of lateral forces during platelets activation Figure 3.7D. However, one drawback of MFM is that the image acquisition is time-consuming (~3.6 s), which makes it more suitable to measure the orientation of relatively less dynamic forces.

As more and more of the mysteries of receptor mechanics and its role in cell biology have been solved, recent research has been pushing imaging resolution to the nanometer level for force mapping. Wang and coworkers reported a cellular force nanoscopy (CFN) method in 2020, achieving 50 nm resolution in integrin force mapping.[53] This method utilizes a modified version of a conventional TGT, comprised of a DNase-resistant PNA/DNA duplex rather than a DNA/DNA duplex. The TGT is labeled with a fluorophore–quencher pair (Cy5–BHQ2); thus the Cy5 is dequenched once force is applied. This technique is straightforward and easy to implement with a TIRF microscope by following these steps: First, the dequenched Cy5 is bleached in 0.5–1 s; second, newly ruptured TGT molecules are imaged in the next frame; and third, the just imaged TGTs are bleached. By repeating this cycle, 50 nm resolution integrin force mapping is achieved in both migratory and stationary cells. One pitfall of this technique is that around 3% of the signal is false positive signal, coming from the dissociation of the TGTs during image acquisition. Another recent paper describes a relatively more complicated but powerful method to achieve super-resolved tension mapping. Leveraging DNA-points accumulation for imaging in nanoscale topography (DNA-PAINT) technique, the resolution is further improved to 25 nm.[54] The authors presented a real-time DNA tension probe featuring a strain-free force-sensing region with a cryptic docking site that is advantageous for a complementary imager strand to form a transient binding interaction (Figure 3.7E). By imaging the imager sampling of the mechanically opened probes, a super-resolved tension map can be reconstructed. This method is also adaptable to the TGTs for mapping accumulative tension. Similar to CFN, one of the limitations of tPAINT is the temporal resolution. The other disadvantage is that tPAINT undersamples the mechanical events due to the imager on-rate. However, it is still superior given that it provides super-resolved tension mapping in both real time and the accumulative history.

3.4 Conclusion and Future Perspectives

In this chapter, we discussed receptor mechanics and the following signal transduction inside the cell using integrins and TCR as examples. The common methods of studying mechanobiology were briefly introduced, and the application of DNA as a force-sensing material was highlighted as a new direction in molecular mechanobiology. The force-sensing mechanisms of different basic DNA structures were discussed, followed by the introduction of powerful DNA-based tools to investigate the receptor mechanics from detecting the molecular receptor force to manipulating mechanical signaling.

Despite the encouraging advances that have been made, new techniques are needed to depict the receptor mechanics and its regulation. For example, mapping the cell mechanics inside and outside cells simultaneously can be powerful in elucidating how force is transmitted throughout different components within the cellular machinery. Coupling the DNA tension probes with protein-based FRET probes or measurement of actin retrograde flow is a good starting point for such investigations. Other interesting topics to explore further include the mechanical forces of virus particle entry, mechanics of engineered CAR-T cells, cell nuclear mechanosensing, differences in mechanics in healthy *versus* diseased cells, *etc.* To summarize, elucidating the relationship between tension and function needs extensive work, and there are many uncharted territories in the field of mechanobiology.

In addition to the mechanobiological questions, there are also some future directions for advancing DNA-based force detection. For example, nuclease-resistant DNA tension probes will allow force mapping for cells in culture, which will enable tracking of the mechanical footprint of stem cell differentiation. Moreover, integrating tension probes into 3D matrices or cell–cell junctions is critical, as it better mimics the biological environment than a 2D planar surface. The next generation of tools based on DNA nanotechnology that are used to study cell mechanics should also be tried to deconvolute the effects of force and work when evaluating the cell mechanics. Likewise, another domain that is forgotten is time. Specifically, interesting questions can be asked to further characterize the receptor mechanics in regard to the loading rate of the force, as well as the output power of a single receptor. Furthermore, tools that can detect molecular pushing or protrusive forces are equally important, as the cells not only mechanically sense the environment but also respond to it, especially at the cell–cell junction. We envision that a force-sensing/manipulating tool box based on DNA nanotechnology will greatly aid the field of mechanobiology as it continues to assist research on cell mechanics from many different angles.

Acknowledgements

K.S. would like to acknowledge support from the National Institutes of Health (NIH) through R01GM124472 and 1R01GM131099, as well as the National Science Foundation (NSF) through CHE 2004126.

References

1. D. Mohammed, M. Versaevel, C. Bruyere, L. Alaimo, M. Luciano, E. Vercruysse, A. Proces and S. Gabriele, *Front. Bioeng. Biotechnol.*, 2019, **7**, 162.
2. T. Iskratsch, H. Wolfenson and M. P. Sheetz, *Nat. Rev. Mol. Cell Biol.*, 2014, **15**, 825.
3. Y. Chen, L. Ju, M. Rushdi, C. Ge and C. Zhu, *Mol. Biol. Cell*, 2017, **28**, 3134.
4. Y. Zhang, Y. Qiu, A. T. Blanchard, Y. Chang, J. M. Brockman, V. P.-Y. Ma, W. A. Lam and K. Salaita, *Proc. Natl. Acad. Sci. U. S. A.*, 2018, **115**, 325.
5. Y. Liu, L. Blanchfield, V. P.-Y. Ma, R. Andargachew, K. Galior, Z. Liu, B. Evavold and K. Salaita, *Proc. Natl. Acad. Sci. U. S. A.*, 2016, **113**, 5610.
6. K. M. Spillane and P. Tolar, *J. Cell Biol.*, 2017, **216**, 217.
7. X. Wang and T. Ha, *Science*, 2013, **340**, 991.
8. Z. Sun, M. Costell and R. Fassler, *Nat. Cell Biol.*, 2019, **21**, 25.
9. A. Banno and M. H. Ginsberg, *Biochem. Soc. Trans.*, 2008, **36**, 229.
10. D. A. Calderwood, I. D. Campbell and D. R. Critchley, *Nat. Rev. Mol. Cell Biol.*, 2013, **14**, 503.
11. L. B. Case and C. M. Waterman, *Nat. Cell Biol.*, 2015, **17**, 955.
12. F. Martino, A. R. Perestrelo, V. Vinarsky, S. Pagliari and G. Forte, *Front. Physiol.*, 2018, **9**, 824.
13. K. A. Beningo, K. Hamao, M. Dembo, Y. L. Wang and H. Hosoya, *Arch. Biochem. Biophys.*, 2006, **456**, 224.
14. C. Le Clainche and M. F. Carlier, *Physiol. Rev.*, 2008, **88**, 489.
15. J. Stricker, Y. Aratyn-Schaus, P. W. Oakes and M. L. Gardel, *Biophys. J.*, 2011, **100**, 2883.
16. P. W. Oakes, Y. Beckham, J. Stricker and M. L. Gardel, *J. Cell Biol.*, 2012, **196**, 363.
17. J. I. Lehtimaki, E. K. Rajakyla, S. Tojkander and P. Lappalainen, *eLife*, 2021, 10.
18. E. Kassianidou, C. A. Brand, U. S. Schwarz and S. Kumar, *Proc. Natl. Acad. Sci. U. S. A.*, 2017, **114**, 2622.
19. A. H. Courtney, W. L. Lo and A. Weiss, *Trends Biochem. Sci.*, 2018, **43**, 108.
20. J. B. Huppa and M. M. Davis, *Nat. Rev. Immunol.*, 2003, **3**, 973.

21. S. Valitutti, S. Muller, M. Cella, E. Padovan and A. Lanzavecchia, *Nature*, 1995, **375**, 148.
22. J. Hellmeier, R. Platzer, A. S. Eklund, T. Schlichthaerle, A. Karner, V. Motsch, M. C. Schneider, E. Kurz, V. Bamieh, M. Brameshuber, J. Preiner, R. Jungmann, H. Stockinger, G. J. Schutz, J. B. Huppa and E. Sevcsik, *Proc. Natl. Acad. Sci. U. S. A.*, 2021, **118**, e2016857118.
23. W. A. Comrie and J. K. Burkhardt, *Front. Immunol.*, 2016, **7**, 68.
24. C. Zhu, W. Chen, J. Lou, W. Rittase and K. Li, *Nat. Immunol.*, 2019, **20**, 1269.
25. R. Basu, B. M. Whitlock, J. Husson, A. Le Floc'h, W. Jin, A. Oyler-Yaniv, F. Dotiwala, G. Giannone, C. Hivroz, N. Biais, J. Lieberman, L. C. Kam and M. Huse, *Cell*, 2016, **165**, 100.
26. E. Cai, K. Marchuk, P. Beemiller, C. Beppler, M. G. Rubashkin, V. M. Weaver, A. Gerard, T.-L. Liu, B.-C. Chen and E. Betzig, *Science*, 2017, **356**, eaal3118.
27. Y. Feng, E. L. Reinherz and M. J. Lang, *Trends Immunol.*, 2018, **39**, 596.
28. V. P. Y. Ma and K. Salaita, *Small*, 2019, **15**, 1900961.
29. A. Stubb, R. F. Laine, M. Miihkinen, H. Hamidi, C. Guzman, R. Henriques, G. Jacquemet and J. Ivaska, *Nano Lett.*, 2020, **20**, 2230.
30. W. J. Polacheck and C. S. Chen, *Nat. Methods*, 2016, **13**, 415.
31. D. R. Stabley, C. Jurchenko, S. S. Marshall and K. S. Salaita, *Nat. Methods*, 2011, **9**, 64.
32. Y. Liu, K. Yehl, Y. Narui and K. Salaita, *J. Am. Chem. Soc.*, 2013, **135**, 5320.
33. Y. Zhang, C. Ge, C. Zhu and K. Salaita, *Nat. Commun.*, 2014, **5**, 5167.
34. S. Cocco, J. Yan, J. F. Leger, D. Chatenay and J. F. Marko, *Phys. Rev. E*, 2004, **70**, 011910.
35. M. T. Woodside, W. M. Behnke-Parks, K. Larizadeh, K. Travers, D. Herschlag and S. M. Block, *Proc. Natl. Acad. Sci. U. S. A.*, 2006, **103**, 6190.
36. M. Mosayebi, A. A. Louis, J. P. Doye and T. E. Ouldridge, *ACS Nano*, 2015, **9**, 11993.
37. K. Hatch, C. Danilowicz, V. Coljee and M. Prentiss, *Phys. Rev. E*, 2008, **78**, 011920.
38. S. Cocco, R. Monasson and J. F. Marko, *Proc. Natl. Acad. Sci. U. S. A.*, 2001, **98**, 8608.
39. X. Zhang, H. Chen, H. Fu, P. S. Doyle and J. Yan, *Proc. Natl. Acad. Sci. U. S. A.*, 2012, **109**, 8103.
40. X. Zhang, Y. Qu, H. Chen, I. Rouzina, S. Zhang, P. S. Doyle and J. Yan, *J. Am. Chem. Soc.*, 2014, **136**, 16073.
41. Z. J. Yang, G. H. Yuan, W. L. Zhai, J. Yan and H. Chen, *Sci. China Phys. Mech.*, 2016, **59**, 680013.
42. B. L. Blakely, C. E. Dumelin, B. Trappmann, L. M. McGregor, C. K. Choi, P. C. Anthony, V. K. Duesterberg, B. M. Baker, S. M. Block, D. R. Liu and C. S. Chen, *Nat. Methods*, 2014, **11**, 1229.
43. Y. Wang, D. N. LeVine, M. Gannon, Y. Zhao, A. Sarkar, B. Hoch and X. Wang, *Biosens. Bioelectron.*, 2018, **100**, 192.

44. F. Chowdhury, I. T. Li, T. T. Ngo, B. J. Leslie, B. C. Kim, J. E. Sokoloski, E. Weiland, X. Wang, Y. R. Chemla, T. M. Lohman and T. Ha, *Nano Lett.*, 2016, **16**, 3892.
45. Y. Wang and X. Wang, *Sci. Rep.*, 2016, **6**, 36959.
46. V. P. Ma, Y. Liu, K. Yehl, K. Galior, Y. Zhang and K. Salaita, *Angew. Chem., Int. Ed.*, 2016, **55**, 5488.
47. P. K. Dutta, Y. Zhang, A. T. Blanchard, C. Ge, M. Rushdi, K. Weiss, C. Zhu, Y. Ke and K. Salaita, *Nano Lett.*, 2018, **18**, 4803.
48. V. P. Ma, Y. Liu, L. Blanchfield, H. Su, B. D. Evavold and K. Salaita, *Nano Lett.*, 2016, **16**, 4552.
49. R. Glazier, J. M. Brockman, E. Bartle, A. L. Mattheyses, O. Destaing and K. Salaita, *Nat. Commun.*, 2019, **10**, 4507.
50. B. Zhao, C. O'Brien, A. Mudiyanselage, N. Li, Y. Bagheri, R. Wu, Y. Sun and M. You, *J. Am. Chem. Soc.*, 2017, **139**, 18182.
51. R. Ma, A. V. Kellner, V. P.-Y. Ma, H. Su, B. R. Deal, J. M. Brockman and K. Salaita, *Proc. Natl. Acad. Sci. U. S. A.*, 2019, **116**, 16949.
52. J. M. Brockman, A. T. Blanchard, V. M. Pui-Yan, W. D. Derricotte, Y. Zhang, M. E. Fay, W. A. Lam, F. A. Evangelista, A. L. Mattheyses and K. Salaita, *Nat. Methods*, 2018, **15**, 115.
53. Y. Zhao, K. Pal, Y. Tu and X. Wang, *J. Am. Chem. Soc.*, 2020, **142**, 6930.
54. J. M. Brockman, H. Su, A. T. Blanchard, Y. Duan, T. Meyer, M. E. Quach, R. Glazier, A. Bazrafshan, R. L. Bender, A. V. Kellner, H. Ogasawara, R. Ma, F. Schueder, B. G. Petrich, R. Jungmann, R. Li, A. L. Mattheyses, Y. Ke and K. Salaita, *Nat. Methods*, 2020, **17**, 1018.

CHAPTER 4

Measurement and Manipulation of Cellular Forces Using Silicone Elastomers

HIROAKI HIRATA*[a,b] AND MASAHIRO SOKABE[a,c]

[a] Mechanobiology Laboratory, Nagoya University Graduate School of Medicine, Nagoya 466-8550, Japan; [b] Department of Applied Bioscience, College of Bioscience and Chemistry, Kanazawa Institute of Technology, Hakusan 924-0838, Japan; [c] Human Information Systems Laboratory, Kanazawa Institute of Technology, Hakusan 924-0838, Japan
*Email: hirata@neptune.kanazawa-it.ac.jp

4.1 Introduction

Cells in our body suffer various kinds of mechanical stimuli from surroundings, such as shear stress from the blood or interstitial flow, tensile and compressive forces generated by neighboring cells/deformed tissues, and rigidity of the surrounding/underlying extracellular matrix. Mechanical forces are also generated inside the cell, which is mainly caused by cytoskeletal activities. A vast body of research in the field of mechanobiology has revealed that these mechanical inputs largely influence fundamental cellular functions including survival, morphogenesis, migration, proliferation, differentiation, and death.[1–3] To study mechanical cue-induced cellular responses and their underlying molecular mechanisms, manipulation and measurement of mechanical parameters at the cellular and molecular levels are essential. In the early days, Harris and coworkers observed that fibroblasts and epithelial cells cultured on a thin silicone film formed wrinkles in the film, which provided the first

Biomaterials Science Series No. 12
Material-based Mechanobiology
Edited by Jun Nakanishi and Koichiro Uto

Published by the Royal Society of Chemistry, www.rsc.org

evidence for contractile force generation in single nonmuscle cells, even though this observation was rather qualitative.[4] Afterward, various methods have been developed and applied for measurement and manipulation of mechanical forces acting on cells and subcellular organelles/molecules, which include traction force microscopy,[5] atomic force microscopy,[6,7] optical and magnetic tweezers,[8,9] and molecular tension sensors.[10] Together with the progress in imaging technologies for cells and molecules, these methodological developments have provided driving forces for research in the mechanobiology field.

As a major approach to develop novel tools for force measurement/manipulation at the cellular and subcellular levels, device fabrication using chemical polymers has been widely employed. Among the polymer materials, the silicone elastomer poly(dimethylsiloxane) (PDMS) has been used most commonly for cell biological and mechanobiological research, since PDMS has several desirable properties.[11,12] PDMS, which is chemically inert and biocompatible, is low cost and easy to fabricate with the high efficiency of mold pattern transfer. The transparent and low-autofluorescence nature of PDMS makes it suitable for optical and microscopic observations. Furthermore, PDMS is highly flexible and elastic, which is ideal for constructing devices for measurement and manipulation of the low magnitude of force that acts on cells and subcellular structures. In this chapter, we review silicone elastomer-based approaches for the measurement and manipulation of cellular and subcellular forces.

4.2 Mechanical Measurement Using Silicone Elastomers

4.2.1 Measurement of Force at Cell–ECM Adhesion

Cell adhesion to the extracellular matrix (ECM) is essential for the survival and growth of normal cells. At the site of cell–ECM adhesion, called focal adhesion, the integrin family of transmembrane receptors for ECM proteins are clustered, and the actin cytoskeleton is connected to the cytoplasmic side of the integrin cluster through various adaptor proteins.[13,14] Thus contractile force generated by the actomyosin motor activity in the actin cytoskeleton is transmitted to FA and ECM. The force exertion to ECM is pivotal to drive cell morphogenesis and migration.

4.2.1.1 Traction Force Microscopy

Exertion of cell-generated force to ECM was qualitatively visualized by Harris *et al.*,[4] as described in the Introduction. Quantitative measurement of the exerted force was first achieved by Dembo and Wang, who developed traction force microscopy (TFM).[5] TFM is based on the usage of elastic polyacrylamide hydrogel in which fluorescent beads are embedded (Figure 4.1A). The surface of the polyacrylamide gel is covalently conjugated with the ECM protein (typically collagen or fibronectin), and cells are seeded and grown on it. The

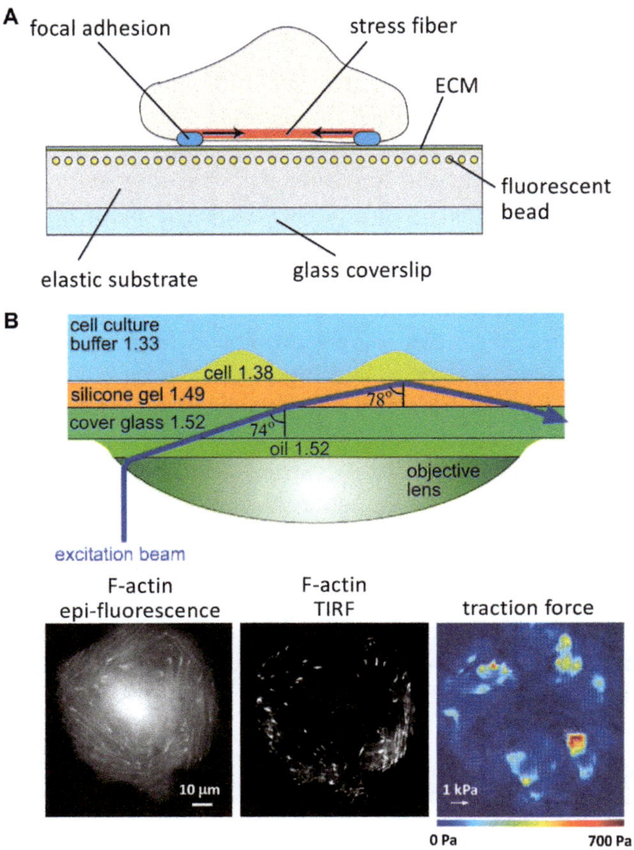

Figure 4.1 Traction force microscopy (TFM): (A) Schematic of a typical TFM platform. (B) Upper schematic: schematic diagram of silicone elastomer-based TFM combined with total internal reflection fluorescence (TIRF) microscopy. Lower panels: epifluorescence (left) and TIRF (middle) images of F-actin in a HUVEC cell expressing Lifeact-EGFP, and a vector map of traction forces in the same field (right). Reproduced from ref. 20, https://doi.org/10.1371/journal.pone.0023807, under the terms of the CC BY 4.0 license https://creativecommons.org/licenses/by/4.0/.

cell exerts traction force to the polyacrylamide gel, which causes distortion of the gel. The deformation field in the gel, which is obtained by tracking tracer beads in the gel, is then converted into the traction stress field by solving the inverse Boussinesq problem.[5,15] A prototype of TFM was constructed using a silicone film, which produced quantitative traction stress maps in fast-moving fish keratocytes.[16] However, the silicone film used in this prototype was too stiff to measure traction force exerted by mammalian cells. By contrast, stiffness of polyacrylamide gels can be easily tuned in the range of 0.1–100 kPa;[15] Young's modulus of the acrylamide gel used by Dembo and Wang was ∼6 kPa.[5] More recently, softer silicone elastomers (3–8 kPa) have been introduced into TFM, which enables the measurement of traction stresses in mammalian cells.[17,18] Compared with the acrylamide hydrogel, silicone elastomers have higher refractive indexes. Therefore, TFM using soft silicone elastomers can be combined with total internal reflection fluorescence (TIRF) observation of cells[19,20] (Figure 4.1B).

With this approach, detailed dynamics of actin cytoskeleton and focal adhesion molecules in the vicinity of the extracellular substrate surface have been observed simultaneously with monitoring time-dependent changes in traction force exertion.[19,21]

To extract the deformation field in the TFM substrate for calculation of traction forces, an image of the substrate under traction has to be compared with the reference image showing the undeformed substrate. In conventional TFM, the reference image is typically obtained by detaching cells from the substrate (*e.g.* by treating cells with trypsin) after taking an image under the cell traction stress. However, this "cell removal" step can be skipped by using micropatterned silicone elastomer as a TFM substrate. Silicone elastomers are well compatible with microfabrication, and micron- or submicron-scaled regular patterns can be created on the elastomer surface. Thus, given the elastomer stiffness that is measured separately, traction forces can be calculated from distortion of the micropattern on the substrate surface without removal of cells.[22]

TFM is one of the most successful and widely used techniques for cellular force measurement. This technology has brought numerous findings in cell mechanobiology, which includes the architectural basis for force generation and propagation in the actin cytoskeleton network,[23] the molecular friction mechanism for force transmission from the retrograding actin cytoskeleton to focal adhesions and ECM,[24,25] the role of traction force in focal adhesion development,[22] the existence of distinct classes of focal adhesions in traction force exertion,[26,27] and a close correlation between metastatic potential and traction force generation in cancer cells.[28] Especially, since stiffness of the TFM substrate can be easily altered, TFM has extensively contributed to unveiling molecular and biophysical mechanisms for rigidity-dependent regulations of actomyosin contractility and cell–ECM adhesions.[29–34]

4.2.1.2 Micropillar Arrays

TFM has been widely used and is still being subjected to technological advancements, such as improvement of spatial resolution and accuracy.[35,36] However, conversion of the deformation field into the traction stress field in an elastic continuum is an ill-posed problem, and its mathematical bases are not so simple and easy for biological researchers. To avert this difficulty in TFM, another class of methods for traction force measurement has been developed, wherein arrays of elastic micropillars (also called microposts) are employed.[37] When cells are grown on the micropillar arrays, which are typically made of PDMS, traction force exertion by cells causes deflection of micropillars (Figure 4.2). It is of note that each micropillar independently acts as a simple elastic cantilever. Thus local traction forces can be quantified just by measuring deflections of individual micropillars. To enable cells to adhere to the

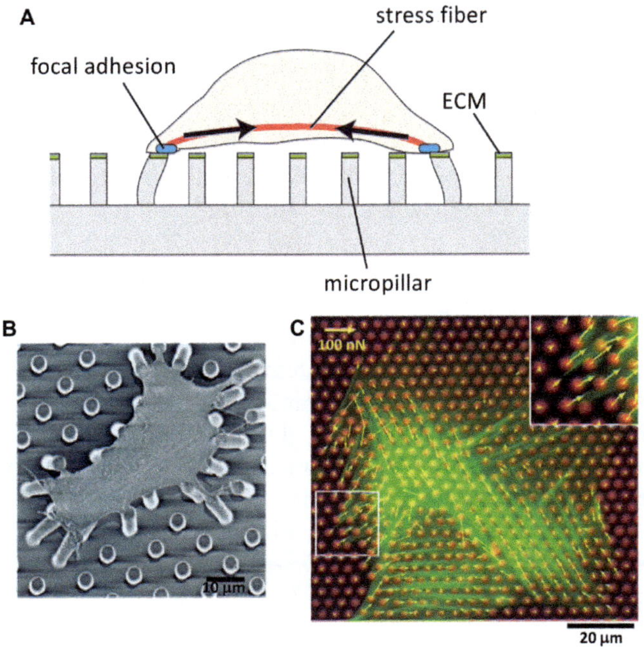

Figure 4.2 Micropillar arrays for traction force measurement: (A) Schematic of traction force measurement using micropillar arrays. (B) Scanning electron microscopy (SEM) image of a smooth muscle cell cultured on the array of micropillars whose tips were coated with fibronectin. (C) Force map (yellow arrows) and an epifluorescence image of F-actin (green) in a fibroblast cultured on the array of micropillars whose tips were coated with fluorescently labeled fibronectin (red). The inset shows a magnified image of the outlined region. B was reproduced from ref. 37 with permission from National Academy of Sciences, Copyright 2003. C was reproduced from ref. 41 with permission from John Wiley & Sons, Copyright © 2014 The Authors.

hydrophobic surface of PDMS micropillars, the micropillars have to be coated with ECM proteins. When the entire surface of the micropillar substrate is coated with ECM, cells tend to spread down the length of micropillars and adhere not only to micropillar tips but also to the side of micropillars and the surface of the substrate base,[37] which makes it difficult to evaluate actual traction forces. This problem can be overcome by introducing the microcontact printing technique, which allows selective coating of micropillar tips with ECM proteins.[37,38] Micropillar deflection caused by traction force exertion from cells is typically measured by either bright-field observation of the optical contrast at tips of micropillars[39] or fluorescence observation of micropillar tips that are coated with fluorescently labeled ECM proteins.[40] With these approaches using micropillars, traction forces exerted from individual focal adhesions and connected single stress fibers have been successfully quantified.[37,41]

There are two major geometrical parameters of micropillars: radius and height. The radius of micropillars largely affects the formation and organization of focal adhesions, as well as the magnitude and the direction of traction force.[42] The actual cue causing these cellular responses on micropillars with different radii is not clear because a difference in the pillar radius alters not only the pillar tip area available for focal adhesion formation but also the bending stiffness of the pillar. To discriminate the effect of the bending stiffness of micropillars on cell behaviors, micropillars with varied heights have been used; taller pillars have lower bending stiffness.[42–45] Thus, using micropillars with different heights, it has been revealed that traction force exertion at focal adhesions increases with the stiffness of micropillars, suggesting existence of a mechanical feedback mechanism for regulation of actomyosin contractility.[42–46]

Spatial anisotropy of extracellular substrate stiffness can be created by using elliptical micropillars. Such pillars have larger bending stiffness along the long axis of the pillar ellipse compared with that along the short axis. On micropillars with anisotropic stiffness, cells elongate and stress fibers orient along the stiffer axis, causing the generation of traction forces oriented along this axis, leading to cell migration along its length.[44,47] Importantly, these cellular behaviors were not observed when cells were seeded onto the glass surface on which the ECM pattern with the same elliptical shape was microprinted.[44,47] These results reveal that cells can detect local anisotropy of extracellular substrate stiffness at cell–ECM adhesion sites.

Micropillars with micron-scaled diameters (typically 1–2 μm) have been used for traction force measurement. In this case, a single micropillar is associated with one focal adhesion or a few, and therefore the traction force is measured at the single focal adhesion level. The spatial resolution of traction force measurement has been improved by using pillars with submicron diameters and submicron interpillar spacing. When cells were grown on such "nanopillars", individual focal adhesions were formed across multiple pillars.[42] Thus it became possible to measure the traction force distribution within a single focal adhesion.[42]

4.2.2 Measurement of Force at Cell–Cell Adhesion

Cells adhere not only to ECM but also to neighboring cells under the crowded condition. When cells are densely packed in a monolayer, cells should feel a laterally compressive force. At the same time, monolayered cells form several types of intercellular adhesion structures. Among them, adherens junctions provide adhesion sites to adjacent cells through the intercellular binding of the adhesion receptors, cadherins. On the cytoplasmic side of adherens junctions, actomyosin cables are connected to clustered cadherin through the actin-binding protein α-catenin, forming a complex with β-catenin that binds to the cytoplasmic domain of cadherin.[48] Therefore, adherens junctions are subjected to actomyosin-generated tensile force from neighboring cells. Direct measurement of force acting at the cell–cell boundary is not easy because, compared with cell–ECM adhesion sites, cell–cell adhesion sites are relatively difficult to externally access for force measurement. However, significant efforts have been made to achieve quantification of the force at cell–cell adhesion.

A simple approach for this direction is modifying the methods of traction force measurement at cell–ECM adhesion. The elastomer surface or the tips of micropillars are coated with recombinant ectodomains of cadherin molecules, instead of ECM proteins. When cells are sparsely seeded onto the surfaces coated with cadherin ectodomains, cells adhere to the surfaces by forming complexes of adherens junction components at the boundary between the cell and the substrate surface.[49] This "pseudo" adherens junction is connected to the actomyosin fiber, which is similar to the case of focal adhesion formed on the ECM substrate. Thus traction forces are exerted through the "pseudo" adherens junctions to the cadherin-coated substrate, which can be quantified by measuring the distortion of the elastomer substrate or deflection of micropillars.[49–51]

Force measurement on elastic surfaces coated with cadherin ectodomains is relatively simple and easy. However, the lack of cell adhesion to ECM in this assay causes various defects in cell behaviors such as integrin-mediated intracellular signaling and establishment of the apico-basal polarity. To avert such disadvantages of this assay, several methods for the quantitative evaluation of forces that act at the actual cell–cell boundary have been developed. These include measurement of the recoiling velocity of the cell–cell junction after laser ablation of the junction[52,53] and theoretical inference of tension at the cell–cell boundaries based on individual cell shapes in a cell monolayer.[54,55] However, these two approaches cannot derive the absolute value of force at the cell–cell boundary. By contrast, Liu *et al.* developed a method that allowed the measurement of the absolute value of force acting at the boundary of two contacting cells by using micropillars with the micro-printed ECM pattern[56] (Figure 4.3). When a single cell is confined in the area of a cell-sized ECM pattern printed onto micropillar arrays, the sum of traction force vectors exerted from this nonmoving cell should be zero. However, when two confined cells are in contact with each other, the sum of traction force vectors in each cell becomes not zero. This nonzero force in a cell corresponds to the force loaded from the contacting cell. With this principle, Liu *et al.* have revealed the quantitative relationship between the intercellular tugging force and the size of the cell–cell junction.[56] This strategy for quantification of cell–cell boundary force has been extended and applied to larger clusters of cells,[57,58] which has enabled the analysis of the role of intercellular mechanical interaction in collective dynamics of cells in a cell monolayer.[58–60]

4.2.3 Cell Stiffness Measurement Using Microfluidics

In addition to forces actively generated in cells, another mechanical property of cells, cellular deformability, is also highly related to cellular activities and functions such as cell motility, cell cycle progression, cell differentiation, and cancer cell malignancy.[61–65] While atomic force microscopy, magnetic tweezers, and micropipette aspiration have long been used for measuring the

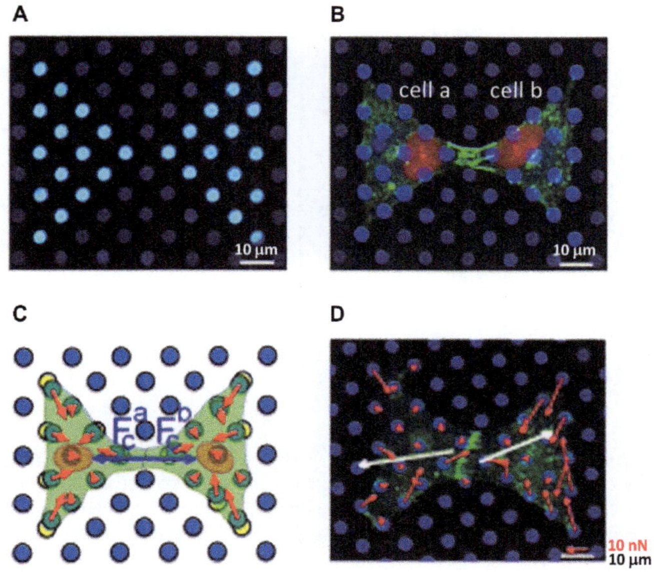

Figure 4.3 Measurement of force acting at the cell–cell boundary: (A) A bowtie pattern of fibronectin (*cyan*) was microcontact printed onto the array of micropillars. (B) Two cells confined in the bowtie pattern on the micropillar array were contacting each other and formed adherens junctions (marked by β-catenin in green) at the cell–cell boundary. (C) For each of two contacting cells, intercellular tugging force (F_c^a and F_c^b) is equal in magnitude and opposite in direction to the vector sum of traction forces in the cell. (D) Traction forces (red arrows) and intercellular tugging forces (white arrows) in two contacting cells confined in the bowtie pattern on the micropillar array. Adherens junctions were marked by β-catenin (green). Reproduced from ref. 56 with permission from National Academy of Sciences.

deformability of cells, microfluidic techniques have been increasingly employed, which highly improves the throughput of the deformability measurement (reviewed in ref. 66). Using PDMS, microfluidics with a variety of designs can be easily created. For the quantitative measurement of cell elasticity, microfluidics with constrictions smaller than cell diameter are typically used (Figure 4.4A). A cell in a constriction is deformed by the fluid pressure. If the cell is purely elastic, it is deformed in proportion to the applied pressure without any delay after being trapped in the constriction. However, since the cell behaves as a viscoelastic body, cell strain changes with time (Figure 4.4B). This creep behavior is described by a power-law rheology, as

$$\varepsilon(t) = \frac{\Delta P}{E}\left(\frac{t}{t_0}\right)^{\beta}. \tag{4.1}$$

Here, ε is cell strain over time t, ΔP is the effective pressure that the cell experiences, and E is the cell's elastic modulus, which is evaluated at an

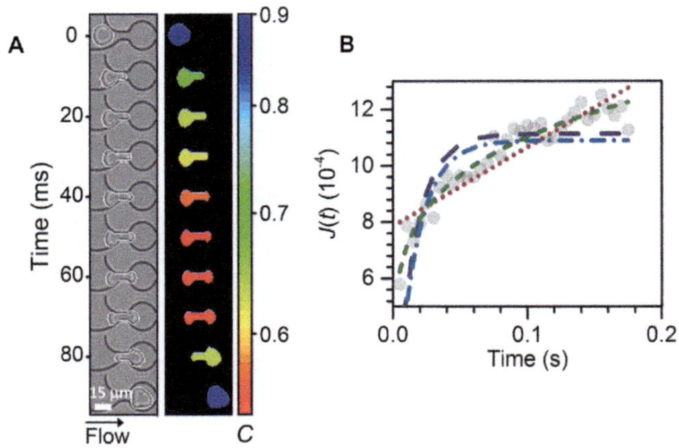

Figure 4.4 Microfluidic measurement of cellular viscoelasticity: (A) Time sequences of a cell transiting through a microfluidic constriction. The color overlay indicates the circularity, C, of the cell. (B) Creep trajectory for a cell at the constriction (gray dots). $J(t)$ is the creep function ($= \varepsilon(t)/\Delta P$ in eqn (4.1)). The trajectory is better fitted with a power-law rheology model (green short-dashed line) than with Maxwell (red dotted line), Kelvin–Voigt (purple long-dashed line) and standard linear solid (blue dot-dashed line) models. Reproduced from ref. 68 with permission from Elsevier, Copyright 2017.

arbitrary reference time t_0. The power-law exponent β reflects stress dissipation and thus provides a measure of the fluidity of the cell; $\beta = 0$ for a purely elastic solid, and $\beta = 1$ for a Newtonian fluid. Based on this simple equation, the elasticity and the fluidity of cells can be quantitatively extracted by tracking time-dependent changes of the cell shape in the constriction. The measured elastic modulus varied from ~0.06 to ~2 kPa depending on cell types, and the power-law exponent fell in the range of 0.2–0.4.[67–69] Both the elasticity and the fluidity of cells are largely affected by the actin cytoskeleton; disruption of the actin cytoskeleton reduces the elasticity of cells and fluidizes them.[67,68] Notably, narrower constrictions and higher pressures give rise to an increase in the elasticity of cells,[68] which is consistent with the strain-stiffening behavior observed in a variety of biopolymer networks and some synthetic polymer networks.[70,71] Due to such nonlinear nature of the cell elasticity, we need to consider the possibility that slight difference in experimental conditions of different studies may largely affect the reported values of the cell elasticity.

4.3 Mechanical Manipulation by Stretching Silicone Elastomer Membranes

Among methods for mechanical manipulation of cells, stretching cells cultured on silicone elastomer membranes has been broadly utilized. This approach is easy to control numerical parameters of manipulation and applicable to both small-scale (*e.g.* live imaging of a single cell) and bulk (*e.g.* biochemical assays for a large number of cells) analyses of cells. As membrane stretching is simple and versatile operation, numerous devices for this purpose have been developed so far (reviewed in ref. 72 and 73).

4.3.1 Distinct Strain Types

Uniaxial distension is the simplest case of membrane stretching. Two opposite sides of a rectangular piece of ECM-coated silicone membrane, on which cells are cultured, are gripped, and the membrane is extended between the grips by displacing one grip (or two grips in the counterdirection). It is ideal if the membrane is extended uniformly in a single direction. However, due to the Poisson effect, the uniaxial distention of the membrane induces compressive strain in the direction perpendicular to the stretch axis (Figure 4.5A). In such a situation, cells on the membrane experience both tensile and compressive stimuli, making the obtained results difficult to interpret. This problem can be overcome, at least in part, by attaching thick elastic walls to the free sides of the membrane; the elastic walls are extended along the stretch axis but prevent the compressive deformation of the membrane in the perpendicular direction.[74] Uniaxial stretching of membranes is typically performed manually or by using motorized devices. As an alternative way to stretch the membrane uniaxially, vacuum chambers were introduced into a microfluidics-based stretch device, in which sides of the membrane were attached to deformable walls of vacuum chambers[75,76] (Figure 4.5B). When vacuum was applied, the chamber walls were deformed, causing extension of the attached membrane. Compared with motorized devices, this PDMS-fabricated device is compact in size, thus being suitable for placement on a microscope stage.

Cellular responses to uniaxial stretching largely depend on the relative orientation of cells against the stretch axis.[41,77,78] Since cells cultured under the subconfluent condition are randomly oriented, a result obtained by uniaxial stretching reflects the mixture of different responses of cells. In order to reduce heterogeneity in cellular responses against stretching, Norton *et al.* developed a device in which an elastic membrane was pulled in perpendicular directions.[79] With this device, biaxial strain was provided in the central region distant from the gripped portions (Figure 4.5C). Alternative approaches to obtain the isotropic strain field include circular stretching of membranes. Typically, a circular membrane whose periphery is tethered is expanded by applying negative or positive pressure to the undersurface of the membrane (Figure 4.5D, E). In this case, the radial component of membrane strain is nearly uniform.[80] However, since the tethered periphery cannot be extended in the circumferential direction, circumferential strain drops quadratically from the maximum at the membrane center to zero at the periphery.[80] Furthermore, the membrane also suffers out-of-plane bending. As such, circular stretching of periphery-tethered membrane is not ideal to provide homogeneous stimuli to cells. These limitations in the usage of peripherally tethered circular membrane have been overcome by introducing a cylindrical plunger. When the center of a tethered circular membrane is pushed up by a cylindrical plunger whose diameter is smaller than that of the membrane, the membrane portion in contact with the upper surface of the plunger suffers in-plane extension (Figure 4.5F). Alternatively, the circular membrane is deformed not by moving up the plunger but by applying vacuum from below to pull down the annular portion of the membrane outside the plunger (Figure 4.5G). Remarkably, both methods achieve generation of a uniform, equibiaxial strain field in the membrane portion overlying the upper surface of the cylindrical plunger.[72,81,82] However, these methods have a problem in application to cell biology experiments; cells

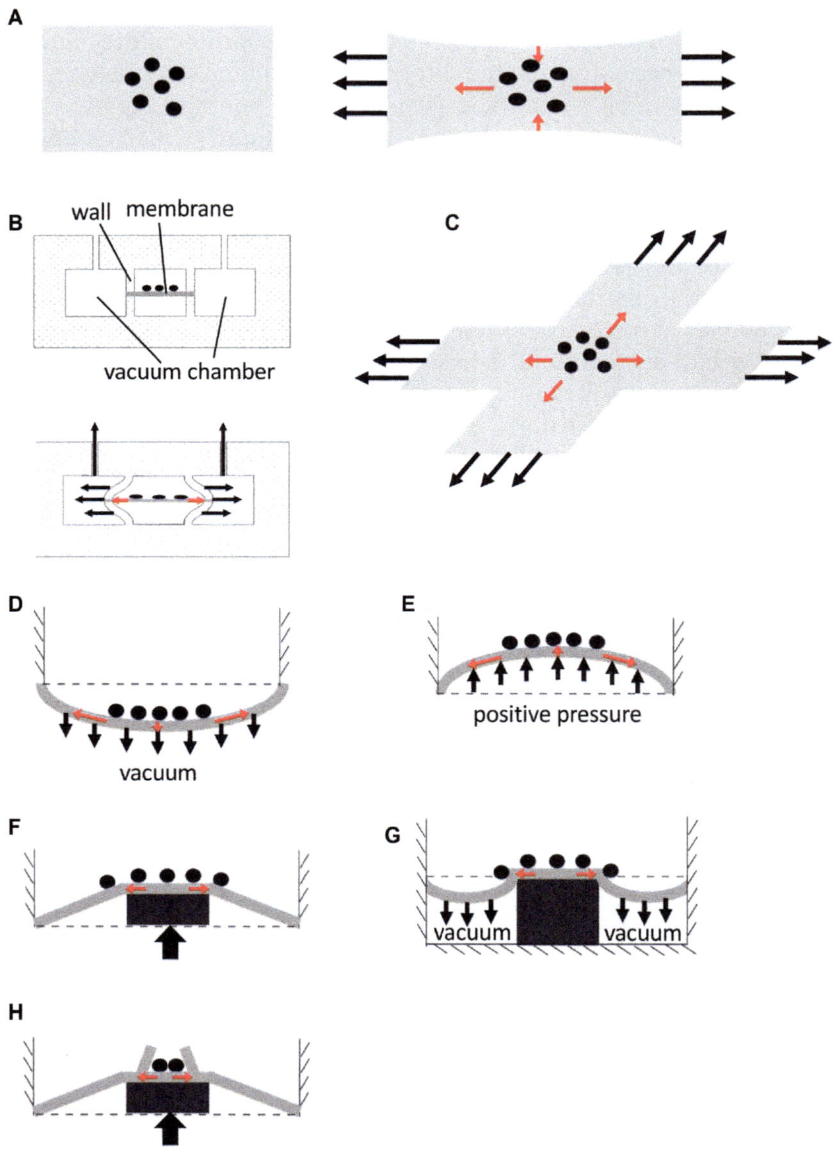

residing in the membrane portion outside the plunger are subjected to undesired, heterogeneous stretches. This problem could be solved by confining cells in the central region of the membrane. To this end, a stretchable circular wall was attached to the membrane, which created a cell culture chamber in the membrane center[83] (Figure 4.5H). By optimizing height and thickness of the wall, an eventually uniform and equibiaxial strain field was obtained in the membrane within the chamber.[83]

4.3.2 Sustained or Cyclic

Eventually all tissue cells sustain prestress (isometric tension) because contractile force generated by the intracellular actomyosin machinery is transmitted to and counterbalanced by extracellular substances.[84] At the same time, tissues in our body are repeatedly deformed by physical activities in the body including heartbeat, breathing, and skeletal muscle contraction. For instance, arteries are cyclically expanded by the periodic change in the blood pressure, which leads to cyclic stretches of artery endothelial cells in the circumferential direction of the vessel. Thus forces with different temporal modes are acting on cells in tissues. To mimic such mechanical environments of tissue cells, two major temporal modes of membrane stretching have been employed in *in vitro* studies: sustained and cyclic. When endothelial cells cultured on a silicone elastomer membrane are cyclically (typically at 1 Hz) stretched in a single direction, cells are reoriented perpendicularly to the stretch direction, which is associated with remodeling of actin stress fibers to realign along the longitudinal axis of the cells.[85] This cell reorientation perpendicular to the stretch axis is consistent with the *in vivo* situation in which endothelial cells are aligned in a longitudinal direction of the vessel. By contrast, when cells are subjected to sustained uniaxial stretch, cells and their stress fibers and focal adhesions tend to polarize along the stretch axis.[41,86] Furthermore, cyclic stretching of

Figure 4.5 Distinct ways to stretch membranes: (A) Uniaxial stretch. Simple extension of a membrane between two gripped sides causes compressive strain in the direction perpendicular to the stretch axis due to Poisson effect (right). (B) Cross section of a microfluidic device integrating a stretchable membrane and vacuum chambers. When vacuum is applied, the walls of the vacuum chambers are distorted, inducing uniaxial stretch of the attached membrane (lower diagram). (C) Biaxial stretch. (D and E) Out-of-plane stretch of a periphery-tethered circular membrane. Vacuum (D) or positive pressure (E) is applied from below to extend the membrane. (F and G) Stretching of a circular membrane using a cylindrical plunger. The membrane is extended either by moving up the plunger (F) or by applying vacuum from below (G). The central region of the membrane, which is in contact with the upper surface of the plunger, suffers in-plane, equibiaxial stretch. (H) Stretchable circular wall attached to the circular membrane provides a cell culture chamber to confine cells in the central region of the membrane. In all panels, red arrows represent deformation in the membranes.

cells at the same strain magnitude but different frequencies differentially affects orientation responses of cells and stress fibers.[87] Thus cells can distinguish different temporal modes of stretch stimuli.

When an elastic object is cyclically stretched, tensile stress is developed repeatedly in the object in phase with strain (Figure 4.6A). By contrast, if the object is totally viscous, stress developed in the object is in proportion to the deformation rate of the objective. Therefore, when the viscous object is cyclically stretched in a sinusoidal fashion, tensile and compressive stresses appear alternately (Figure 4.6B). Since cells and intracellular structures

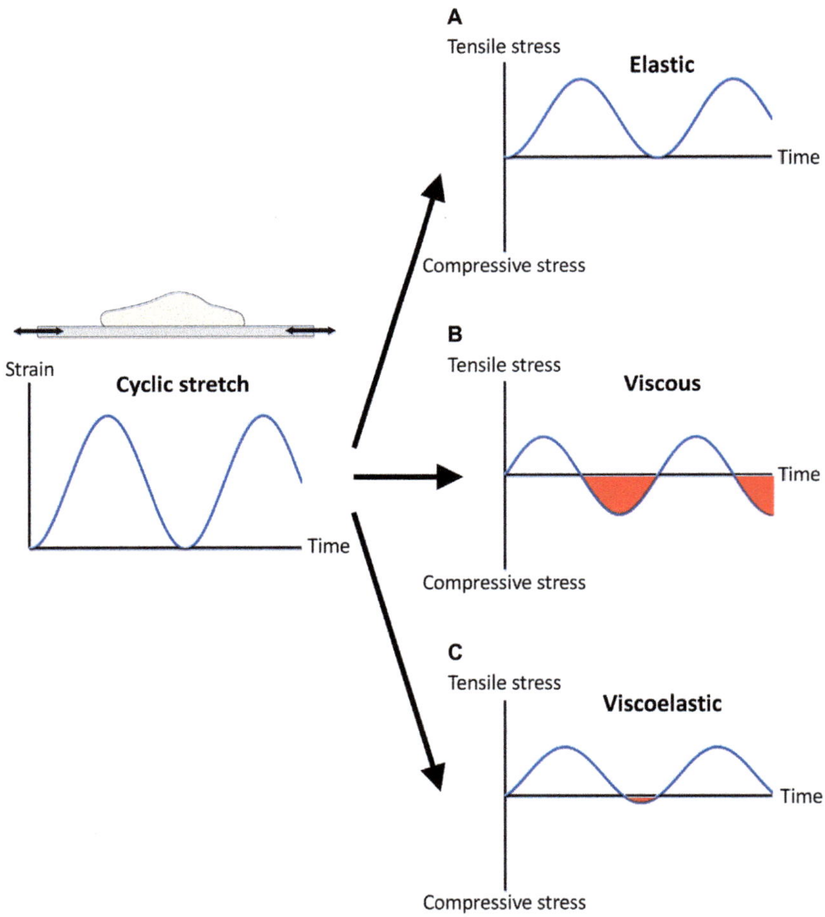

Figure 4.6 Elastic, viscous, and viscoelastic responses against cyclic stretch: (A) When sinusoidal stretch is applied, tensile stress is developed repeatedly in an elastic body in phase with strain. (B) By contrast, tensile and compressive stresses are developed alternately in a viscous body with a $\pi/2$ phase lead over strain. (C) Viscoelastic response in stress development against cyclic stretch is the superimposition of elastic and viscous responses. Compressive stresses are developed in viscous and viscoelastic bodies upon cyclic stretching (red regions).

including the actin cytoskeleton behave as viscoelastic bodies,[88] their stress responses against cyclic stretch are the superimposition of elastic and viscous responses (Figure 4.6C). Thus compressive stress is potentially developed in cells and the cytoskeleton during cyclic stretching (red in Figure 4.6C). Such compressive stress would reduce actomyosin-dependent prestress in stress fibers, which may be responsible for cyclic stretch-induced disassembly of stress fibers through severing of unloaded actin filaments by cofilin.[89] In contrast to cyclic stretching, sustained stretching exclusively develops tensile stress. Therefore, sustained stretch can substitute for actomyosin-generated tensile force and hence restore tension-dependent molecular events (*e.g.* molecular recruitment and phosphorylation of signaling molecules) at focal adhesions and stress fibers in actomyosin-inhibited cells.[34,41,78,90]

4.4 Combination of Mechanical Manipulation and Measurement

As described in the above sections, silicone elastomers are highly useful for both mechanical measurement and manipulation in cell biological research. As a natural direction for the next step, several attempts have been made to combine mechanical manipulation with quantitative assessment of mechanical statuses of cells and intracellular structures.

Many adherent cell types, including fibroblasts, neurons, and smooth and skeletal muscle cells, have polarized morphologies that are associated with the polarized organization of internal cytoskeletons. Materials composed of polymers with oriented arrangement exhibit anisotropy in their mechanical properties. Consistent with this, when cells with polarized cytoskeletons are subjected to uniaxial stretch, their responses depend on orientation of cells/cytoskeletons against the stretch axis.[41,86] Particularly, in the case of linear stress fibers whose tips are anchored to the extracellular substrate at focal adhesions, relative stress development in the fibers upon sustained uniaxial stretch is orientation dependent and can be estimated simply based on their orientation.[41,91] As shown in Figure 4.7A, we assume that a stress fiber is oriented with an angle θ with respect to the stretch axis and that the length of the fiber is changed from l to l' in response to stretching of the extracellular substrate membrane. An a% stretch of the membrane extends the parallel component of the stress fiber from x to $(1+a/100)x$, whereas the perpendicular component (y) remains unchanged. Thus the resultant strain (S) along the stress fiber length can be expressed as

$$S \equiv \frac{l'-l}{l} = \frac{\sqrt{(1+a/100)^2 x^2 + y^2}}{\sqrt{x^2+y^2}} - 1. \tag{4.2}$$

Since y is expressed as $x \times \tan\theta$, eqn (4.2) is reexpressed as

$$S = \frac{\sqrt{(1+a/100)^2 + \tan^2\theta}}{1+\tan^2\theta} - 1 = \sqrt{(1+a/100)^2 \cos^2\theta + \sin^2\theta} - 1.$$

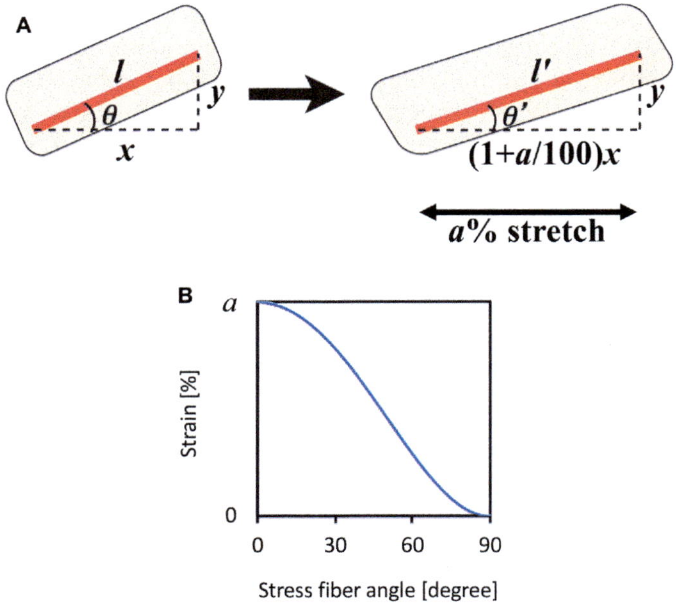

Figure 4.7 Orientation dependency of stress fiber strain under sustained uniaxial stretch of the extracellular substrate: (A) Stress fiber (with the length l and the orientation angle θ against the stretch axis), whose tips are anchored to the extracellular substrate through focal adhesions, suffers a% uniaxial stretch of the substrate. (B) Dependency of stretch-induced stress fiber strain against the orientation angle of the stress fiber.

Thus strain in the stress fiber monotonically decreases with the angle θ, and stress fibers oriented perpendicularly to the stretch axis do not suffer strain (Figure 4.7B). Since developed stress in the fiber is proportional to the strain, it has the same dependency on the angle θ. This method for the estimation of stretch-induced stress development in stress fibers has been applied to the analysis of mechanotransduction on stress fibers. When sustained uniaxial stretch was applied to cells pretreated with the myosin II inhibitor to exclude the effect of actomyosin-dependent prestress, phosphorylation of the MAP kinase ERK on stress fibers increased linearly with developed stress, revealing that stress fibers act as a platform for tension-dependent ERK activation.[41] This simple approach is just based on geometric calculation of object deformation and has a limitation that absolute values of developed tension or tensile force cannot be obtained.

To measure the stress responses of cells against stretching in a more direct way, micropillar array-based force measurement has been combined with elastic membrane stretching.[92,93] Herein, PDMS-fabricated micropillar arrays are bonded onto a PDMS elastic membrane. After coating micropillar tips with ECM proteins by a microcontact printing technique, cells are cultured on the micropillar array–integrated membrane. The membrane is then stretched uniaxially[92] or equi-biaxially.[93] Importantly, even though the membrane stretching alters interpillar spacing, it does not affect the shape and the bending stiffness of individual pillars.[93] By monitoring the deflection of each micropillar during the stretch experiment, time-dependent changes in tensile force under stretch can be measured in subcellular regions. Accordingly, in addition to a sudden, passive change in cellular tensile stress upon the onset of stretch, active, long-term changes in cell contractility were observed during the sustained stretch; the contractile force generated in a cell gradually increased in the first ∼20 min after the onset of stretch, which was followed by a gradual decrease in the contractile force generation to the initial level or lower by ∼60 min after the onset.[93] Furthermore, it has been revealed that these responses of contractile force against stretching are spatially heterogeneous in a cell.[92,93] Thus the combination of membrane stretching with micropillar-based force measurement is useful for spatiotemporal analyses of biomechanical properties and responses in a cell.

As an alternative, a silicone elastomer–based tool for combined manipulation and measurement of local force in cells, magnetically functionalized micropillars, has been developed. PDMS micropillars, in which nanowires/nanoparticles of cobalt, nickel, or iron are embedded, are actuated by applying a magnetic field while measuring force from pillar deflection.[94–96] By optimizing the wire/particle density in a PDMS base or by patterned casting of a wire/particle-containing PDMS, micropillar arrays with a subpopulation of magnetized micropillars can be prepared. With this approach, it was shown that mechanical force applied locally to a cell through one or a few focal adhesions induced both sudden and gradual changes in traction force at focal adhesions located distant from the site(s) of force application.[94] Such responses of traction force were heterogeneous among focal adhesions in a cell, suggesting that spatially discrete mechanical or biochemical connections between focal adhesions are involved in traction force regulation under mechanical perturbations. Another advantage of using the magnetic field is that oscillatory forces with different frequencies are easily generated by using an electrical magnet. With this advantage, magnetized micropillars have been used for the dynamic viscoelasticity measurement of cells. A single pillar whose tip was adhered to a cell was actuated with a range of frequencies (0.1–100 Hz), and the motion of the pillar was recorded using a high-speed camera (>100 frames per s), which revealed that cell mechanics at focal adhesion sites exhibited power-law viscoelasticity against the frequency; $|k(\omega)| \propto \omega^{\beta}$, where k is the cellular stiffness, ω is the actuation frequency, and β is the power-low exponent.[96]

4.5 Concluding Remarks

Due to their versatile natures, silicone elastomers have been widely utilized in the fields of cell biomechanics and mechanobiology. In this chapter, we have briefly reviewed only core technologies for mechanical measurement and manipulation based on silicone elastomers. Compared with specialized instruments such as atomic force microscopy, silicone-based devices are inferior in their sensitivity and resolution of force measurement; even though atomic force microscopy can measure piconewton force at the nanometer spatial resolution in aqueous solution,[97] micropillar arrays typically have micrometer interpillar spacing and nanonewton sensitivity. Nevertheless, since silicone elastomers are highly compatible with various fabrication techniques, a large variety of derivative devices with modified designs have been developed for the mechanical measurement/manipulation of cells. Another large class of silicone-based devices, microfluidic devices, are not included in this chapter. Although we have briefly presented an application of microfluidics to the measurement of mechanical properties of cells in Section 4.2.3, microfluidic devices have been widely employed for applying hydrodynamic forces to cells, as well as trapping or sorting cells based on mechanical characteristics of cells (reviewed in ref. 98 and 99). Microfluidics have also been integrated with other types of PDMS-based technologies, which include membrane stretching[75] and traction force measurement by micropillar arrays.[100] As fundamental tools that drive biomechanical/mechanobiological research and its application to biomedical fields, the spectrum of silicone elastomer devices will continue to expand.

Acknowledgements

This work was supported by JSPS KAKENHI grants (19K23748 and 20K12596 to HH).

References

1. N. I. Petridou, Z. Spiró and C. P. Heisenberg, *Nat. Cell Biol.*, 2017, **19**, 581.
2. K. H. Vining and D. J. Mooney, *Nat. Rev. Mol. Cell Biol.*, 2017, **18**, 728.
3. T. Panciera, L. Azzolin, M. Cordenonsi and S. Piccolo, *Nat. Rev. Mol. Cell Biol.*, 2017, **18**, 758.
4. A. K. Harris, P. Wild and D. Stopak, *Science*, 1980, **208**, 177.
5. M. Dembo and Y. L. Wang, *Biophys. J.*, 1999, **76**, 2307.
6. C. Rotsch, K. Jacobson and M. Radmacher, *Proc. Natl. Acad. Sci. U. S. A.*, 1999, **96**, 921.
7. A. Elosegui-Artola, I. Andreu, A. E. M. Beedle, A. Lezamiz, M. Uroz, A. J. Kosmalska, R. Oria, J. Z. Kechagia, P. Rico-Lastres, A. L. Le Roux, C. M. Shanahan, X. Trepat, D. Navajas, S. Garcia-Manyes and P. Roca-Cusachs, *Cell*, 2017, **171**, 1397.
8. N. Wang, J. P. Butler and D. E. Ingber, *Science*, 1993, **260**, 1124.
9. D. Choquet, D. P. Felsenfeld and M. P. Sheetz, *Cell*, 1997, **88**, 39.
10. C. Grashoff, B. D. Hoffman, M. D. Brenner, R. Zhou, M. Parsons, M. T. Yang, M. A. McLean, S. G. Sligar, C. S. Chen, T. Ha and M. A. Schwartz, *Nature*, 2010, **466**, 263.
11. Y. Xia and G. M. Whitesides, *Angew. Chem., Int. Ed.*, 1998, **37**, 550.

12. J. C. McDonald and G. M. Whitesides, *Acc. Chem. Res.*, 2002, **35**, 491.
13. P. Kanchanawong, G. Shtengel, A. M. Pasapera, E. B. Ramko, M. W. Davidson, H. F. Hess and C. M. Waterman, *Nature*, 2010, **468**, 580.
14. H. Hirata, M. Sokabe and C. T. Lim, *Prog. Mol. Biol. Transl. Sci.*, 2014, **126**, 135.
15. R. W. Style, R. Boltyanskiy, G. K. German, C. Hyland, C. W. MacMinn, A. F. Mertz, L. A. Wilen, Y. Xu and E. R. Dufresne, *Soft Matter*, 2014, **10**, 4047.
16. T. Oliver, M. Dembo and K. Jacobson, *Cell Motil. Cytoskeleton*, 1995, **31**, 225.
17. A. F. Mertz, Y. Che, S. Banerjee, J. M. Goldstein, K. A. Rosowski, S. F. Revilla, C. M. Niessen, M. C. Marchetti, E. R. Dufresne and V. Horsley, *Proc. Natl. Acad. Sci. U. S. A.*, 2013, **110**, 842.
18. S. R. K. Vedula, H. Hirata, M. H. Nai, A. Brugués, Y. Toyama, X. Trepat, C. T. Lim and B. Ladoux, *Nat. Mater.*, 2014, **13**, 87.
19. Y. Iwadate and S. Yumura, *Biotechniques*, 2008, **44**, 739.
20. E. Gutierrez, E. Tkachenko, A. Besser, P. Sundd, K. Ley, G. Danuser, M. H. Ginsberg and A. Groisman, *PLoS One*, 2011, **6**, e23807.
21. S. J. Han, Y. Oak, A. Groisman and G. Danuser, *Nat. Methods*, 2015, **12**, 653.
22. N. Q. Balaban, U. S. Schwarz, D. Riveline, P. Goichberg, G. Tzur, I. Sabanay, D. Mahalu, S. Safran, A. Bershadsky, L. Addadi and B. Geiger, *Nat. Cell Biol.*, 2001, **3**, 466.
23. T. Vignaud, C. Copos, C. Leterrier, M. Toro-Nahuelpan, Q. Tseng, J. Mahamid, L. Blanchoin, A. Mogilner, M. Théry and L. Kurzawa, *Nat. Mater.*, 2021, **20**, 410.
24. M. L. Gardel, B. Sabass, L. Ji, G. Danuser, U. S. Schwarz and C. M. Waterman, *J. Cell Biol.*, 2006, **183**, 999.
25. Y. Aratyn-Schaus and M. L. Gardel, *Curr. Biol.*, 2010, **20**, 1145.
26. K. A. Beningo, M. Dembo, I. Kaverina, J. V. Small and Y. L. Wang, *J. Cell Biol.*, 2001, **153**, 881.
27. S. Munevar, Y. L. Wang and M. Dembo, *Mol. Biol. Cell*, 2001, **12**, 3947.
28. C. M. Kraning-Rush, J. P. Califano and C. A. Reinhart-King, *PLoS One*, 2012, 7, e32572.
29. C. E. Chan and D. J. Odde, *Science*, 2008, **322**, 1687.
30. S. V. Plotnikov, A. M. Pasapera, B. Sabass and C. M. Waterman, *Cell*, 2012, **151**, 1513.
31. A. I. Yip, K. Iwasaki, C. Ursekar, H. Machiyama, M. Saxena, H. Chen, I. Harada, K. H. Chiam and Y. Sawada, *Biophys. J.*, 2013, **104**, 19.
32. A. Elosegui-Artola, E. Bazellières, M. D. Allen, I. Andreu, R. Oria, R. Sunyer, J. J. Gomm, J. F. Marshall, J. L. Jones, X. Trepat and P. Roca-Cusachs, *Nat. Mater.*, 2014, **13**, 631.
33. A. Elosegui-Artola, R. Oria, Y. Chen, A. Kosmalska, C. Pérez-González, N. Castro, C. Zhu, X. Trepat and P. Roca-Cusachs, *Nat. Cell Biol.*, 2016, **18**, 540.

34. H. Hirata, W. C. Ku, A. K. Yip, C. P. Ursekar, K. Kawauchi, A. Roy, A. K. Guo, S. R. K. Vedula, I. Harada, K. H. Chiam, Y. Ishihama, C. T. Lim, Y. Sawada and M. Sokabe, *J. Cell Sci.*, 2016, **129**, 3574–3582.
35. B. Sabass, M. L. Gardel, C. M. Waterman and U. S. Schwarz, *Biophys. J.*, 2008, **94**, 207.
36. H. Colin-York, D. Shrestha, J. H. Felce, D. Waithe, E. Moeendarbary, S. J. Davis, C. Eggeling and M. Fritzsche, *Nano Lett.*, 2016, **16**, 2633.
37. J. L. Tan, J. Tien, D. M. Pirone, D. S. Gray, K. Bhadriraju and C. S. Chen, *Proc. Natl. Acad. Sci. U. S. A.*, 2003, **100**, 1484.
38. M. Gupta, L. Kocgozlu, B. R. Sarangi, F. Margadant, M. Ashraf and B. Ladoux, *Methods Cell Biol.*, 2015, **125**, 289.
39. O. Du Roure, A. Saez, A. Buguin, R. H. Austin, P. Chavrier, P. Silberzan and B. Ladoux, *Proc. Natl. Acad. Sci. U. S. A.*, 2005, **102**, 2390.
40. M. Gupta, B. R. Sarangi, J. Deschamps, Y. Nematbakhsh, A. Callan-Jones, F. Margadant, R. M. Mège, C. T. Lim, R. Voituriez and B. Ladoux, *Nat. Commun.*, 2015, **6**, 7525.
41. H. Hirata, M. Guputa, S. R. K. Vedula, C. T. Lim, B. Ladoux and M. Sokabe, *EMBO Rep.*, 2015, **16**, 250.
42. S. Ghassemi, G. Meacci, S. Liu, A. A. Gondarenko, A. Mathur, P. Roca-Cusachs, M. P. Sheetz and J. Hone, *Proc. Natl. Acad. Sci. U. S. A.*, 2012, **109**, 5328.
43. A. Saez, A. Buguin, P. Silberzan and B. Ladoux, *Biophys. J.*, 2005, **89**, L52.
44. M. Ghibaudo, A. Saez, L. Trichet, A. Xayaphoummine, J. Browaeys, P. Silberzan, A. Buguin and B. Ladoux, *Soft Matter*, 2008, **4**, 1836.
45. J. Fu, Y. K. Wang, M. T. Yang, R. A. Desai, X. Yu, Z. Liu and C. S. Chen, *Nat. Methods*, 2010, **7**, 733.
46. L. Trichet, J. Le Digabel, R. J. Hawkins, S. R. K. Vedula, M. Gupta, C. Ribrault, P. Hersen, R. Voituriez and B. Ladoux, *Proc. Natl. Acad. Sci. U. S. A.*, 2012, **109**, 6933.
47. A. Saez, M. Ghibaudo, A. Buguin, P. Silberzan and B. Ladoux, *Proc. Natl. Acad. Sci. U. S. A.*, 2007, **104**, 8281.
48. R. M. Mège and N. Ishiyama, *Cold Spring Harbor Perspect. Biol.*, 2017, **9**, a028738.
49. B. Ladoux, E. Anon, M. Lambert, A. Rabodzey, P. Hersen, A. Buguin, P. Silberzan and R. M. Mège, *Biophys. J.*, 2010, **98**, 534.
50. A. Ganz, M. Lambert, A. Saez, P. Silberzan, A. Buguin, R. M. Mège and B. Ladoux, *Biol. Cell.*, 2006, **98**, 721.
51. A. Chopra, E. Tabdanov, H. Patel, P. A. Janmey and J. Y. Kresh, *Am. J. Physiol. Heart Circ. Physiol.*, 2011, **300**, H1252.
52. M. Rauzi, P. Verant, T. Lecuit and P. F. Lenne, *Nat. Cell Biol.*, 2008, **10**, 1401.
53. R. Fernandez-Gonzalez, S. D. M. Simoes, J. C. Röper, S. Eaton and J. A. Zallen, *Dev. Cell*, 2009, **17**, 736.
54. K. K. Chiou, L. Hufnagel and B. I. Shraiman, *PLoS Comput. Biol.*, 2012, **8**, e1002512.
55. S. Ishihara and K. Sugimura, *J. Theor. Biol.*, 2012, **313**, 201.

56. Z. Liu, J. L. Tan, D. M. Cohen, M. T. Yang, N. J. Sniadecki, S. A. Ruiz, C. M. Nelson and C. S. Chen, *Proc. Natl. Acad. Sci. U. S. A.*, 2010, **107**, 9944.
57. V. Maruthamuthu, B. Sabass, U. S. Schwarz and M. L. Gardel, *Proc. Natl. Acad. Sci. U. S. A.*, 2011, **108**, 4708.
58. D. T. Tambe, C. C. Hardin, T. E. Angelini, K. Rajendran, C. Y. Park, X. Serra-Picamal, E. H. Zhou, M. H. Zaman, J. P. Butler, D. A. Weitz, J. J. Fredberg and X. Trepat, *Nat. Mater.*, 2011, **10**, 469.
59. E. Basellières, V. Conte, A. Elosegui-Artola, X. Serra-Picamal, M. Bintanel-Morcillo, P. Roca-Cusachs, J. J. Muñoz, M. Sales-Pardo, R. Guimerà and X. Trepat, *Nat. Cell Biol.*, 2015, **17**, 409.
60. M. Uroz, S. Wistorf, X. Serra-Picamal, V. Conte, M. Sales-Pardo, P. Roca-Cusachs, R. Guimerà and X. Trepat, *Nat. Cell Biol.*, 2018, **20**, 646.
61. J. Guck, S. Schinkinger, B. Lincoln, F. Wottawah, S. Ebert, M. Romeyke, D. Lenz, H. M. Erickson, R. Ananthakrishnan, D. Mitchell, J. Käs, S. Ulvick and C. Bilby, *Biophys. J.*, 2005, **88**, 3689.
62. V. Swaminathan, K. Mythreye, E. T. O'Brien, A. Berchuck, G. C. Blobe and R. Superfine, *Cancer Res.*, 2011, **71**, 5075.
63. O. Otto, P. Rosendahl, A. Mietke, S. Golfier, C. Herold, D. Klaue, S. Girardo, S. Pagliara, A. Ekpenyong, A. Jacobi, M. Wobus, N. Töpfner, U. F. Keyser, J. Mansfeld, E. Fischer-Friedrich and J. Guck, *Nat. Methods*, 2015, **12**, 199.
64. Q. Luo, D. Kuang, B. Zhang and G. Song, *Biochim. Biophys. Acta*, 2016, **1860**, 1953.
65. J. Lin, D. Kim, H. T. Tse, P. Tseng, L. Peng, M. Dhar, S. Karumbayaram and D. Di Carlo, *Microsyst. Nanoeng.*, 2017, **3**, 17013.
66. H. Ito and M. Kaneko, *Robomech J.*, 2020, **7**, 3.
67. L. R. Lange, J. Steinwachs, T. Kolb, L. A. Lautscham, I. Harder, G. Whyte and B. Fabry, *Biophys. J.*, 2015, **109**, 26.
68. K. D. Nyberg, K. H. Hu, S. H. Kleinmam, D. B. Khismatullin, M. J. Butte and A. C. Rowat, *Biophys. J.*, 2017, **113**, 1574.
69. Z. Chen, Y. Zhu, D. Xu, M. M. Alam, L. Shui and H. Chen, *Lab Chip*, 2020, **20**, 2343.
70. C. Storm, J. J. Pastore, F. C. Mackintosh, T. C. Lubensky and P. A. Janmey, *Nature*, 2005, **435**, 191.
71. K. A. Erk, K. J. Henderson and K. R. Shull, *Biomacromolecules*, 2010, **11**, 1358.
72. T. D. Brown, *J. Biomech.*, 2000, **33**, 3.
73. H. Kamble, M. J. Barton, M. Jun, S. Park and N. T. Nguyen, *Lab Chip*, 2016, **16**, 3193.
74. K. Naruse, T. Yamada and M. Sokabe, *Am. J. Physiol.*, 1998, **274**, H1532.
75. D. Huh, B. D. Matthews, A. Mammoto, M. Montoya-Zavala, H. Y. Hsin and D. E. Ingber, *Science*, 2010, **328**, 1662.
76. Y. Huang and N. T. Nguyen, *Biomed. Microdevices*, 2013, **15**, 1043.

77. J. H. C. Wang, G. Yang, Z. Li and W. Shen, *J. Biomech.*, 2004, **37**, 573.
78. H. Hirata, H. Tatsumi and M. Sokabe, *J. Cell Sci.*, 2008, **121**, 2795.
79. L. A. Norton, K. L. Andersen, D. Arenholt-Bindslev, L. Andersen and B. Melsen, *Arch. Oral Biol.*, 1995, **40**, 863–872.
80. J. L. Williams, J. H. Chen and D. M. Belloli, *J. Biomech. Eng.*, 1992, **114**, 377.
81. C. T. Hung and J. L. Williams, *J. Biomech.*, 1994, **27**, 227.
82. J. L. Schaffer, M. Rizen, G. J. L'Italien, A. Benbrahim, J. Megerman, L. C. Gerstenfeld and M. L. Gray, *J. Orthop. Res.*, 1994, **12**, 709.
83. C. P. Ursekar, S. K. Teo, H. Hirata, I. Harada, K. H. Chiam and Y. Sawada, *PLoS One*, 2014, **9**, e90665.
84. D. E. Ingber, *FASEB J.*, 2006, **20**, 811.
85. V. P. Shirinsky, A. S. Antonov, K. G. Birukov, A. V. Sobolevsky, Y. A. Romanov, N. V. Kabaeva, G. N. Antonova and V. N. Smirnov, *J. Cell Biol.*, 1989, **109**, 331.
86. Y. Chen, A. M. Pasapera, A. P. Koretsky and C. M. Waterman, *Proc. Natl. Acad. Sci. U. S. A.*, 2013, **110**, E2352.
87. H. J. Hsu, C. F. Lee and R. Kaunas, *PLoS One*, 2009, **4**, e4853.
88. D. Stamenović, *Cell. Mol. Life Sci.*, 2008, **65**, 3592.
89. K. Hayakawa, H. Tatsumi and M. Sokabe, *J. Cell Biol.*, 2011, **195**, 721.
90. H. Hirata, H. Tatsumi, C. T. Lim and M. Sokabe, *Am. J. Physiol.: Cell Physiol.*, 2014, **306**, C607.
91. H. Hirata, M. Guputa, S. R. K. Vedula, C. T. Lim, B. Ladoux and M. Sokabe, *Methods Mol. Biol.*, 2017, **1487**, 223.
92. K. Nagayama, A. Adachi and T. Matsumoto, *J. Biomech.*, 2011, **44**, 2699.
93. J. M. Mann, R. H. W. Lam, S. Weng, Y. Sun and J. Fu, *Lab Chip*, 2012, **12**, 731.
94. N. J. Sniadecki, A. Anguelouch, M. T. Yang, C. M. Lamb, Z. Liu, S. B. Kirschner, Y. Liu, D. H. Reich and C. S. Chen, *Proc. Natl. Acad. Sci. U. S. A.*, 2007, **104**, 14553.
95. K. Nagayama, T. Inoue, Y. Hamada and T. Matsumoto, *J. Biomech.*, 2017, **65**, 194.
96. Y. Shi, C. L. Porter, J. C. Crocker and D. H. Reich, *Proc. Natl. Acad. Sci. U. S. A.*, 2019, **116**, 13839.
97. A. Ikai, R. Afrin, M. Saito and T. Watanabe-Nakayama, *Semin. Cell Dev. Biol.*, 2018, **73**, 132.
98. G. Velve-Casquillas, M. Le Berre, M. Piel and P. T. Tran, *Nano Today*, 2010, **5**, 28.
99. P. K. Chaudhuri, M. E. Warkiani, T. Jing, Kenry and C. T. Lim, *Analyst*, 2016, **141**, 504.
100. R. H. W. Lam, Y. Sun, W. Chen and J. Fu, *Lab Chip*, 2012, **12**, 1865.

CHAPTER 5

Geometric Cues for Directing Cell Fate

SARA ROMANAZZO[a] AND KRISTOPHER A. KILIAN*[a,b]

[a] School of Chemistry, Australian Centre for NanoMedicine, University of New South Wales, Sydney, NSW, Australia; [b] School of Materials Science and Engineering, University of New South Wales, Sydney, NSW, Australia
*Email: k.kilian@unsw.edu.au

5.1 Introduction

Tissue form and function arise from a myriad of signals that coax cells to adopt defined arrangements that dictate cell state and associated activities. These proscribed activities are at the heart of tissue-level processes that ultimately give rise to important macroscopic properties underlying tissue and organ function. The most widely studied signals during tissue formation are soluble cues that are secreted by cells and diffuse with well-defined kinetics in the tissue microenvironment to be received by neighboring cells. The downstream signal transduction cascades ultimately regulate gene expression, which guides cell fate decisions. These dynamic changes give rise to complex transformations, including migration and proliferation, mesenchymal and epithelial transitions, and other morphogenetic processes.[1–3]

Another important aspect of the cell and tissue microenvironment that underlies all functional activity is the biophysical and biochemical attributes of the surrounding insoluble matrix.[4,5] Precise peptide motifs in natural biopolymers engage cell surface receptors to catalyze cytoplasmic signal transduction; the physical properties of the engaged extracellular matrix (ECM) play a vital role in how these insoluble peptide signals are interpreted

Biomaterials Science Series No. 12
Material-based Mechanobiology
Edited by Jun Nakanishi and Koichiro Uto

Published by the Royal Society of Chemistry, www.rsc.org

by the cells. Biomaterials scientists have probed these matrix cues through a wide array of synthetic and natural systems that serve to deconstruct the complex interplay between the surrounding support material and gradients of soluble factors in tissue. For instance, the use of designer hydrogels has demonstrated the impact of matrix viscoelasticity on cell decision making,[6,7] and nano- and microstructuring materials in 2D[8–10] and 3D[11] have demonstrated how the context in which cells are cultured is decisive in lineage determination.

A popular technique that has demonstrated how geometry and topography dictate cellular decision making involves microengineering of substrates and materials to recreate tissue-like patterns that are present *in vivo*.[12] Micropatterning is the process by which molecules are deposited onto a substrate with microscale resolution. It has been widely demonstrated that, by manipulating micropattern geometry, adherent cells adapt their cytoskeletal architecture and associated signaling to the shape of the microenvironment. In fact, positioning cells on a substrate by controlling size and spatial confinement, can influence cell–substrate and cell–cell interaction. Cell actin filaments remodeling and microtubule networks are now known to take part in the adaptation process of cell polarity with respect to external constraints. These modifications further impact cell migration, growth, and differentiation.

Thus micropatterning can be used *in vitro* to understand relationships between cells and the surrounding extracellular environment, by replicating as closely as possible the complex architecture that emerges *in vivo*. Whilst the first micropatterning techniques were developed over 40 years ago,[13,14] they became applicable as an important tool in cell biology and have been commercialized only recently. Information gathered by these observations can be used not only to conduct fundamental biological studies but also to perform drug testing experiments, to understand biochemical processes, and to design microfluidic devices for medical applications[15] (see Figure 5.1). Substrates commonly used for micropatterning are rigid surfaces, like glass, silicon, and substrates, with deposited thin layers of coinage metals[16] and those composed of polymers, such as polydimethylsiloxane (PDMS), polymethylmethacrylate (PMMA), cyclic olefin copolymer (COC), and polyimide (PI).[17,18] With rapid advances in microfabrication and 3D printing technologies, virtually any material can be structured at the microscale to control the cell- and tissue-level environment.

Constraining the cell adhesion footprint through micropatterning can influence diverse processes, including cell shape, division, polarity and migration, and lineage determination. Early work using soft lithography of alkanethiolate self-assembled monolayers on gold demonstrated how the adhesive footprint will dictate cell life or death.[19] Subsequent studies demonstrated how cells captured on anisotropic geometries would influence polarity,[20] and the further development of dynamic surface chemistry facilitated the triggered release of polarized cells to explore migratory phenotypes.[21,22] In the subsequent years, micropatterning has been used to dissect signaling within single cells and across pattern-confined multicellular

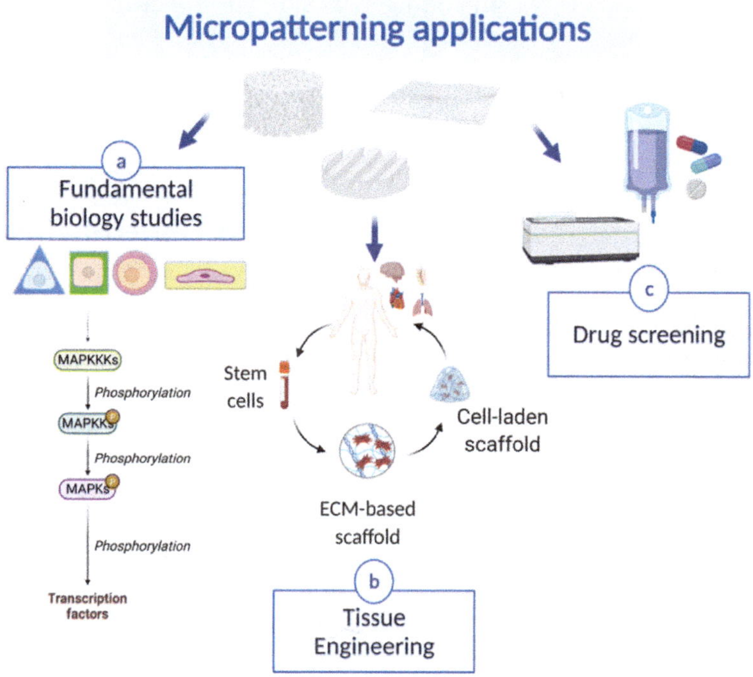

Figure 5.1 Micropatterning applications in biomedical fields.

aggregates. Cell motility is critical for many functions within human physiology, such as embryonic development, wound healing, angiogenesis, regulation of the immune system, stem cell differentiation, which are all processes relying on the ability of cells to migrate within the extracellular matrix (ECM).[23]

The advantage of using micropatterned materials in biological research is the exquisite control over the shape and organization of cells across (and within) a material. In contrast to cell culture on planar glass or plastic dishes, controlling cell and tissue shape with well-defined matrices for adhesion facilitates the investigation of complex processes while keeping shape, adhesion, cell–cell and cell–matrix connectivity, *etc.* constant within an experiment. This enables the exploration of more sophisticated biological questions that underlie physiological and pathological contexts. For instance, cells adhere to one another through cadherin proteins and to the ECM *via* integrin receptors.[4,24] Biophysical aspects of the matrix, including geometry, topography, and viscoelasticity, are converted to signaling cascades through nano- and microscale focal adhesions, which propagate through cytoskeletal filaments to the nucleus.[1,25,26] By controlling cell shape and topography at the boundary of multicellular aggregates, downstream signal transduction can be probed and related to biophysical constraints that may be imposed by the surrounding ECM. The numerous signaling cascades downstream of shape-mediated mechanotransduction will dictate complex cellular processes, including adhesion, proliferation, migration, and differentiation.[27]

Here we provide an overview of popular micropatterning techniques used for cell micropatterning, along with a description of how they are applied on a variety of cell and biomaterial types, for the investigation of cell–matrix structure–function relationships.

5.2 Microengineering Techniques to Control the Shape of Cells and Tissue

All micropatterning techniques have the common goal to achieve a controlled spatial architecture to reproducibly dictate cell adhesion. This can be achieved on different scales, from single cells to tissue level patterns, based on the technique used. All techniques include three basic steps: (1) generation of a pattern of controlled surface chemistry/adhesiveness, (2) seeding a cell population targeted to adhere to the patterned regions, and (3) washing to remove excess cells and reveal the pattern. Single-cell pattern dimensions can range from one to hundreds of microns and from hundreds to thousands for multicellular aggregates.[17,28–31] Here we describe the most widely used micropatterning techniques for studying how cell and tissue geometry/topography influences biological processes (see Table 5.1 and Figure 5.2).

5.2.1 Photolithography

Photolithography is one of the first techniques used to pattern cells and substrates, and utilizes materials that harden or soften in response to light irradiation.[32] A thin film of photopolymerizable material, such as a photoresist (positive or negative) or a polymerizable polymer hydrogel, *e.g.*, poly (ethylene glycol) diacrylate (PEGDA), is deposited uniformly onto a substrate. The material is polymerized through a photomask of the desired pattern, which either hardens or softens after light exposure. The unpolymerized material is then washed away to leave microscale features in relief that can be used either to directly culture cells or as a master to mold and shape 3D scaffolds, elastomeric stamps for soft lithography, or microfluidic devices. This technique can give a pattern with a resolution typically ~2–500 μm of the desired pattern dictated by the photomask. However, photolithography processes usually require clean room facilities, which are expensive and not easily accessible. In addition, many of the solvents used to process the photoresist can denature biological molecules and are cytotoxic. Once created, the photoresist can be reused for several experiments, but changing patterns or design requires fabrication of new photoresist masters, which can be cost limiting.

5.2.2 Soft Lithography

Whitesides and colleagues developed a patterning process termed soft lithography, which overcomes many of the shortcomings of photolithography.[33]

Table 5.1 Micropatterning techniques, advantages, and disadvantages.

Technique	Resolution	Advantages	Disadvantages
Photolithography	2–500 μm	Often directly applicable to proteins and cells	– Intrinsically expensive – It gives limited control over surface properties – Time from design to prototype can be long – Techniques unfamiliar and inaccessible to most biologists
Soft-lithography	20–50 μm	– Low cost – Rapid prototyping – The stamp can be used multiple times with little degradation	– Microfluidic printing: limited spacing between microchannels – Difficulty in controlling the ink and the surface robustness
Inkjet cell printing	50 μm	– Moderate cost – Good controllability	– Droplet formation – Requires an external power source
Optical fabrication	<100 nm	– Remote and large-scale manipulation – High precision due to the intrinsic charge and dielectric properties of cells	– Cell viability might be affected by laser wavelength, laser power, and duration of exposure to the trap – Requires an external power source – Limited manipulation area which
Dielectrophoresis	100 mm	– Combine electrokinetic forces with hydrodynamic effects – High-resolution patterning – Large-scale parallel manipulation	– Requires an external power source – Dielectric force decreases due to the separation distance of electrodes
Other approaches:Electron beam lithography	10 nm	– Higher resolution than lithography	– Very expensive and complex with high maintenance cost

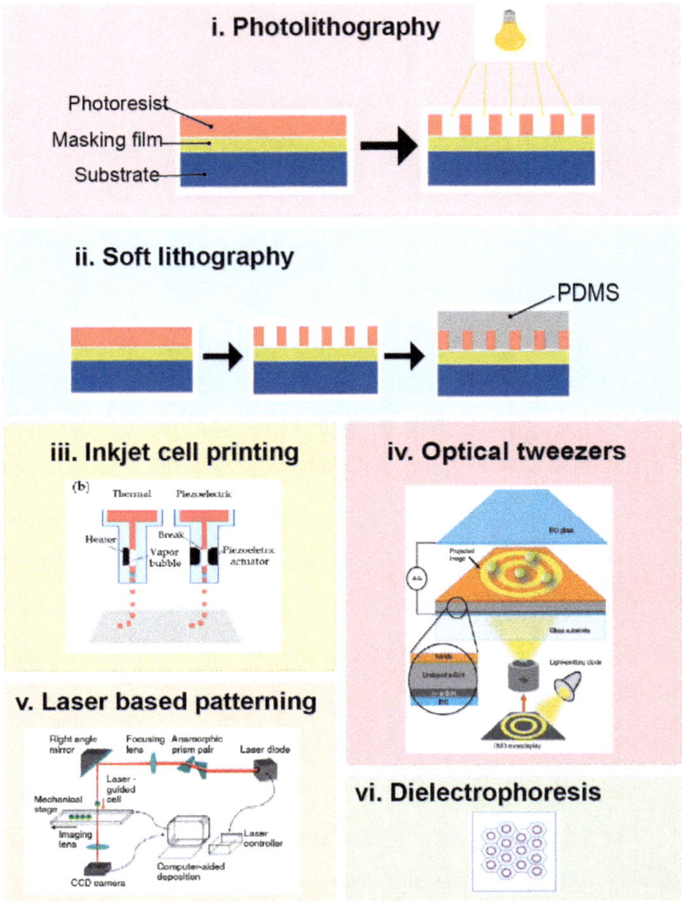

Figure 5.2 Micropatterning techniques: (i) Photolithography. (ii) Soft lithography. (iii) inkjet printing. Reproduced from ref. 54, https://doi.org/10.3390/bioengineering4010010, under the terms of the CC BY 4.0 license https://creativecommons.org/licenses/by/4.0. (iv) Optoelectronic tweezers. Reproduced from ref. 57 with permission from Springer Nature, Copyright 2005. (v) Laser-based patterning. Reproduced from ref. 60 with permission from Springer Nature, Copyright 2006; (vi) dielectrophoresis.

There are different types of soft lithography techniques, such as microcontact and microfluidic printing, but the commonality among all of them is that they use "soft" materials (polymers and organic molecular layers) rather than the rigid inorganic materials used in microelectronic systems.[22,34] The most common material used for the lithographic transfer is polydimethylsiloxane (PDMS). PDMS is a liquid prepolymer at room temperature that can be readily cured to form an elastomeric material that deforms on contact to transfer at the nanoscale to the macroscale.[35] At the same time, it is a moderately stiff elastomer (Young's modulus 1 MPa).[22] To fabricate PDMS stamps, the prepolymer is mixed with a curing agent, poured onto a template, and cured to crosslink the polymer. Once a PDMS master is formed, it can be used over an extended period of time with little degradation.[36,37]

Among the numerous micropatterning techniques, soft lithography—also known as microcontact printing (μCP)—has become a versatile technique in the study of cell and tissue processes. The general procedure for μCP involves a stamp that is coated with a solution of the substance to be printed, such as a specific chemistry to be conjugated to a substrate, a polymer or ECM-derived peptide or protein.[21] The ligand density can vary from experiment to experiment since transfer efficiency from the stamp can vary. To increase the fidelity of the deposited pattern, Whitesides and colleagues adapted this technique to self-assembled monolayer chemistry, where the transferred material is a solution of alkanethiolates that will self-assemble into a highly uniform region on the surface of gold or silver.[38,39] These self-assembled monolayers (SAMs) form discrete patterned regions, and the remaining background is passivated with a chemistry that resists protein adsorption (for instance, oligo(ethylene oxide) layers). Addition of matrix-derived peptides or proteins will then physically adsorb, or react covalently with distal conjugation motifs, to form a cell–adhesive layer where cells can adhere. The antifouling nature of the surrounding diluent provides a robust platform to restrict cells from leaving the patterned regions. Several other surface chemistries have been investigated using similar principles, including μCP alkylsilanes on glass and silicon,[40,41] phosphonate monolayers on metal oxides,[42,43] and end-functionalized organic monomers across diverse hard materials.[44] There are also examples of mixed SAMs of alkanethiols on metals, such as gold, composed of aliphatic and polar alkanethiols.[45]

To push the resolution of soft lithography, Mirkin and colleagues developed an array of pyramid-shaped microstructures through a so-called polymer or beam pen lithography technique, where a PDMS mold and the arrayed surface micropyramids were initially negatively replicated from a master silicon mold through a wet etching process and subsequently coated with a thin metal layer to define the grid of pyramidal objects. With controlled compression of the pyramidal PDMS, chemistry can be patterned at the nanoscale to define subcellular adhesion points.[46,47]

PDMS materials casted on photolithography masters are also used in microfluidic devices and in the technique of microfluidic printing, where microchannels formed by the bonding of PDMS to glass substrates can be used to selectively deliver monolayer chemistry,[48,49] ECM proteins,[50,51] or cells onto the tissue culture substrate through laminar flow delivery.[52] A block of elastomeric polymer, *e.g.* PDMS, containing channels is molded from the template made by photolithography, crosslinked, and removed from the template. The obtained channels are then fixed to a substrate (usually glass), and a series of solutions are perfused through the channels to generate substrate patterning. Channels are then removed before cell seeding. In contrast to μCP, which divides the culture surface into two regions (adhesive and nonadhesive), microfluidics allows creation of a multicomponent surface, where different solutions can be chosen for each channel.[28]

5.2.3 Inkjet Cell Printing

Inkjet cell printing consists in depositing droplets of cells of the ink loaded in the printer. This technique is encompassed by three types of methodologies known as continuous inkjet (CIJ) printing, drop-on-demand (DOD) printing, and electrohydrodynamic jet printing[17,53] (e-jet printing), which uses a rapid nozzle-free jet process called pyroelectrohydrodynamic jet (p-jet), because it uses a pyroelectric effect that modifies the bioink fluid, modulating the dot sizes from 200 μm down to 0.5 μm.[54] With the inkjet printing technique developed by Yusof *et al.*, single-cell printing in microarray plates allowed single-cell arrays for patterning different cell types.[55] Mecozzi *et al.* used jet-printing to deposit either a fibronectin bioink or 3-aminopropyl triethoxysilane (APTES), a coating able to provide a SAM with a distal amine group, to achieve a stable ionic bonding with the negative charge of the cell membrane, on glass coverslips coated with a thin layer of cytophobic fluorinated product, named Fluorolink S10. Both inks were printed with sizes ranging from 50 to 200 μm. The aligned dots generated by printing the inks were then used to deposit cells.[56] With the diversity of bioinks and the capability of fabricating complex structures, inkjet bioprinting appears as an effective tool for the automatic and reproducible creation of multiple formats and cell types.

5.2.4 Optical Tweezers and Optoelectronic Tweezers

Optical fabrication of micropatterning consists in using electrical forces to move cells. Optical tweezers use radiation pressure emitted either by a laser beam or by other use of infrared lasers.[57] Due to the laser power and exposure duration to the trap, cell viability might be affected by this technique. In order to overcome the radiation forces used in optical tweezers, optoelectronic ones can be used to micropattern cells by trapping them in specific positions. However, thermal effect and photodamage of cells are still concerning when using these two techniques.[17] In a study by Wu and colleagues, cells were placed between an upper indium tin oxide-coated glass and lower multiple layers of photosensitive surfaces. In this way, it was possible to distinguish dead cells from live cells, based on the image obtained.[57] Using similar strategies with surface chemistry, a wide array of materials systems may be used with this technique.

5.2.5 Laser-based Cell Patterning

This method uses a laser to transfer cells from one source film (donor, ribbon, or target) to a receiving or acceptor substrate. It could be divided as follows: laser-induced forward transfer (LIFT), absorbing film-assisted laser-induced forward transfer (AFA-LIFT), biological laser processing (BioLP), matrix-assisted pulsed laser evaporation direct writing (MAPLE DW), and laser-guided direct writing (LGDW).[17] In the LIFT process, cells are suspended in either a culture medium or a hydrogel and uniformly spread onto the bottom side of the print ribbon, typically made of quartz of glass and coated with a metal.[58] Cell suspension is then vaporized with a laser pulse focused onto the metal layer, which acts to move cells from the print ribbon to a receiving substrate.[59] LGDW is a variation technique of the commonly used optical trap (laser tweezers, optical tweezers), capable of depositing cells on different ECM matrices by using optical forces of a laser (700–1000 nm range, above the wavelength absorption of most proteins).[32] Nahmias *et al.* demonstrated LGDW by using a simple set of optics that can be mounted on any standard inverted microscope. Endothelial cells were deposited by simply pushing them toward a collagen-coated surface for adhesion.[60] LDW was also utilized by Chrisey and colleagues to deposit breast cancer cells and fibroblasts into spatially defined patterns on cultured rat mesenteric tissues and used as an *ex vivo* platform to study cancer cell dynamics.[61]

5.2.6 Dielectrophoresis

Dielectrophoresis (DEP) is a phenomenon in which polarizable objects are modified due to their interaction with a nonuniform electric field.[36] DEP has been applied on cells to align different types of cells in complex tissues, for tissue engineering purposes.[62,130] As an example, Ho *et al.* applied dielectrophoresis to pattern two types of cells. Endothelial and hepatic cells are initially randomly distributed in a microfluidic chamber, containing concentric stellate-tip electrodes that resemble the classic hepatic lobule structure in which the two cell types are distributed in a specific pattern. The positive DEP force applied showed an attraction for cells to the region of the local electric field maximum, providing the desired cell positioning.[63]

5.2.7 Other Techniques

Among other micropatterning techniques, we have (1) the masked ion beam lithography, which has been used to create patterns in nonstandard microfabrication materials, such as polymethylmethacrylate (PMMA), giving an additional option to silicon-based manufacturing techniques;[64] (2) electron beam lithography, which is similar to photolithography but exposes the resistance through a finely focused and precisely controlled beam of electrons, giving a resolution down to 10 nm. This technique is highly precise, but it can be used only with materials suitable for subsequent clean room etching techniques.[65] In a recent study, electron microscopy grids were patterned by contactless and mask-free photomicropatterning for site-specific deposition of ECM-related proteins. Precise cell positioning for micromachining was obtained by cryofocused ion beam milling. The complex micropatterns could be related to intracellular organization, allowing direct correlation between cell architecture and 3D intracellular structural characterization of the underlying molecular machinery.[66]

5.3 Controlling Tissue Geometry in 3D

It is important to design and engineer cellular microenvironments to direct cells *in vitro* with functional similarities to living tissues and organs. Simulating the native environment's 3D structures is essential. Although micropatterning techniques can now accurately reproduce microstructures on surfaces, 3D scaffolds with micropatterns have not yet been easily achieved in the same way. For this reason, 3D micropatterning fabrication has been explored through a variety of techniques (See Figure 5.3).

When transitioning a material system to 3D, a useful approach is the microfabrication of 3D microwells in a planar material, akin to creating a master *via* photolithography but where the cells can be integrated within the 3D "well". One of the first approaches using this methodology was to cast a polymer or hydrogel material onto a silicon master fabricated with photolithography.[67] Human embryonic stem cells have been cultured in microwells,[68,69] Khademhosseini and colleagues demonstrated how cardiac and endothelial cell differentiation from embryonic stem cells (ESCs) would depend on microwell geometry and associated paracrine signals in the microenvironment,[70] and Lee and colleagues showed how germ layer specification in ESCs could be influenced in concave microwells.[71] Adult stem cells were also shown to be influenced by the geometry of microwells, with fibroblastic morphology on planar substrates, compared to a morphology in 3D that corresponds to the geometry of the well.[72] In all of these cases, it should be noted that the adhesion and associated cytoskeleton were significantly affected by the transition from 2D to 3D, directly influencing the propagation of mechanochemical signals to direct differentiation programs.

Beyond this lithographic templating approach, Hoffman *et al.* used two-photon laser scanning lithography (TP-LSL), to create micropatterns in poly(ethylene glycol) (PEG) hydrogels with PEG-RGDS.[73] Applegate *et al.* used multiphoton absorption (MPA) of light to generate voids of 5 μm in diameter within silk fibroin hydrogels. Silk has been used widely in the tissue engineering field as it is cytocompatible, biodegradable, and able to stabilize labile compounds such as enzymes and drugs[74] (See Figure 5.3A.). In both of the studies mentioned, 3D micropatterns were created to study the cell migration process. Roberts *et al.* developed 3D microvessels with a diameter of the order of 100 μm, to improve endothelial cell culture *in vitro*. A mask of the network pattern was realized using high-resolution printing, from which PDMS was then patterned and used to inject collagen hydrogel at 0.6% and incubated at 37 °C to allow gelation. Once collagen has crosslinked, the PDMS molds are removed, and the top and bottom housing jigs are screwed together to enclose the network[75] (see Figure 5.3C).

Ma *et al.* proposed a maskless 3D micropatterning approach to generate 3D nanofibrous matrices with highly organized tubular architecture for tissue regeneration.[76] Gelatin nanofibers were fabricated by electrospinning and then micropatterned through a laser-guided technique using Leica LMD 7000 system. The obtained micropatterned matrix of tubular pore density of 20 000 mm^{-2} and

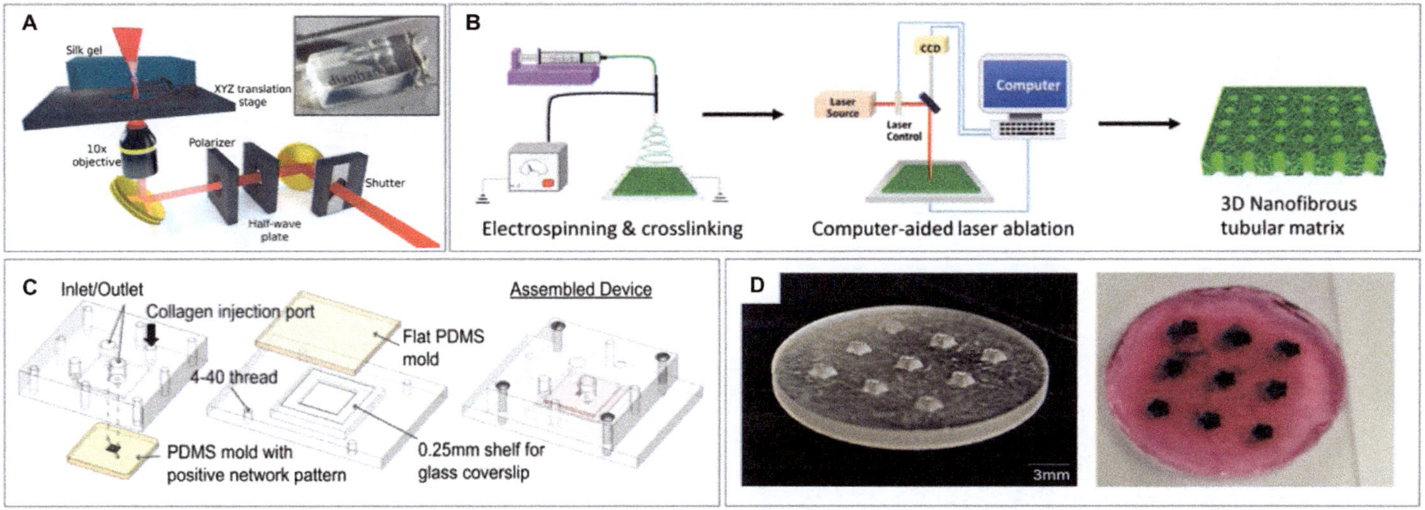

Figure 5.3 Examples of 3D micropatterning techniques: (A) Multiphoton absorption (MPA) for hydrogels patterning. Reproduced from ref. 74 with permission from National Academy of Sciences. (B) Fibronectin functionalized micropattern by multiphoton lasers. Reproduced from ref. 76 with permission from John Wiley & Sons, Copyright © 2017 WILEY-VCH Verlag GmbH & Co. KGaA, Weinheim. (C) 3D microvessels (100 μm diameter) realized with a mask of network pattern through high-resolution printing. Reproduced from ref. 75 with permission from JoVE, Copyright 2016. (D) Replica molding approach with 3D printing to create patterns for tumors *in vitro* and implant *in vivo*. Reproduced from ref. 78 with permission from John Wiley & Sons, Copyright © 2020 WILEY-VCH Verlag GmbH & Co. KGaA, Weinheim.

a tubular diameter of less than 5 μm, resembled the natural dentinal matrix that the researchers were trying to recreate in order to use the obtained material as a scaffold for tubular dentin regeneration (see Figure 5.3B).

Jeon and Alsberg were able to spatially pattern growth factors to bind to heparin, by making dual-crosslinkable hydrogels consisting of methacrylated alginate hydrogels photofunctionalized with methacrylated heparin. Heparin-binding growth factors, such as FGFs, VEGF, TGF-βs, and BMPs, were then able to be 3D micropatterned in the hydrogels.[77] Our group recently 3D micropatterned poly(ethylene glycol) (PEGDM) and gelatin-based hydrogels, by using 3D printing technique to generate a resin inverse topology, in which the biocompatible PEGDM is then added into the mold and allowed to crosslink[78] (see Figure 5.3D). Details of application of these hydrogels are given in the next section. An important advantage of photoresponsive biomaterials is the capability of spatiotemporal tunability, which allows site-specific drug or other therapeutic delivery *in vivo* and biochemical and biophysical modulation of synthetic cell culture, in order to mimic the dynamic 4D environment present in the human body.[79]

5.4 Deconstructing Biological Behavior Through Micropatterning

In this section we will explore how microengineering techniques have made an impact on elucidating the properties of the microenvironment that directs diverse processes in human physiology and pathology. Early work by Watt and colleagues[80] and Whitesides, Ingber, and colleagues[81] demonstrated how adhesive area guides DNA synthesis, thereby directing terminal differentiation in keratinocytes or apoptosis in fibroblasts, respectively. Using micropatterned materials, subsequent work demonstrated how geometry will also influence cell growth and migration[82–85] and cell adaptation and lineage determination.[86–92] Changing the nonadhesive space under cells with the same triangular area, was shown to direct focal adhesion formation and the contractility of tethered actomyosin structures,[37] and intermediate filaments were shown to be guided by the interfacial boundary of adhesive islands.[93] In addition to filaments, cell polarity and protrusive features can be directed by the geometry of single micropatterned cells,[94] which corresponds to migratory phenotypes.[95,96] See Figure 5.4A for examples of how geometry directs adhesion and cytoskeletal elements.

In addition to these works, numerous fundamental studies of cell and tissue systems have used micropatterned materials, including signaling and morphogenesis in endothelial cells,[97,98] liver cells,[15] fibroblasts,[99] and skeletal muscle cells.[40]

5.4.1 Teasing out Mechanotransduction During Stem Cell Differentiation

While these pioneering studies with micropatterning focused on questions related to control of adhesive and motility features, the most studied

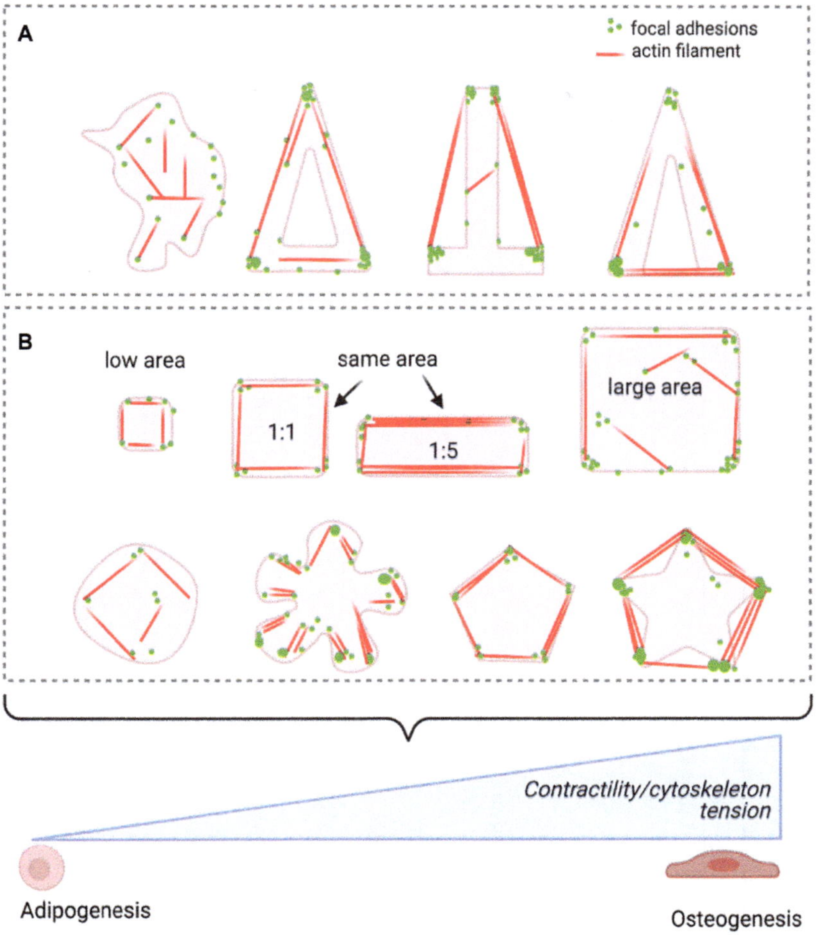

Figure 5.4 Schematical representations of (A) how geometry directs cell adhesion and cytoskeleton, and (B) how cell shape and related contractility influence cell differentiation.

biological areas in recent years are the use of micropatterns for stem cell mechanobiology, lineage determination, and investigations into cancer signaling.[27] Geometric confinement is known to be a critical aspect in guiding embryogenesis, where pluripotent stem cells start differentiating to create specific tissue patterns that ultimately make up organs in the developing embryo.[100] Recent studies have started to use micropatterning techniques to control pluripotent stem cell differentiation. For instance, Table 5.2 lists examples of micropatterned substrates that have been used for studying adult and pluripotent stem cells. Warmflash and colleagues used commercially available micropatterned glass to culture ESCs and determine how the cells in the population spatially distribute during germ layer differentiation.[101] Pagliari *et al.* showed that mesoderm specification of induced pluripotent stem cells (iPSCs) can be obtained by using specific

Table 5.2 Examples of patterning for stem cell studies.[a]

Material	Patterning technique	Purpose	Reference
Glass	Micropattern on glass obtained by photolithography. Glass coverslip was embedded in a chamber system.	Micropatterned chamber for quantitative parallel analysis of neural stem cell proliferation and differentiation	104
Glass	Printing proteins with noncontact arrayer (*BCA arrayer*, Perkin–Elmer)	Neural precursor differentiation	105
Photoreactive PVA-coated TCP	UV lithography	MSC transfection	107
Discontinuous PDTEC substrates	Phase separation in a mixture containing substrate (PDTEC) and sacrificial (PS) polymers	Stem cell differentiation (osteogenic and adipogenic differentiation)	131
Glass	Multiphoton	MSC focal adhesion maturation studies	113
Titanium	Combination of photolithography and ICP-based dry etching	MSC osteogenic differentiation	112
PLGA	Microcontact printing	aMSC, bmMSC, fMSC myogenesis	111
Polytetrafluoroethylene (PTFE)	Spraying natural microscopic droplets that are adsorbed on biomaterials surface	Enhancing endothelialization of bovine aortic endothelial cells	98
PA hydrogels	Microcontact printing	MSC differentiation	132
Glass coverslips (Arena CYTOO chips)	CYTOO chips obtained by photolitography Matrigel-coated CYTOO chips	Controlling iPSC mesoderm specification	102
	Laminin-coated CYTOO chips	Controlling human embryonic stem cells spatial organization	133
	Laminin-coated CYTOO chips	Study of gastrulation process in mouse iPSC	134

[a]MSC: mesenchymal stem cells; PVA: polyvinyl alcohol; TCP: tissue culture polystyrene; PDTEC: poly(desaminotyrosyl-tyrosine carbonate); ICP: inductively coupled plasma; PLGA: poly(lactic-*co*-glycolic acid); PDMS: polydimethylsiloxane; PA: polyacrylamide; aMSC: adipose-derived MSC; bmMSC: bone marrow–derived fMSC: fetal tissue–derived MSC; iPSC: induced pluripotent stem cells.

micropatterns and identified YAP-TEAD1 transcriptional activity responsible for the mechanotransduction process.[102]

During embryogenic nervous system development, the axonal pathfinding is crucial for cell differentiation into neural precursor cells (NPC).[103] NPC differentiation has been studied using micropatterned glass surfaces.[104–106] Yang *et al.* prepared micropatterned surfaces by UV lithography with photomasks and used them to control cell spreading area and elongation of hMSCs. These cell parameters were used to assess transfection efficiency and cellular uptake of complexes and DNA synthesis activity. Their study demonstrated that highly spread and elongated cells having an aspect ratio of 8 : 1 had higher transfection efficiency compared to other groups and well organized actin filaments.[107]

Single epidermal stem cells have been captured on micropatterned surfaces to study differentiation, where it was demonstrated that constraining cell area would augment terminal differentiation in keratinocytes.[108] McBeath, Chen, and colleagues used micropatterned glass substrates to show how controlling the area of single cells would influence cytoskeletal tension and lineage commitment in MSCs. When exposed to media supplements for adipogenesis, myogenesis, chondrogenesis, and osteogenesis, small round cells would prefer to undergo adipogenesis[25] or chondrogenesis,[109] while large flat cells would prefer to undergo osteogenesis[25] or myogenesis.[109] Further work by Kilian, Mrksich, and colleagues demonstrated how changing aspect ratio or the degree of curvature at the periphery of single patterned MSCs would similarly influence contractility and integrated soluble signaling pathways.[110] As aspect ratio or the degree of negative curvature at the cell edge increases, there are larger contractile bundles between the points, which translates into increased osteogenic gene expression. See Figure 5.4B for examples of shape-contractility-differentiation relationships.

Poly(lactic-*co*-glycolic acid) (PLGA) micropatterned films were used to explore MSCs from different tissue sources, such as adipose, fetal, and bone marrow tissues, for their ability to undergo myogenesis.[111] Zhu *et al.* focused their study on MSC osteogenic differentiation and found that microgrooved titanium substrates have no influence on cell proliferation, but they dramatically influence cell osteogenic differentiation *in vitro*. After screening 34 microgrooved patterns, the one with 3 mm ridge width, 7 mm groove width, and 2 mm depth showed maximum focal adhesion formation and related osteogenic differentiation.[112] MSC mature adhesion process, where cells form fibrillar adhesion (FBA) and 3D matrix adhesion (3DMA), has been studied by using fibronectin micropatterned glass substrates at different concentrations and volumes, in order to obtain a gradient of stiffness that would mimic the native ECM in which cells adhere. This study demonstrated that fibronectin functionalized patterns with high anisotropy benefit the FBA maturation. Other characteristics of the stem cell niches, such as higher bending of cells seemed to have a positive effect on the maturation of cell–matrix adhesion.[113]

5.4.2 Controlling Shape and Interfacial Topography in Cancer Cultures

One of the major areas in biology studied using micropatterned surfaces is cancer cell behavior (Table 5.3). Much early work was devoted to studying how shape can influence cancer cell adhesion and migratory features.[94,114,115] As hyaluronic acid (HA) is the ECM component known to facilitate tumor progression, Dickinson and Gerecht analyzed the interaction between HA and cancer cells by using micropatterned surfaces and discovered that cell adhesion to HA occurs with CD44 receptor.[116,117] HA and CD44 correlation was already proposed to be a key regulator of cancer cell proliferation, migration, and invasion and thus associated with patient outcome.[118]

Weihs and colleagues showed that cancer cells and benign cells differ in the way they adhere long-term to microchannels, where cancer cells prefer to adhere on high-curvature channel bottoms as single cells, whereas benign cells form tissue-like cell sheet in narrow channels, suggesting a way to distinguish the two cell types and to target them accordingly.[119] Zhong *et al.* used m-Eraser strategy to micropattern cells on either PLGA or tissue culture plate (TCP) substrates. The technique consisted in connecting a PDMS stamp to a mechanical device for stable and repeated moving. When cell culture medium was removed, the PDMS stamp was pressed onto a confluent layer of cells to induce their lysis and then their removal by PBS wash. In this way, one type of cells was patterned. At this point, the second cell type was seeded to obtain coculture micropatterns that were used to evaluate the efficacy of anticancer drugs, such as doxorubicin.[120]

Microengineered hydrogels can control cell and tissue geometry as well as matrix mechanics independently and together. Gomez, Nelson, Bissel, and colleagues demonstrated how regions of curvature within collagen microislands would augment epithelial to mesenchymal transition in breast epithelial

Table 5.3 Cancer cells micropatterning.[a]

Material	Patterning technique	Purpose	Reference
(PEGDM) (10 kDa)	Stereolithography	Microtumors modeling	78
Glass-coated with collagen	Laser engraver (Epilog)-based patterning	Role of mitochondrial activities in breast cancer model	135
TCP and PLGA nanofiber sheets	m-Eraser strategy	Lung cancer bone metastasis modeling	120
Titanium grids on SiO_2 films	Cryopatterning	Shape control of HeLa cells	66
GelMA	Two-step photolithography	Breast cancer migration modeling	129
Glass patterned with hyaluronic acid	Microcontact printing	Study of cancer cells' interaction with hyaluronic acid	116
PLGA gels	Soft lithography	Differences between benign and cancer cells	119

[a]PEGDM: poly(ethylene glycol) dimethacryloyl; TCP: tissue culture polystyrene; PLGA: poly(lactic-*co*-glycolic acid); GelMA: gelatin methacrylate.

cells and reveal geometry-invasion relationships.[121–123] Using polyacrylamide conjugated with matrix protein, we demonstrated how confined populations of melanoma cells on hydrogels of varying stiffness would undergo epithelial to mesenchymal transition at the interface coinciding with reprogramming to a cancer stem cell-like state[124] (see Figure 5.5C). Following on from this work, it was revealed that the microscale topography of a tumor could enhance

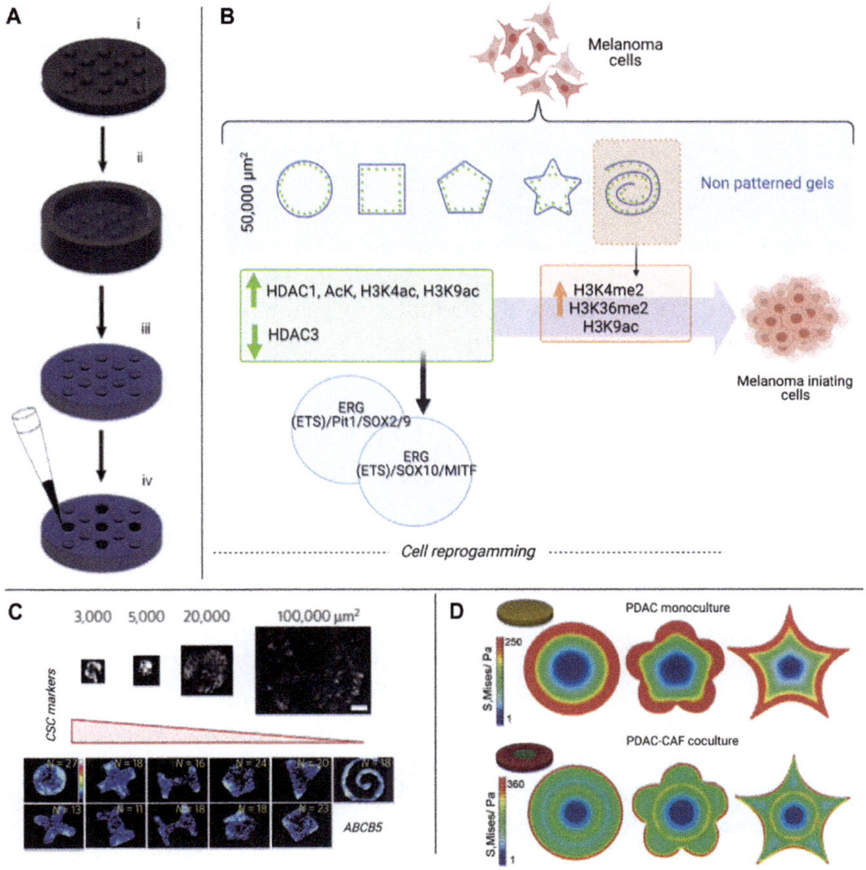

Figure 5.5 (A) Schematical representation of replica moulding method used to fabricate 3D micropatterned hydrogels. Reproduced from ref. 78 with permission from John Wiley & Sons, Copyright © 2020 WILEY-VCH Verlag GmbH & Co. KGaA, Weinheim. (B) illustration of epigenetic factors involved in cell reprogramming when using micropatterned geometric cues for cell culture. (C) With decreasing pattern size, individual cells occupy longer arcs that correspond to higher cancer stem cell (CSC) markers, such as ABCB5. Reproduced from ref. 124 with permission from Springer Nature, Copyright 2016. (D) Computational modeling representation of stress induced on cancerous pancreatic cells (pancreatic ductal adenocarcinoma, PDAC) cultures when cultured on PA hydrogels with different shape micropatterns, either in mono (top) or in coculture with noncancerous pancreatic cells (cancer-associated fibroblasts, CAF) (bottom). Reproduced from ref. 128 with permission from John Wiley & Sons, Copyright © 2021 Wiley-VCH GmbH.

stemness through mechanotransduction *via* signal transduction pathways coinciding with hypoxia signaling.[125] It is known that the malignant transformation of a healthy cell is a process mediated by the combination of genetic abnormalities and epigenetic alterations.[126] In order to better understand the relationship between geometry and cancer state, we investigated a series of epigenetic markers controlling oncogenic genes activation, to see whether they may regulate cancer stem-like behavior in melanoma.[127] Using a soft lithography micropatterning approach that primes stem fractions at the interface,[124] soft hydrogels were prepared across a range of shapes or without patterns. When melanoma cells were cultured for at least 5 days on each substrate, results showed that cells displayed elevated histone 3 methylation, such as H3K4me2 and H3K36me2, compared to cells on nonpatterned surfaces. Moreover, cells at the periphery of each pattern showed higher histone 3 acetylation, which corresponds to chromatin accessibility and gene regulation. H3K9ac in particular was significantly elevated in cells at the periphery of each shape, compared to cells in the center, which corresponds to the stemness phenotype. Figure 5.5B shows a schematic of main findings of epigenetic markers influenced by cell geometry. Recently, using the same platform we showed how cocultures of two or more cells can be integrated within micropatterned substrates.[128] By adjusting the seeding density, pancreatic cancer cells coordinate their spatial assembly through stromal cell activity and biophysical cues from geometry and stiffness, with evidence for bidirectional crosstalk facilitating invasive signaling (see Figure 5.5D). Together, these studies demonstrate how micropatterned hydrogels can be used to deconstruct dynamic cancer processes like invasion and heterogeneity.

Toward true 3D microenvironments, Nikkhah and colleagues developed a 3D tumor model to study breast cancer migratory systems. For this purpose, they used both breast cancer cell lines (MDA-MB-231, MCF7) and non-tumorigenic mammary epithelial cells (MCF10A) and embedded separately in the tumor model. To briefly describe the 3D model, it was generated by a two-step photolithography technique where high density array of circular constructs were surrounded by lower-stiffness hydrogels. This was obtained by different crosslinking incubation time. Increased exposure to the crosslinking method resulted in obtaining higher-stiffness GelMA hydrogels, located in the circular constructs. Each cell type was embedded in the gel forming those circular constructs. As a result, cancer cell line MDA-MB-231 showed great migratory ability and invaded the surrounding matrix; MCF7 or MCF10A could only form clusters and remained confined within the circular regions where they were initially positioned.[129]

Molley and colleagues developed a different 3D model using poly(ethylene glycol) and gelatin-based hydrogels (PEGDM) where melanoma cells could be not only spatially localized but also in different shapes, including hemisphere, cube, pyramid, and star shapes. A stereolithographic (SLA) 3D printer was used to fabricate resin molds that were subsequently filled with prehydrogel solution consisting of PEGDM. The hydrogel was then allowed to crosslink and be removed from the mold. At this point, ECM proteins, such as fibronectin, were physically absorbed at the interface by protein deposition in the semidried microwells, followed by cell adhesion (see Figure 5.5A). Through this method, it was possible to generate tumor cell aggregates *in vitro* that could be then implanted *in vivo* and gave the opportunity to study the relationship between geometry and invasion in the progression of a tumor.[78] Importantly, geometric structuring led to enrichment of cells adopting a stem cell-like phenotype at periphery features similar to that observed in 2D studies.[124]

5.5 Conclusion and Future Perspectives

From the early embryo to the adult organism, form underlies function, with complex spatiotemporal organization of cells. The development of micropatterned systems has provided a better understanding of how the geometry and topography of the ECM impacts on cellular physiology. Going forward, through advances in microengineering techniques, in particular in the 3D and 4D platforms, it will be possible to generate *in vitro* models that resemble accurately the structure of the native environment, and thereby facilitate elucidation of cell and tissue function.

Abbreviations

2D	two-dimensional
3D	three-dimensional
CAF	cancer-associated fibroblasts
COC	cyclic olefin copolymer
ECM	extracellular matrix
ESCs	embryonic stem cells
hMSCs	human mesenchymal stem cells
iPSCs	induced pluripotent stem cells
MSCs	mesenchymal stem cells
NPC	neural progenitor cells
PA	polyacrylamide
PDAC	pancreatic ductal adenocarcinoma
PDMS	polydimethylsiloxane
PEGDA	poly (ethylene glycol) diacrylate
PI	polyimide
PLGA	poly(lactic-*co*-glycolic acid)
PMMA	polymethylmethacrylate
TCP	tissue culture plate

References

1. C. M. Nelson and J. P. Gleghorn, *Annu. Rev. Biomed. Eng.*, 2012, **14**, 129–154.
2. C. Dahmann, A. C. Oates and M. Brand, *Nat. Rev. Genet.*, 2011, **12**, 43–55.
3. F. Fagotto, *Development*, 2014, **141**, 3303–3318.

4. D. A. C. Walma and K. M. Yamada, *Development*, 2020, 147.
5. F. Guilak, D. M. Cohen, B. T. Estes, J. M. Gimble, W. Liedtke and C. S. Chen, *Cell Stem Cell*, 2009, **5**, 17–26.
6. A. J. Engler, S. Sen, H. L. Sweeney and D. E. Discher, *Cell*, 2006, **126**, 677–689.
7. D. E. Discher, P. Janmey and Y.-L. Wang, *Science (80-.)*, 2005, **310**, 1139–1143.
8. M. J. Dalby, N. Gadegaard, R. Tare, A. Andar, M. O. Riehle, P. Herzyk, C. D. W. Wilkinson and R. O. C. Oreffo, *Nat. Mater.*, 2007, **6**, 997–1003.
9. C. S. Chen, *Science (80-.)*, 1997, **276**, 1425–1428.
10. L. E. McNamara, R. J. McMurray, M. J. P. Biggs, F. Kantawong, R. O. C. Oreffo and M. J. Dalby, *J. Tissue Eng.*, 2010, **1**, 1–13.
11. H. Donnelly, M. Salmeron-Sanchez and M. J. Dalby, *J. R. Soc., Interface*, 2018, 15.
12. J. Laurent, G. Blin, F. Chatelain, V. Vanneaux, A. Fuchs, J. Larghero and M. Théry, *Nat. Biomed. Eng.*, 2017, **1**, 939–956.
13. C. H. O'Neill, P. N. Riddle and P. W. Jordan, *Cell*, 1979, **16**, 909–918.
14. A. Harris, *Exp. Cell Res.*, 1973, **77**, 285–297.
15. G. H. Underhill and S. R. Khetani, *CMGH*, 2018, **5**, 426–439, e1.
16. D. Cuvelier, O. Rossier, P. Bassereau and P. Nassoy, *Eur. Biophys. J.*, 2003, **32**, 342–354.
17. A. Martinez-Rivas, G. K. González-Quijano, S. Proa-Coronado, C. Séverac and E. Dague, *Micromachines*, 2017, **8**, 12.
18. T. C. McDevitt, K. A. Woodhouse, S. D. Hauschka, C. E. Murry and P. S. Stayton, *J. Biomed. Mater. Res., Part A*, 2003, **66**, 586–595.
19. C. S. Chen, M. Mrksich, S. Huang, G. M. Whitesides and D. E. Ingber, *Science (80-.)*, 1997, **276**, 1425–1428.
20. M. Théry, V. Racine, M. Piel, A. Pépin, A. Dimitrov, Y. Chen, J. B. Sibarita and M. Bornens, *Proc. Natl. Acad. Sci. U. S. A.*, 2006, **103**, 19771–19776.
21. M. Mrksich, L. E. Dike, J. Tien, D. E. Ingber and G. M. Whitesides, *Exp. Cell Res.*, 1997, **235**, 305–313.
22. G. M. Whitesides, E. Ostuni, S. Takayama, X. Jiang and D. E. Ingber, *Annu. Rev. Biomed. Eng.*, 2001, **3**, 335–373.
23. H. Do Kim and S. R. Peyton, *Integr. Biol.*, 2012, **4**, 37–52.
24. R. A. Desai, L. Gao, S. Raghavan, W. F. Liu and C. S. Chen, *J. Cell Sci.*, 2009, **122**, 905–911.
25. R. McBeath, D. M. Pirone, C. M. Nelson, K. Bhadriraju and C. S. Chen, *Dev. Cell*, 2004, **6**, 483–495.
26. J. Swift, I. L. Ivanovska, A. Buxboim, T. Harada, P. C. D. P. Dingal, J. Pinter, J. D. Pajerowski, K. R. Spinler, J.-W. Shin and M. Tewari, *Science (80-.)*, 2013, **341**, 1240104.
27. M. Théry, *J. Cell Sci.*, 2010, **123**, 4201–4213.
28. E. D'Arcangelo and A. P. McGuigan, *Biotechniques*, 2015, **58**, 13–23.
29. T. Vignaud, H. Ennomani and M. Théry, *Methods Cell Biol.*, 2014, **120**, 93–116.

30. M. Ermis, E. Antmen and V. Hasirci, *Bioact. Mater.*, 2018, **3**, 355–369.
31. D. Falconnet, G. Csucs, H. M. Grandin and M. Textor, *Biomaterials*, 2006, **27**, 3044–3063.
32. L. R. Trump, G. Timp and L. B. Schook, *Stem Cells Revasc. Ther.*, 2011, 107–130.
33. A. Kumar and G. M. Whitesides, *Appl. Phys. Lett.*, 1993, **63**, 2002–2004.
34. Y. Xia and G. M. Whitesides, *Angew. Chem., Int. Ed.*, 1998, **37**, 550–575.
35. S. Alom Ruiz and C. S. Chen, *Soft Matter*, 2007, **3**, 168–177.
36. F. L. Yap and Y. Zhang, *Biosens. Bioelectron.*, 2007, **22**, 775–788.
37. M. Théry, A. Pépin, E. Dressaire, Y. Chen and M. Bornens, *Cell Motil. Cytoskeleton*, 2006, **63**, 341–355.
38. M. Mrksich, C. S. Chen, Y. Xia, L. E. Dike, D. E. Ingber and G. M. Whitesides, *Proc. Natl. Acad. Sci. U. S. A.*, 1996, **93**, 10775–10778.
39. J. C. Love, D. B. Wolfe, M. L. Chabinyc, K. E. Paul and G. M. Whitesides, *J. Am. Chem. Soc.*, 2002, **124**, 1576–1577.
40. K. Y. Vajanthri, R. K. Sidu and S. K. Mahto, *IRBM*, 2020, **41**, 48–57.
41. N. L. Jeon, K. Finnie, K. Branshaw and R. G. Nuzzo, *Langmuir*, 1997, **7463**, 3382–3391.
42. E. Hoque, J. A. DeRose, P. Hoffmann, H. J. Mathieu, B. Bhushan and M. Cichomski, *J. Chem. Phys.*, 2006, **124**, 174710.
43. O. Acton, D. Hutchins, L. Árnadóttir, T. Weidner, N. Cernetic, G. G. Ting, T.-W. Kim, D. G. Castner, H. Ma and A. K.-Y. Jen, *Adv. Mater.*, 2011, **23**, 1899–1902.
44. A. Perl, D. N. Reinhoudt and J. Huskens, *Adv. Mater.*, 2009, **21**, 2257–2268.
45. J. L. Tan, J. Tien and C. S. Chen, *Langmuir*, 2001, **21205**, 2002–2004.
46. F. Huo, G. Zheng, X. Liao, L. R. Giam, J. Chai, X. Chen, W. Shim and C. A. Mirkin, *Nat. Nanotechnol.*, 2010, **5**, 637–640.
47. M. D. Cabezas, B. Meckes, C. A. Mirkin and M. Mrksich, *ACS Nano*, 2019, acsnano.9b03937.
48. L. J. Millet, M. E. Stewart, R. G. Nuzzo and M. U. Gillette, *Lab Chip*, 2010, **10**, 1525–1535.
49. J. T. Koepsel and W. L. Murphy, *Langmuir*, 2009, **25**, 12825–12834.
50. R. A. Desai, S. B. Gopal, S. Chen and C. S. Chen, *J. R. Soc., Interface*, 2013, **10**, 1–9.
51. S. A. Ruiz and C. S. Chen, *Stem Cells*, 2008, **26**, 2921–2927.
52. J. Lee, A. A. Abdeen and K. A. Kilian, *Sci. Rep.*, 2014, **4**, 5188.
53. M. Zhang, S. Krishnamoorthy, H. Song, Z. Zhang and C. Xu, *J. Appl. Phys.*, 2017, **121**, 124904.
54. H. G. Yi, H. Lee and D. W. Cho, *Bioengineering*, 2017, **4**, 10.
55. A. Yusof, H. Keegan, C. D. Spillane, O. M. Sheils, C. M. Martin, J. J. O'Leary, R. Zengerle and P. Koltay, *Lab Chip*, 2011, **11**, 2447–2454.
56. L. Mecozzi, O. Gennari, R. Rega, L. Battista, P. Ferraro and S. Grilli, *Macromol. Biosci.*, 2017, **17**, 1600307.
57. P. Y. Chiou, A. T. Ohta and M. C. Wu, *Nature*, 2005, **436**, 370–372.
58. E. Fogarassy, C. Fuchs, F. Kerherve, G. Hauchecorne and J. Perriere, *J. Mater. Res.*, 1989, **4**, 1082–1086.

59. N. R. Schiele, D. T. Corr, Y. Huang, N. A. Raof, Y. Xie and D. B. Chrisey, *Biofabrication*, 2010, **2**, 1–14.
60. Y. Nahmias and D. J. Odde, *Nat. Protoc.*, 2006, **1**, 2288–2296.
61. T. B. Phamduy, R. S. Sweat, M. S. Azimi, M. E. Burow, W. L. Murfee and D. B. Chrisey, *Integr. Biol.*, 2015, **7**, 1068–1078.
62. Z. X. Zhou, F. Buchanan, A. Lennon and N. Dunne, *Key Eng. Mater.*, 2015, **631**, 306–311.
63. C. T. Ho, R. Z. Lin, W. Y. Chang, H. Y. Chang and C. H. Liu, *Lab Chip*, 2006, **6**, 724–734.
64. J. L. Charest and W. P. King, *BioNanoFluidic MEMS*, 2008, pp. 251–278.
65. A. N. Broers, A. C. F. Hoole Andrew and J. M. Ryan, *Microelectron. Eng.*, 1996, **32**, 131–142.
66. M. Toro-Nahuelpan, I. Zagoriy, F. Senger, L. Blanchoin, M. Théry and J. Mahamid, *Nat. Methods*, 2020, **17**, 50–54.
67. M. Charnley, M. Textor, A. Khademhosseini and M. P. Lutolf, *Integr. Biol.*, 2009, **1**, 625–634.
68. J. C. Mohr, J. J. de Pablo and S. P. Palecek, *Biomaterials*, 2006, **27**, 6032–6042.
69. H.-C. Moeller, M. K. Mian, S. Shrivastava, B. G. Chung and A. Khademhosseini, *Biomaterials*, 2008, **29**, 752–763.
70. Y.-S. Hwang, B. G. Chung, D. Ortmann, N. Hattori, H.-C. Moeller and A. Khademhosseini, *Proc. Natl. Acad. Sci. U. S. A.*, 2009, **106**, 16978–16983.
71. Y. Y. Choi, B. G. Chung, D. H. Lee, A. Khademhosseini, J.-H. Kim and S.-H. Lee, *Biomaterials*, 2010, **31**, 4296–4303.
72. X. Xu, W. Wang, K. Kratz, L. Fang, Z. Li, A. Kurtz, N. Ma and A. Lendlein, *Adv. Healthcare Mater.*, 2014, **3**, 1991–2003.
73. J. C. Hoffmann and J. L. West, *Integr. Biol.*, 2013, **5**, 817–827.
74. M. B. Applegate, J. Coburn, B. P. Partlow, J. E. Moreau, J. P. Mondia, B. Marelli, D. L. Kaplan, F. G. Omenetto and D. A. Weitz, *Proc. Natl. Acad. Sci. U. S. A.*, 2015, **112**, 12052–12057.
75. M. A. Roberts, S. S. Kotha, K. T. Phong and Y. Zheng, *J. Vis. Exp*, 2016, **2016**, 54457.
76. C. Ma, T. Qu, B. Chang, Y. Jing, J. Q. Feng and X. Liu, *Adv. Healthcare Mater.*, 2018, **7**, 3.
77. O. Jeon, K. Lee and E. Alsberg, *Small*, 2018, **14**, 1800579.
78. T. G. Molley, X. Wang, T. tyng Hung, P. B. Jayathilaka, J. L. Yang and K. A. Kilian, *Adv. Biosyst.*, 2020, **4**, 1–8.
79. E. R. Ruskowitz and C. A. Deforest, *Nat. Rev. Mater.*, 2018, **3**(17087), 1–14.
80. F. M. Watt, P. W. Jordan and C. H. O'Neill, *Proc. Natl. Acad. Sci. U. S. A.*, 1988, **85**, 5576–5580.
81. C. S. Chen, M. Mrksich, S. Huang, G. M. Whitesides and D. E. Ingber, *Science (80-.)*, 1997, **276**, 1425–1428.
82. P. Roca-Cusachs, J. Alcaraz, R. Sunyer, J. Samitier, R. Farré and D. Navajas, *Biophys. J.*, 2008, **94**, 4984–4995.
83. R. G. Thakar, Q. Cheng, S. Patel, J. Chu, M. Nasir, D. Liepmann, K. Komvopoulos and S. Li, *Biophys. J.*, 2009, **96**, 3423–3432.

84. D. Gospodarowicz, G. Greenburg and C. R. Birdwell, *Cancer Res.*, 1978, **38**, 4155–4171.
85. K. Kandere-Grzybowska, C. J. Campbell, G. Mahmud, Y. Komarova, S. Soh and B. A. Grzybowski, *Soft Matter*, 2007, **3**, 672–679.
86. J. Le Beyec, R. Xu, S.-Y. Lee, C. M. Nelson, A. Rizki, J. Alcaraz and M. J. Bissell, *Exp. Cell Res.*, 2007, **313**, 3066–3075.
87. C. M. Nelson, D. Khauv, M. J. Bissell and D. C. Radisky, *J. Cell. Biochem.*, 2008, **105**, 25–33.
88. N. Jain, K. V. Iyer, A. Kumar and G. V. Shivashankar, *Proc. Natl. Acad. Sci.*, 2013, **110**, 11349–11354.
89. J. Lee, A. A. Abdeen, T. H. Huang and K. A. Kilian, *J. Mech. Behav. Biomed. Mater.*, 2014, **38**, 209–218.
90. X. Yao, R. Peng and J. Ding, *Biomaterials*, 2013, **34**, 930–939.
91. B. M. Spiegelman and C. A. Ginty, *Cell*, 1983, **35**, 657–666.
92. M. J. Bissell, D. Farson and A. S. Tung, *J. Supramol. Struct.*, 1977, **6**, 1–12.
93. S. H. Shabbir, M. M. Cleland, R. D. Goldman and M. Mrksich, *Biomaterials*, 2014, **35**, 1359–1366.
94. J. James, E. D. Goluch, H. Hu, C. Liu and M. Mrksich, *Cell Motil. Cytoskeleton*, 2008, **65**, 841–852.
95. X. Jiang, D. A. Bruzewicz, A. P. Wong, M. Piel and G. M. Whitesides, *Proc. Natl. Acad. Sci. U. S. A.*, 2005, **102**, 975–978.
96. G. Mahmud, C. J. Campbell, K. J. M. Bishop, Y. A. Komarova, O. Chaga, S. Soh, S. Huda, K. Kandere-Grzybowska and B. A. Grzybowski, *Nat. Phys.*, 2009, **5**, 606–612.
97. D. E. J. Anderson and M. T. Hinds, *Ann. Biomed. Eng.*, 2011, **39**, 2329–2345.
98. L. Gagne and G. Laroche, *Adv. Mater. Res.*, 2007, **2**, 77–82.
99. J. A. Park, H. Lee, S. Park and S. Jung, *Adv. Biosyst.*, 2020, **4**, 2070054.
100. G. Blin, D. Wisniewski, C. Picart, M. Thery, M. Puceat and S. Lowell, *Stem Cells Regener.*, 2018, **145**, dev166025.
101. A. Deglincerti, F. Etoc, M. C. Guerra, I. Martyn, J. Metzger, A. Ruzo, M. Simunovic, A. Yoney, A. H. Brivanlou, E. Siggia and A. Warmflash, *Nat. Protoc.*, 2016, **11**, 2223–2232.
102. S. Pagliari, V. Vinarsky, F. Martino, A. R. Perestrelo, J. Oliver De La Cruz, G. Caluori, J. Vrbsky, P. Mozetic, A. Pompeiano, A. Zancla, S. G. Ranjani, P. Skladal, D. Kytyr, Z. Zdráhal, G. Grassi, M. Sampaolesi, A. Rainer and G. Forte, *Cell Death Differ.*, 2020, 1–15.
103. B. A. Reynolds and S. Weiss, *Science (80-.)*, 1992, **255**, 1707–1710.
104. V. I. Chin, P. Taupin, S. Sanga, J. Scheel, F. H. Gage and S. N. Bhatia, *Biotechnol. Bioeng.*, 2004, **88**, 399–415.
105. Y. Soen, A. Mori, T. D. Palmer and P. O. Brown, *Mol. Syst. Biol.*, 2006, **2**, 37.
106. P. García-Parra, F. Cavaliere, M. Maroto, L. Bilbao, I. Obieta, A. L. de Munain, J. I. Álava and A. Izeta, *Front. Cell. Neurosci.*, 2012, **6**, 1–12.
107. Y. Yang, X. Wang, X. Hu, N. Kawazoe, Y. Yang and G. Chen, *ACS Appl. Mater. Interfaces*, 2019, **11**, 1932–1941.

108. J. T. Connelly, J. E. Gautrot, B. Trappmann, D. Wei-Min Tan, G. Donati, W. T. Huck and F. M. Watt, *Nat. Cell Biol.*, 2010, **12**, 711–718.
109. L. Gao, R. McBeath and C. S. Chen, *Stem Cells*, 2010, **28**, 564–572.
110. K. A. Kilian, B. Bugarija, B. T. Lahn and M. Mrksich, *Proc. Natl. Acad. Sci. U. S. A.*, 2010, **107**, 4872–4877.
111. T. Yu, C. K. Chua, C. Y. Tay, F. Wen, H. Yu, J. K. Y. Chan, M. S. K. Chong, D. T. Leong and L. P. Tan, *Macromol. Biosci.*, 2013, **13**, 799–807.
112. M. Zhu, H. Ye, J. Fang, C. Zhong, J. Yao, J. Park, X. Lu and F. Ren, *ACS Appl. Mater. Interfaces*, 2019, **11**, 43888–43901.
113. J. Ma, C. Li, N. Huang, X. Wang, M. Tong, A. H. W. Ngan and B. P. Chan, *ACS Appl. Mater. Interfaces*, 2017, **9**, 29469–29480.
114. A. I. Baba and C. Câtoi, *Tumor Cell Morphology*, The Publishing House of the Romanian Academy, 2007.
115. G. P. Gupta and J. Massagué, *Cell*, 2006, **127**, 679–695.
116. L. E. Dickinson and S. Gerecht, *J. Vis. Exp.*, 2010, **46**, 2413.
117. L. E. Dickinson, C. C. Ho, G. M. Wang, K. J. Stebe and S. Gerecht, *Biomaterials*, 2010, **31**, 5472–5478.
118. M. Götte and G. W. Yip, *Cancer Res.*, 2006, **66**, 10233–10237.
119. M. B. Alvarez-Elizondo, C. W. Li, A. Marom, Y. T. Tung, G. Drillich, Y. Horesh, S. C. Lin, G. J. Wang and D. Weihs, *Med. Eng. Phys.*, 2020, **75**, 5–12.
120. H. Zhong, L. Xuan, D. Wang, J. Zhou, Y. Li and Q. Jiang, *RSC Adv.*, 2017, 7, 21837.
121. E. Boghaert, J. P. Gleghorn, K. Lee, N. Gjorevski, D. C. Radisky and C. M. Nelson, *Proc. Natl. Acad. Sci. U. S. A.*, 2012, **109**, 19632–19637.
122. E. W. Gomez, Q. K. Chen, N. Gjorevski and C. M. Nelson, *J. Cell. Biochem.*, 2010, **110**, 44–51.
123. H. Mori, N. Gjorevski, J. L. Inman, M. J. Bissell, C. M. Nelson and K. Yamada, *Proc. Natl. Acad. Sci. U. S. A.*, 2009, **106**, 14890–14895.
124. J. Lee, A. A. Abdeen, K. L. Wycislo, T. M. Fan and K. A. Kilian, *Nat. Mater.*, 2016, **15**, 856–862.
125. J. Lee, A. A. Abdeen, J. Hedhli, K. L. Wycislo, I. T. Dobrucki, T. M. Fan, L. W. Dobrucki and K. A. Kilian, *C A N C E R Melanoma Topology Reveals a Stem-like Phenotype that Promotes Angiogenesis*, 2017.
126. M. Esteller, *Advances in Experimental Medicine and Biology*, Kluwer Academic/Plenum Publishers, 2003, vol. 532, pp. 39–49.
127. J. Lee, T. G. Molley, C. H. Seward, A. A. Abdeen, H. Zhang, X. Wang, H. Gandhi, J. L. Yang, K. Gaus and K. A. Kilian, *Commun. Biol.*, 2020, **3**, 1–9.
128. S. Nemec, J. Lam, J. Zhong, C. Heu, P. Timpson, Q. Li, J. Youkhana, G. Sharbeen, P. A. Phillips and K. A. Kilian, *Adv. Biol.*, 2021, 2000525.
129. N. Peela, F. S. Sam, W. Christenson, D. Truong, A. W. Watson, G. Mouneimne, R. Ros and M. Nikkhah, *Biomaterials*, 2016, **81**, 72–83.
130. W. N. W. Yahya, N. A. Kadri and F. Ibrahim, *Sensors*, 2014, **14**, 11714–11734.

131. S. L. Vega, V. Arvind, P. Mishra, J. Kohn, N. Sanjeeva Murthy and P. V. Moghe, *Acta Biomater.*, 2018, **76**, 21–28.
132. K. A. Kilian, B. Bugarija, B. T. Lahn and M. Mrksich, *Proc. Natl. Acad. Sci. U. S. A.*, 2010, **107**, 4872–4877.
133. A. Deglincerti, F. Etoc, M. C. Guerra, I. Martyn, J. Metzger, A. Ruzo, M. Simunovic, A. Yoney, A. H. Brivanlou, E. Siggia and A. Warmflash, *Nat. Protoc.*, 2016, **11**, 2223–2232.
134. S. M. Morgani, J. J. Metzger, J. Nichols, E. D. Siggia and A. K. Hadjantonakis, *eLife*, 2018, **7**, e32839.
135. H. M. Begum, H. P. Ta, H. Zhou, Y. Ando, D. Kang, K. Nemes, C. F. Mariano, J. Hao, M. Yu and K. Shen, *Sci. Rep.*, 2019, **9**, 11187.

CHAPTER 6

Dynamic Interfaces for Mechanobiological Studies

JUN NAKANISHI*[a,b,c] AND KAZUHIRO TATEMATSU[a,b]

[a] Research Centre for Functional Materials, National Institute for Materials Science (NIMS), 1-1 Namiki, Tsukuba 305-0044, Japan; [b] Graduate School of Advanced Science and Engineering, Waseda University, 3-4-1 Okubo, Shinjuku-ku, Tokyo 169-8555, Japan; [c] Graduate School of Advanced Engineering, Tokyo University of Science, 6-3-1 Niijuku, Katsushika-ku, Tokyo 125-8585, Japan
*Email: NAKANISHI.Jun@nims.go.jp

6.1 Introduction

In living tissues, cells are surrounded by extracellular matrices (ECMs), where they receive a wide variety of biochemical and physical stimuli that determine cellular functions and fates. From a cellular viewpoint, their intracellular activities are controlled by signals coming from the outside; thereby this modality is called as the *outside-in* regulation (Figure 6.1A). In conventional cell culture in the laboratory, native ECMs are substituted for plastic culture dishes as well as proteins adsorbed on their surfaces. However, after Engler's epoch-making report on the dependence of stem cell lineage on matrix elasticity,[1] researchers have become aware of the critical role of the mechanical properties of ECMs in the regulation of a wide variety of cellular activities. This work has encouraged materials scientists to develop mechanically tuneable materials, including viscoelastic hydrogels and elastomers, to recapitulate the mechanical properties of the native ECM or to identify the mechanical properties of the substrates that the cells are really

Biomaterials Science Series No. 12
Material-based Mechanobiology
Edited by Jun Nakanishi and Koichiro Uto

Published by the Royal Society of Chemistry, www.rsc.org

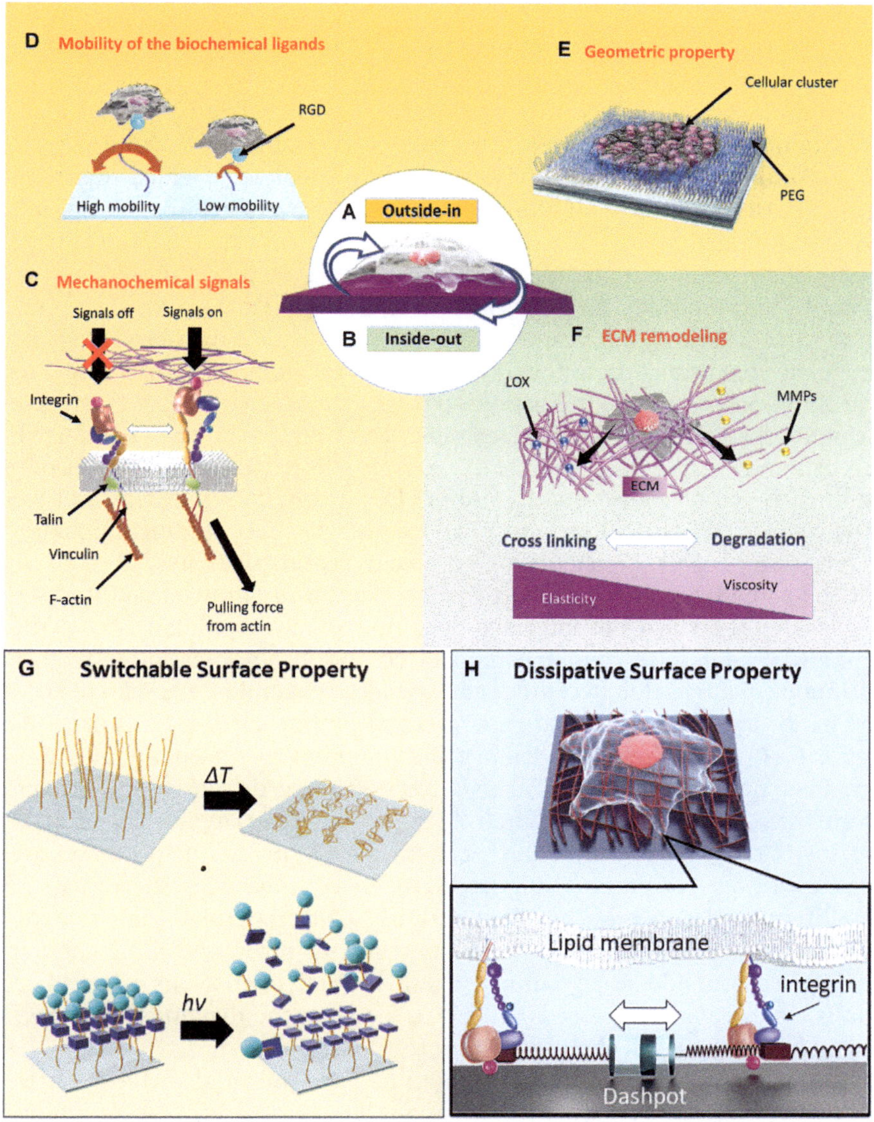

Figure 6.1 Schematic representation of the dynamic interaction of cells with extracellular matrices (ECMs) at various spatial and temporal scales: (A) Outside in regulation. (B) Inside-out regulation. (C–E) Various extracellular factors that regulate cellular mechanosensing: (C) molecules mediating cell-ECM interactions, (D) mobility of ECM ligands, (E) cell and cluster geometry. (F) ECM remodelling: biochemical remodelling through the digestion and crosslinking of pre-existing ECMs. (G,H) Basic concepts of (G) switchable interfaces and (H) dissipative interfaces.

sensing. On the other hand, such mechanical interactions between cells and the ECM are transmitted through the direct binding of ECM proteins and membrane receptor molecules, where their mechanotransduction efficiencies and dynamics are highly dependent on the molecules mediating the cell–ECM interactions (Figure 6.1C).[2–8] From this perspective, biointerfacial design is crucial for understanding materials-based mechanobiology, which this book addresses. Here, the scope of biointerfacial design covers not only the species and the amount (surface density) of proteins that mediate cell–ECM interactions but also their spatial distribution (patterns) at subcellular scales, topography, and mobility (Figure 6.1D). These characteristics affect the degree of integrin clustering[6] and adaptor proteins[9,10] at the juxta membrane regions to regulate force transmission and transduction (Figure 6.1C). Moreover, by controlling cell–substrate interactions at larger scales, comparable to the size of single cells or much larger, researchers can manipulate the geometry of each cell and cell cluster (Figure 6.1E), which affects the stress distribution within each cell or across multiple rows of cells, eventually altering cell polarity, cytoskeletal structures, and phenotypes.[11,12] These findings highlight the crosstalk between biochemical and mechanical regulations.[13,14]

At the same time, it is noteworthy that, even in *in vitro* cell culture conditions, the cell–substrate interface does not remain unchanged from the originally set chemical and mechanical conditions. Rather, it undergoes continuous changes due to cellular activities. For example, cells secrete ECM proteins or the proteins that digest or crosslink pre-existing ECM proteins (Figure 6.1F).[15] Furthermore, from a mechanobiology viewpoint, it is important to note that cellular intrinsic traction force driven by actomyosin contractility causes conformational changes and rearrangements in ECM proteins.[16] These biochemical and mechanical ECM modelling processes are called *inside-out* regulations (Figure 6.1B), in contrast with the *outside-in* signalling just described. They alter various ECM properties, such as composition, mechanics, topography, and porosity, in turn controlling cellular activities *via* outside-in signalling. These mutual regulations proceed at various time scales; hence, it is crucial to focus on the dynamics and time-dependent processes at the cell–ECM interfaces.

In this chapter, we introduce dynamic interfacial materials, which can be used to investigate and control dynamic cell–ECM interactions at various spatial and temporal scales. The word "dynamic" will be used in two meanings here: (1) the interfaces whose properties change in response to external stimuli (Figure 6.1G) and (2) those bearing intrinsically mobile ligands or those that are compliant against cellular traction force (Figure 6.1H). The former can also be called switchable or smart interfaces,[17] where interfacial properties can be controlled on demand to manipulate outside-in signalling (Figure 6.1A). On the other hand, the latter can be called dissipative or adaptive interfaces, which can be used to investigate inside-out mechanical regulation owing to their ability to respond to cellular traction forces (Figure 6.1G). There are a wide variety of reports on dynamic materials in general, but here we tried to focus on the functional interfaces mainly used for mechanobiological applications, while excluding three-dimensional hydrogels and topographical materials to avoid overlaps with Chapter 9, 11, or 13. Readers can also check other comprehensive reviews about switchable materials[18–21] and dissipative mechanobiology.[22–24] The first half of this chapter is devoted to switchable dynamicity, starting from the initially studied cell adhesion control in a larger area for cell migration studies, going down to smaller subcellular and single-molecule scales and eventually discussing their new direction for application to dynamic mechanobiology. The second half discusses dissipative dynamicity, covering non-covalent bond-based functionalised interfaces and fluid interfaces. We will also mention a cutting-edge example of an actuating dynamic interface at the end of this section. It is also noteworthy that there are unique features in the interfaces, regarding their controllability with respect to voltage and, as well as the critical involvement of biointerfacial sciences related to the concept of bioinertness.

6.2 Switchable Interfaces

The history of switchable interfaces goes back to the early 1990s, when Okano *et al.* succeeded in controlling cell adhesion by taking advantage of the temperature-dependent phase-separatable feature of poly(*N*-isopropylacrylamide) (pNIPAAm).[25] Upon decreasing the temperature, the pNIPAAm-functionalised interface swelled, resulting in the dissociation of the adsorbed ECM proteins together with the cell sheets. Owing to the strong impact of their initial work and their useful applications to produce tissue substitutes for regenerative medicines, researchers are still seeking new materials and methodologies for the mild harvesting of adhered cells.[26]

6.2.1 Switchable Interfaces for the Spatiotemporal Control of Cell Adhesion

On the other hand, switchable interfaces, whose cell adhesiveness changes from non-cell-adhesive to cell-adhesive, can be used for dynamic cell patterning. These dynamic interfaces are more popular in mechanobiology research because they allow researchers to investigate cell migration, where mechanical force plays an essential role not only at the cell–ECM interface but also in cell–cell junctions.[27,28] Moreover, as discussed in Chapters 5 and 7, the manipulation of cellular geometry as well as the degree of spreading can change the stress distribution and intensity in each cell and cell collective. Therefore, switchable interfaces capable of altering the cell adhesion geometry *via* external stimuli should provide a versatile platform for the mechanobiological investigation of the migration of single cells and cell collectives,[29,30]

Especially in the early periods, self-assembled monolayers (SAMs) bearing electroactive groups have been investigated for creating switchable interfaces. The first example was reported by Yousaf *et al.*, which was based on a redox-active hydroquinone/quinone group. The oxidised quinone group can react with a cyclopentene bearing a cell-adhesive RGD peptide through the Diels–Alder

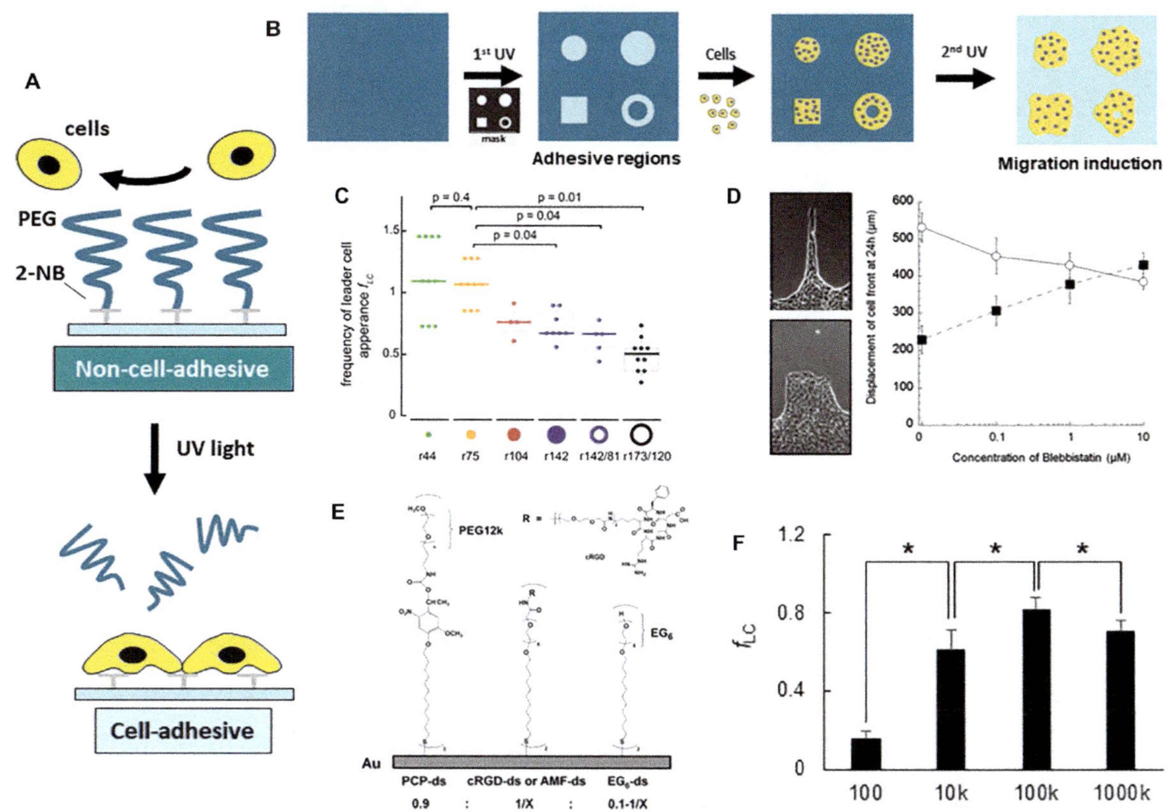

Figure 6.2 Photoactivatable interfaces for cell migration studies: (A) Basic strategy for switching the interface from non-cell-adhesive to cell-adhesive in response to light. (B) Procedure to induce collective cell migration from any given geometry cluster. Adapted from ref. 41 with permission from Elsevier, Copyright 2012. (C) Effect of initial cluster geometry on leader cell appearance. Reproduced from ref. 41 with permission from Elsevier, Copyright 2012. (D) Migration path width–dependent different activation of blebbistatin on collective migration speed. Adapted from ref. 42 with permission from the Royal Society of Chemistry. (E) Mixed self-assembled monolayers to manipulate RGD density in photoactivatable interfaces. (F) Impact of RGD density on leader cell formation (f_{LC}) plotted against the dilution factor. Reproduced from ref. 45 with permission from Elsevier, Copyright 2018.

reaction, thereby making the interface switchable for cell adhesion depending on the applied voltage.[31] The same hydroquinone/quinone chemistry has also been applied to reversibly conjugate oxyamine ligands.[32] An alternative approach has been reported based on the direct electrochemical desorption of the SAM. A cell-repellent tri(ethylene glycol) (EG_3)-terminated thiol can be dissociated from a gold surface with a reducing potential to change the surface, making it cell-adhesive, thereby allowing cells to migrate into the desorbed regions.[33] This surface has been utilised to investigate the impact of the asymmetric shapes of single cells on cell polarity and the eventual determination of the direction of cell migration.[29] To mimic migrating cells, teardrop-shaped cells were prepared, where the cells organised their centrosome–nuclear axis as if the blunt and sharp ends were front and rear, respectively. Although this was not clearly discussed in this paper, there was no question about the role of mechanical forces in controlling the directed cell migration in geometrically controlled cells. In addition, the introduction of a redox-active cobalt complex protecting group has also been reported to control click chemistry (Huisgen cycloaddition), where an RGD-bearing azide ligand can be conjugated in response to an oxidation reaction.[34] Owing to the biorthogonal features of these electrochemical and conjugation reactions, these substrates can be activated for cell adhesion in a culture medium and are therefore useful for studying cell migration. Moreover, one of the biggest advantages of the electrochemical approach is its ability to quantitatively and instantaneously control ligand presentation and cell adhesion.[35] However, electrical potentials are usually applied equally onto the substrate surface; therefore, the substrate surface, as an electrode, should be patterned into the electroactive and nonactive regions in advance unless an additional trigger, such as photochemistry,[36] is introduced to create a region-selective electroactive functionality.

In this regard, light offers a more straightforward approach for switching cell adhesion owing to its high spatiotemporal, remote, and bioorthogonal controllability.[37–39] Nakanishi *et al.* have developed a wide variety of photoresponsive interfaces by conjugating protein- and cell-repellent poly(ethylene glycol) (PEG) to the surface of materials through a photocleavable 2-nitrobenzyl group and have demonstrated the versatility of the platform for studying cell migration mechanobiology (Figure 6.2A).[40] The substrate surface prevents cell adhesion due to the presence of the PEG brushes, but it changes to allow cell adhesion upon the photocleavage of PEG. The procedure for migration induction is shown in Figure 6.2B. By taking advantage of the remote-controllable nature of light, it becomes possible to precisely control the geometry of the initial cell clusters as well as the paths along which cells migrate. This feature is useful for investigating the dynamic mechanobiological responses in cell migration. For example, Rolli *et al.* have successfully elucidated the impact of the size and curvature of initial cell clusters on the probability of leader cell formation during epithelial sheet migration (Figure 6.2C).[41] Also, Kamimura *et al.* have demonstrated the attenuation of cellular contractility after treatment with blebbistatin, a myosin II inhibitor, which resulted in opposite responses, either decelerating or accelerating, depending on the widths of the migration strips (Figure 6.2D).[42] The same photoactivatable interfaces can also be created on the surface of hydrogels, where the impact of substrate stiffness on collective cell behaviours can be investigated.[43]

As previously discussed, it is physiologically relevant to investigate the synergistic effects of biochemical and mechanical regulation. Therefore, dynamic interfaces with molecularly defined cell–ECM interactions will allow for a deeper understanding of the mechanisms of mechanobiological responses than the mere control of cell–substrate contacting geometries. However, a simple coating of the interface with ECM proteins, like fibronectin and collagen, will not meet such demand because various proteins present in the culture medium can gradually substitute for or cover on the pre-adsorbed proteins through a process called ECM remodelling. In this regard, the concept of bioinertness is critical to allow cells to interact with the substrate specifically through a chemical cue of interest with nonspecific protein adsorption repelling in its background. To this end, Shimizu *et al.* have developed photoactivatable interfaces based on mixed SAMs composed of cRGD, short hexa(ethylene glycol) (EG_6), and photocleavable PEG (Figure 6.2E). Introduction of the EG_6 functionality makes the interface repellent against nonspecific protein adsorption guaranteeing selective cell–substrate interaction through the cRGD ligand.[44] This interfacial design allowed them to investigate the impact of the cRGD ligand on the mechanochemical regulation of epithelial cluster expansion behaviours.[45] This study identified a biphasic response of leader cell appearance against interfacial cRGD density. Moreover, it elucidated how chemical cues controlled actomyosin belts surrounding cell clusters as well as E-cadherin expression at the cell–cell junctional regions, eventually coordinating cell collectives (Figure 6.2F).

A similar concept was developed by Del Campo *et al.* to control the activity of the cRGD ligand by introducing 2-nitrobenzyl ester to the aspartate group of the RGD sequence as a photocleavable protecting group.[46] Because of its ability to introduce this photoresponsive module to various biomaterials, it has been successfully applied not only *in vitro* to investigate the effect of the density and time point of the photoactivation in C2C12 myoblast differentiation[47] but also to control inflammation and vascularisation *in vivo.*[48] To further regulate the activity of integrin-binding motifs with dual ligands (Mg^{2+} and RGD) in a combinatorial fashion, Kang *et al.* developed a dynamic interface based on the metal ion coordination chemistry of bisphosphate (BP) ligands and Mg^{2+} ions.[49] Gold nanoparticles functionalised with BP ligands can be reversibly activated in the presence and absence of Mg^{2+} ions, which are further activated to coordinate the BP-modified RGD peptides. The authors applied this method to induce osteoblast differentiation of mesenchymal stem cells (MSCs) in response to the dual ligands. In addition, a magnetic approach for ligand control has been developed by Khatua *et al.*[50] This group utilised silica-coated superparamagnetic core nanoparticles functionalised with a negatively charged RGD peptide (C<u>DD</u>RG<u>D</u>). The negatively charged magnetic particles were coupled to the positively charged substrate surface *via* electrostatic interactions, enabling the manipulation and condensation of RGD ligands by an externally applied magnetic field. MSCs cultured on these dynamic interfaces exhibited selective enhanced spreading, YAP nuclear localisation, and eventual osteoblastic differentiation in the area exposed to the magnetic field. As seen from these examples, recent mechanobiology studies have shifted their focus from designing dynamic interfaces that enable the simple spatiotemporal control of cell adhesion to investigate molecular-level interactions to resolve mechanical and biochemical coupling.

6.2.2 Switchable Interfaces at Subcellular and Molecular Resolutions

By reducing the cell adhesive area down to the nanoscale, the degree of molecular clustering and location of the stress centre at the cell–substrate interface can be manipulated (Figure 6.3A). This in turn alters cellular activities at the macroscale not only in their adhesion or spreading behaviours but also in their differentiation lineages[51] and collective behaviours,[52] highlighting the importance of nanoarchitectonics in cellular mechanobiology. By introducing the dynamic nature of subcellular patterns, dynamic cellular activities as well as the hysteresis of cellular mechanosensing can be addressed.

Spatz *et al.* developed block co-polymer micellar lithography (BCML), in which gold nanoparticles are arrayed in a quasi-hexagonal arrangement with a given interparticle separation by controlling the conditions of the self-assembly of block co-polymer micelles.[53] The size of the embedded arrays of nanoparticles is smaller than that of integrin heterodimers; therefore, by conjugating a cRGD ligand to the arrayed nanoparticles, the degree of clustering of integrin molecules can be controlled through the interparticle distance, from which universal molecular rulers needed for cell spreading have been identified.[54] The details of this methodology are explained in Chapter 7. To add a dynamic nature to this platform, Shimizu *et al.* co-immobilised photocleavable PEG and the cRGD ligand to the arrayed gold nanoparticles and succeeded in the development of a collective cell migration platform (Figure 6.3B).[52] In this study, the authors identified the involvement of FAK phosphorylation signalling in bridging cellular nanosensing and collective behaviours. Another direction is to develop dynamic interfaces that change subcellular patterns or ligand density in response to an external stimulus. A smart electrochemical approach based on nanoarrays of redox-active polypyrrole has been reported by Jiang *et al.*, where reversible transition at the morphology of the array spots transits between the tube and tip structures depending on the potential applied to the substrate (Figure 6.3C).[55] During the transition of the tip morphologies, the surface wettability changes through the hydrophobic nanotubes and hydrophilic nanotips. MSCs cultured on this platform drastically changed their morphologies with high and low spreading on the nanotube and nanotip states, respectively. More interestingly, the cyclic transition between the two states induced YAP and RUNX2 nuclear localisation, indicating

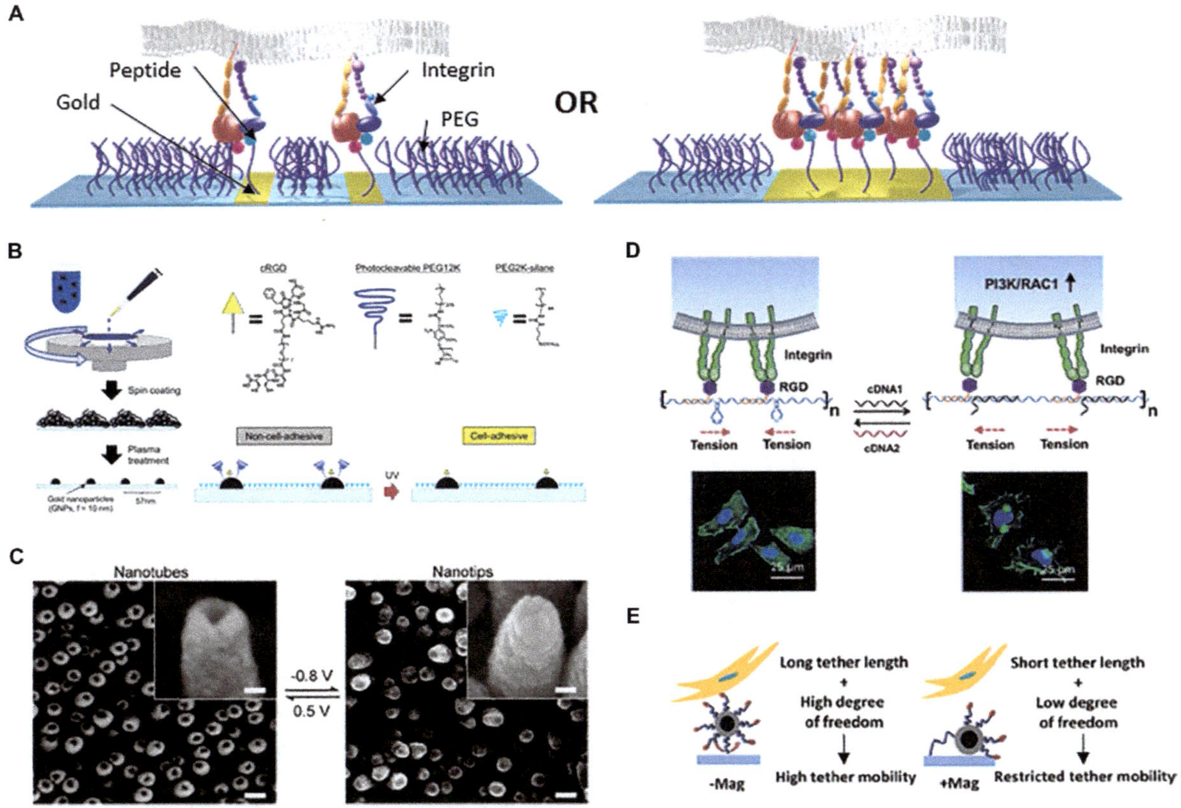

Figure 6.3 Switchable substrates that can resolve mechanobiological events in the subcellular level. (A) Switching area–dependent control of integrin clustering either at an intermolecular distance or at the assembled cluster size. (B) Photoactivatable nanopatterned substrates. Adapted from ref. 52, https://doi.org/10.1371/journal.pone.0091875, under the terms of the CC BY 4.0 license https://creativecommons.org/licenses/by/4.0/. (C) Polypyrrole-based interface with subcellular adhesive patches of electrochemically switchable morphology. Reproduced from ref. 55 with permission from American Chemical Society, Copyright 2017. (D) DNA nanostring–based reversible switching interligand separations. Reproduced from ref. 56 with permission from the Royal Society of Chemistry. (E) Ligand-tethering-tuneable interface based on magnetic nanoparticles. Reproduced from ref. 60 with permission from American Chemical Society, Copyright 2017.

osteogenic differentiation through the switching of mechanotransduction pathways. As an alternative approach, DNA can be a useful platform for controlling ligand density. Zhang *et al.* developed a DNA nanospring based on a long DNA sequence with multiple hairpin structures using a rolling circle amplification reaction (Figure 6.3D).[56] The long DNA scaffold was first hybridised with a short RGD-bearing cDNA sequence; thereafter, the hairpin regions were displaced *via* the addition of cDNA (DNA1) or regenerated *via* the addition of another cDNA (DNA2) complementary to DNA1. The scaffold DNA was extended and contracted through the process, reversibly changing the distance between the RGD ligands. The authors also demonstrated reversible cell morphological changes as well as changes in the expression of integrin-related genes. Recently, Karna *et al.* has extended this concept to a DNA origami bearing a pH-responsive module derived from a C-rich human telomeric region and succeeded in a pH-dependent reversible transformation of the DNA nanospring.[57] On the other hand, Kang *et al.* utilised magnetically responsive CoFe nanocoils, where the pitch of RGD-presenting nanocoils can be magnetically switched without changes in the ligand-presenting area.[58] By using these dynamic interfaces for MSCs, the authors have demonstrated the impact of the dynamic changes of subcellular adhesive spots on focal adhesion maturation, YAP nuclear translocation, and osteogenic differentiation.

Apart from the spatial control of cell–ECM interaction at a subcellular resolution, it has been identified that ECM protein tethering (the length of the ECM fibre segment between two adjacent anchoring points) alters the mechanical feedback against the cellular traction force, thereby changing the fate of MSCs.[59] Taking this into account, Wong *et al.* developed magnetically responsive interfaces based on RGD-modified magnetic beads tethered through flexible linkers, where the degree of tethering can be controlled by applying magnetic field (Figure 6.3E).[60] Using this approach, they demonstrated that higher RGD tether mobility suppressed integrin-mediated mechanotransduction, resulting in reduced osteoblast differentiation of MSCs and enhanced cancer cell stemness.[61]

6.2.3 Switchable Interfaces for Exploring Dynamic Mechanobiology

In the last two sections, we introduced several examples of switchable interfaces for the dynamic control of biochemical cues to cells. These dynamic interfaces are mainly designed to investigate the mechanochemical communications between cells and the ECM, driven by intrinsic cellular contractility, and to understand the mechanical balance between them. However, when we look into physiological and biomedical scenarios, force is applied to cells and tissues in a more dynamic fashion. For example, some cells, such as heart tissues and vocal folds, are always exposed to dynamic force through beating and phonation (Figure 6.4A).[62,63] In hospitals, ultrasonication is a common physical therapy for bone fracture, but the molecular mechanisms of its therapeutic effects have not been elucidated yet.[64] All these provide evidence for the essential role of dynamic forces as well as forces *per se* in our daily life and indicate the significance of its understanding for applied and basic biology, but so much more still needs to be understood.

In this regard, materials and methodologies that can apply dynamic mechanical force or, more specifically, oscillatory force are critical for understanding the molecular mechanisms of its regulation of biological activities. Two modes of vibration can be considered: One is through applying vibration to whole cells (Figure 6.4B), and the other is through applying vibration specifically to integrin molecules (Figure 6.4C). One of the commonly used methods for applying vibration to whole cells is through piezoelectric devices.[65] Nikukar *et al.* demonstrated that MSCs cultured on vibration devices differentiate into osteoblasts, where the best differentiation conditions are at 1 kHz and 22 nm amplitude.[66] This process, named *nanokicking*,[67] was mediated by Rho kinase (ROCK) activation and actomyosin contractility. By comparing the nanovibration approach with the conventional biochemical approach using osteogenic media, they identified cholesterol sulfate as a highly specific metabolite.[68] This shows that nanovibration can be an alternative approach for stem cell engineering over conventional biochemical approaches.

In contrast to these methodologies for shaking entire cells, materials that can apply vibrations selectively to local regions *via* spatially controlled external stimuli or, more specifically, to membrane integrins may provide molecularly resolved information. As an example for local stimulation, Sutton *et al.* developed an array of polymeric microplates, infilled with an NIPAAm-based polymer and gold nanorods (AuNRs) that bend and drag cells in response to NIR irradiation (Figure 6.4D).[69] The hydrogel-actuated integrated responsive system (HAIRS) can apply force to cells at the scale of several microns, resulting in the local elongation of cellular shape due to an increase in intracellular stress.

As an example for molecular-level stimulation, Kadem *et al.* demonstrated a method for applying oscillatory high-frequency mechanical stimuli through cRGD-modified azobenzene interfaces (Figure 6.4E).[70] By taking advantage of the photoisomerisable azobenzene molecule, especially with the rapid relaxing nature of its push–pull type, brings the cRGD molecule front and back for 3.5 Å at a high frequency of up to 10^5 Hz. Such high-frequency vibration significantly increased the accumulation of proteins involved in focal adhesion maturation, such as paxillin, talin, vinculin, and zyxin. This work highlighted the capability of cells to "feel" subnanometer vibrations, and the potential of such photoresponsive dynamic interfaces for the study of cellular responses at high spatiotemporal resolutions. Additionally, Liu *et al.* developed an RGD-functionalised pNIPPAm hydrogel embedded with AuNRs.[71] Absorption of near-IR light by AuNRs increased the local temperature, causing phase transition of the pNIPAAm and gel collapsing, reversibly delivering a piconewton force to the tethered cRGD

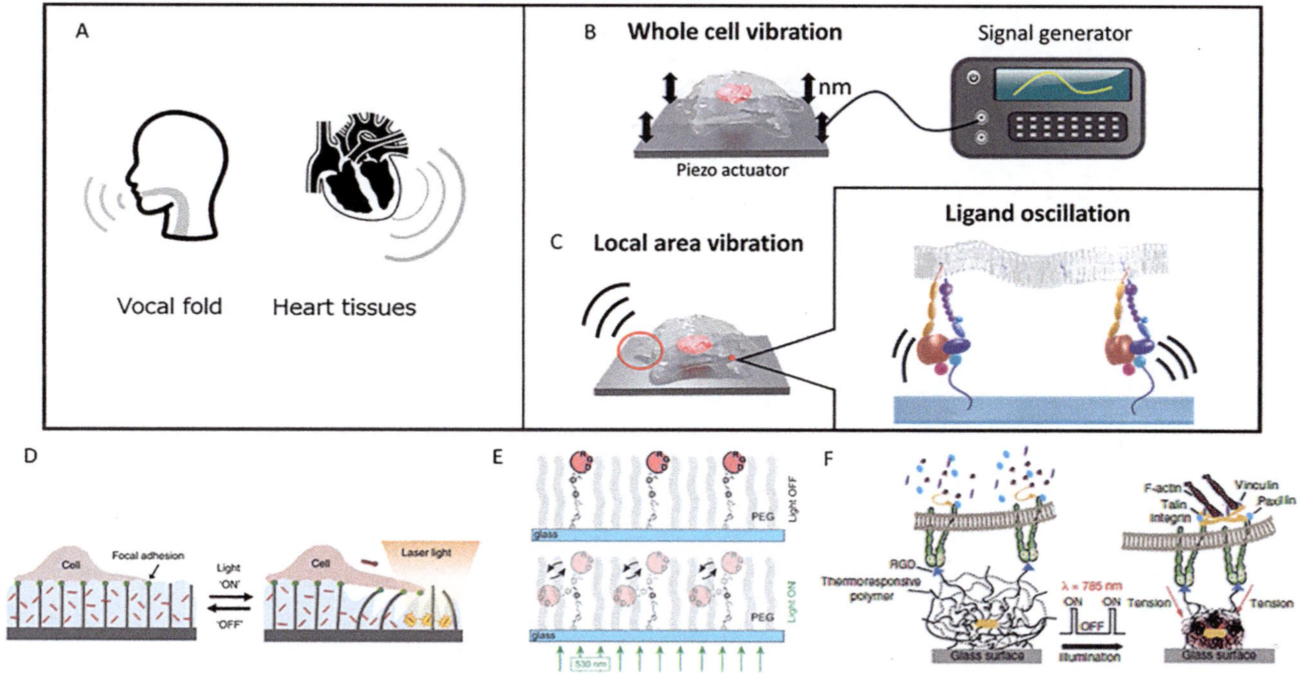

Figure 6.4 Switchable interfaces that can resolve dynamic mechanobiology: (A) Dynamic forces observed in our body. (B, C) Two modes of applying dynamic forces to cells: stimulation to (B) entire cells and (C) subcellular regions, more specifically to molecular levels. In (B), a piezoelectric device–based interface is drawn for the application of mechanical vibration to cultured cells. (D) Hydrogel-actuated integrated responsive system (HAIRS) based on micropillars infilled with a thermoresponsive polymer and gold nanorods. Near-IR irradiation induces aggregation of the polymer, bending the pillars drags the cells. Reproduced from ref. 69, under the terms of the CC BY 4.0 license https://creativecommons.org/licenses/by/4.0/. (E) Push–pull-type azobenzene for oscillatory stimulation at the molecular level. Tension is applied to integrin molecules during the *trans*- to *cis*-photoisomerisation of the azobenzene and its fast relaxation to the *trans* state. Reproduced from ref. 70 with permission from John Wiley & Sons, Copyright © 2017 The Authors. Published by Wiley-VCH Verlag GmbH & Co. KGaA. (F) Near-IR-actuatable interface based on a thermoresponsive polymer and gold nanorods. Tension is applied to integrin molecules in response to swelling/deswelling of the polymer upon cooling/heating. Reproduced from ref. 71 with permission from Springer Nature, Copyright 2017.

ligands (Figure 6.4F). This system allows local vibration at 100 Hz and is exemplified *via* optically controlled cell migration and T cell receptor activation. Moreover, by controlling the regions that received photoirradiation, they demonstrated enhanced myogenesis within an irradiated cluster of C2C12 cells through selective YAP nuclear translocation.[72]

On the other hand, Zhou *et al.* utilised an artificial rotary molecular motor to apply molecular-scale torque at the cell–substrate interface.[73] The concept is based on an overcrowded alkene, which rotates in a unidirectional motion in response to photoirradiation of 365 nm light. MSCs differentiated into osteoblasts on the rotary surface even though they maintained their stemness in the static interface. This unique differentiation was associated with a decrease in the α-helical secondary structure of adsorbed fibronectin, thereby increasing its amount of adsorption. Therefore, this molecular rotational motion is not directly actuating cells; rather it alters the interactions between its molecules and ECM proteins.

6.3 Compliant Interfaces

An initial study in mechanobiology mainly focused on the elastic properties of the ECM, but native ECM is viscoelastic. Intuition tells us that the addition of viscous properties to elastic matter just makes them softer, but this is not always the case in cellular mechanobiological responses. By adding a viscous nature in the matrices whilst maintaining a constant elasticity, numerous counterintuitive cellular behaviours, such as enhanced cell spreading in more relaxing (high loss modulus) hydrogels, have been elucidated.[74–76] These epoch-making studies created a sensation in materials science, especially in the development of dissipative hydrogels, as discussed in Chapter 11. In contrast, this section focuses on interfacial materials with a dynamic and dissipative nature.

6.3.1 Dissipative Interfaces

Seo *et al.* took advantage of supramolecular polyrotaxane, where a PEG chain penetrates through hollow cyclodextrin (CD) molecules to create a mobile cell culture interface (Figure 6.5A).[77] Because the CD molecule is not covalently bonded to the PEG chain but forms a topological complex with it, they can slide along the PEG chain, adding mobility to the interface. Bulky polymers were introduced at both sides of the PEG chain to prevent the slipping off of the CD molecules from PEG. The mobility of CD can be controlled by changing the number of CD molecules complexed with PEG, with less mobility observed for molecules with increased number of CD molecules complexed.[78] The dynamic nature of ECM proteins accelerate their initial interaction with integrins; thereby MSCs exhibited preferential differentiation into adipocytes over osteoblasts due to the dynamic nature of this interface.[79] They also demonstrated enhanced cardiomyogenic differentiation of induced pluripotent stem (iPS) cells on dynamic interfaces.[80]

In contrast to one-dimensional movement in polyrotaxane-based dynamic interfaces, Yu *et al.* demonstrated a methodology for manipulating the two-dimensional mobility of RGD peptides, by controlling the hydrophobicity of the interface (Figure 6.5B).[81] Their approach was based on the modification of a glass surface with alkylsiloxane molecules of different chain lengths (propyl *versus* octyl) and the conjugation of a hydrophobic benzophenone moiety to the RGD peptide. In contrast to observing the simultaneous activation of $\alpha_5\beta_1$ and $\alpha_v\beta_3$ integrins on a constant interface, selective activation of $\alpha_5\beta_1$ integrin on the dynamic interface, where osteogenic differentiation of the MSCs was decreased, was detected. This group further developed a self-strengthening dynamic interface by utilising photochromic spiropyran as a dynamic anchoring group of the cRGD ligand on a hydrophobic surface.[82] The strength of the hydrophobic interaction can be lowered *via* UV irradiation to switch the chromophore into the hydrophilic merocyanine state, which gradually returns to spiropyran in the dark state over a timescale of more than 24 h. MSCs cultured on the self-strengthening dynamic interface exhibited enhanced mechanotransduction and osteogenic differentiation, through the switching of integrin subtypes and Rho GTPases depending on the mobility of the cRGD ligand.

Supported lipid bilayers (SLBs) are another versatile approach to add mobility to interfaces in mechanobiology.[83] They have often been used in the field of bioanalysis to take advantage of membrane receptors as sensing elements. Moreover, the bioinert feature of lipid bilayers is suitable for preventing the nonspecific adsorption of proteins; therefore, they have been widely used to improve the sensitivity of sensors.[84] At the same time, as can be imagined from the concept of the fluid mosaic model, membrane proteins embedded in lipid bilayers are dynamic and mobile and are indispensable for efficient signal transduction across membranes. Owing to the mobile nature of lipid bilayers, we were able to make the cell-adhesive ligand dynamic and examine the effect of dissipative properties on cellular mechanobiological responses. In particular, the lateral mobility of lipid bilayers can be controlled through their phospholipid compositions. Saturated phospholipids, such as dipalmitoylphosphatidylcholine (DPPC), remain in the immobile gel phase, thereby showing lower fluidity. On the other hand, unsaturated phospholipids, such as dioleoylphosphorylcholine (DOPC), are in the disordered fluid phase. Based on this concept, Bennett *et al.* examined the impact of the interfacial mobility of the SLB on cell adhesion behaviours and demonstrated that cellular spreading is highly dependent on the mobility of the SLB; spreading was observed more on the less fluid DPPC, whereas there was no spreading on the more fluid DOPC (Figure 6.5C). Such different cellular spreading behaviours were the consequence of different levels of actin retrograde flow, an indicator of cellular traction force, on the two interfaces. This observation can be explained within the framework of the molecular clutch model.[85] It should be noted that all these studies utilised extremely stiff glass substrates as base materials,

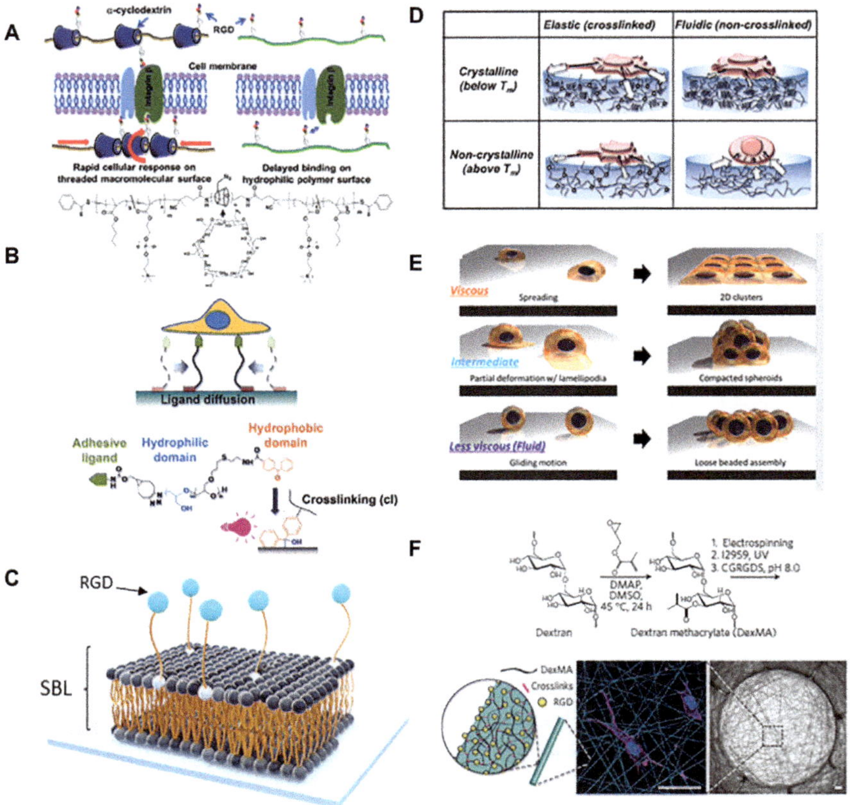

Figure 6.5 Dissipative interfaces. (A) Rotaxane-based interface movable in 1D along PEG. Reproduced from ref. 77 with permission from American Chemical Society, Copyright 2013. (B) Hydrophobic interaction-driven control of ligand diffusion at the 2D interface. Reproduced from ref. 81 with permission from John Wiley & Sons, Copyright © 2020 WILEY-VCH Verlag GmbH & Co. KGaA, Weinheim. (C) Supported lipid bilayers (SLB) with a 2D diffusible feature. (D) P(CL-*co*-DLLA)-based interfaces for the manipulation of elastic and fluid properties *via* crosslinking and controlling their crystalline states. Reproduced from ref. 86 with permission from American Chemical Society, Copyright 2016. (E) Independent tuning of elastic and viscous components on the P(CL-*co*-DLLA) interfaces *via* photocrosslinking determines the initial adhesion behaviours of epithelial cells and their assembly. Reproduced from ref. 90 with permission from Elsevier, Copyright 2021. (F) Nanofibre-based dynamic interfaces made *via* electrospinning. Reproduced from ref. 93 with permission from Springer Nature, Copyright 2015.

but the introduction of interface mobility dramatically changes the cellular adhesion and spreading behaviours on these surfaces.

The effect of interfacial fluidity on cellular adhesion behaviours was also investigated by Uto *et al.* using crosslinked and noncrosslinked poly(ε-caprolactone-*co*-D,L-lactide) (P(CL-*co*-DLLA)) (Figure 6.5D).[86] Even though it is counterintuitive, the system provides higher fluidity and higher elasticity when using noncrosslinked polymers compared to their crosslinked counterparts. The authors demonstrated decreased cell spreading on the noncrosslinked substrate, resulting in the unsuccessful formation of stable focal adhesion on the force-dissipating interface. They also demonstrated the impact of fluidity of the substrate on the susceptibility to anticancer drug,[87] EMT,[88] and senescence.[89] Recently, the system has been further improved to independently manipulate the elastic and viscous components by using photocrosslinkable benzophenone together with P(CL-*co*-DLLA) of different molecular weights.[90] In this study, they elucidated that interfacial relaxation is an essential mechanical parameter for the early adhesion and migration behaviours of epithelial cells and enhanced cell aggregate formation at intermediate levels of interfacial fluidity (Figure 6.5E).

The use of nanofibres is another way of introducing dynamic properties to the cell–substrate interfaces. Nanofibres have been considered as a promising platform for tissue engineering and regenerative medicine in order to mimic the fibrous structure of the native ECM. From a mechanics perspective, noncrosslinked nanofibres showed nonlinear responses, such as that of a shear- or stress-stiffening nature, due to the entanglement between surrounding nanofibres.[91] This feature endows fibrous hydrogels with unique long-range force–transmitting properties, arousing the fundamental question of how far cells can "feel".[92] Looking at the interface of fibrous scaffolds based on electrospun dextran methacrylate fibres, immobilised cell-adhesive ligands were recruited in response to cellular traction forces[93] (Figure 6.5F). Owing to the adaptive nature of fibrous interfaces, stem cell adhesion and proliferation became more efficient in a soft network (2.8 kPa) than in a stiffer one (55 kPa), which was opposite to that commonly observed for normal hydrogels. These results highlight not only the viscoelastic properties of interfaces but also that the dynamics of chemical ligands and rearrangements driven by cellular traction force are critical for the determination of cellular responses at dissipative and adaptive interfaces. As one of the strategies to decouple these two components, Charrier *et al.* synthesised a polyacrylamide gel with independent interpenetrating polymers and conjugated RGD peptides to either of them.[94] In their study, they demonstrated the influence of viscous dissipation on cellular phenotypes. Even though the last example was based on the sophisticated design of adaptive hydrogels, such innovation is also important for understanding dynamic mechanosensing at interfaces.

6.3.2 Liquid–Liquid Interfaces

As discussed in the previous section, cells are sensitive to the viscous nature of interfaces, and the introduction of a very thin lipid bilayer on a glass surface can cause drastic changes in cellular spreading behaviours, where cells can no longer spread on fluidic phospholipids. Based on this, we would expect the liquid–liquid interface between water and immiscible organic solvents to barely permit cell adhesion and spreading because of the fluid nature of the interface, even to its bulk area. The same can also be expected from the fact that the hanging-drop method is commonly used technique for cell spheroid formation. Here, we can consider air as a hydrophobic substrate, facilitating spheroidal formation because cells cannot adhere to the water–air interface but prefer cohesion to one another.[95] In addition, it is known that porous materials infilled with perfluorocarbons exhibit extremely high antibiofouling properties, as bacteria cannot adhere to the interface because of their slippery characteristics.[96] Nevertheless, Rosenberg reported for the first time that mammalian cells can spread and grow in culture media overlaid on fluid perfluorocarbons.[97] What was happening there? It was speculated that cells do not directly adhere to the liquid interface; rather, spreading is mechanically sustained by the denatured protein films formed at the liquid interface. Given this perspective, Keese and Giaever extended their culture method by strengthening protein films through the introduction of poly-L-lysine or by crosslinking them.[98–100] These studies were further followed by several groups that examined the effect of ECM coating or the type of surfactants on cell adhesion behaviours;[101–103] however, most were only phenomenological, except for the determination of minimum interfacial mechanics as measured using a modified viscometer.[104]

More detailed characteristics of the protein films had not been identified until the late 2010s when Kong *et al.* correlated cell adhesion behaviours and the interfacial rheology of protein films.[105] They created a sudden drastic increase in the interfacial shear modulus upon the injection of serum proteins into the aqueous phase, which was further accelerated in the presence of perfluorobenzoyl chloride as a reactive surfactant (Figure 6.6A).[106] Moreover, the protein films have been identified to be 14–19 nm thick based on AFM measurements.[106] Therefore, the introduction of a perfluoroalkyl group to proteins increased their interfacial activity, and hence grafted proteins assembled at the interface. In contrast, Jia *et al.* examined free-standing protein nanolayers formed at the interface without the addition of reactive surfactants.[107] The protein films formed on the perfluorotributylamine interface were robust enough to mature into a solid-like form, brittle enough to be torn after tapping but durable enough to allow the spreading of MSCs (Figure 6.6B).[108] When using bovine serum albumin (BSA), a major component in serum, the drastic loss of secondary structures and enhanced interprotein interactions can be detected *via* infrared (IR) spectroscopy. These indicate that the protein nanofilms were formed through the following procedures: (1) protein conformational change at the interface, (2) interfacial jamming of the denatured proteins driven by interfacial tension, and (3) robust nanoscopic protein thin film formation. When they analysed the protein layers that formed on the perfluorocarbons, allowing cell spreading (perfluorotributylamine, PFTBA) or not (perfluorodecalin, PFD), stiffer and thicker nanofilms were formed on PFTBA (Figure 6.6C). The cellular mechanosensing of the more robust protein nanosheets was also evidenced through the higher nuclear translocation of YAP. However, when they calculated the estimated Young's modulus, the

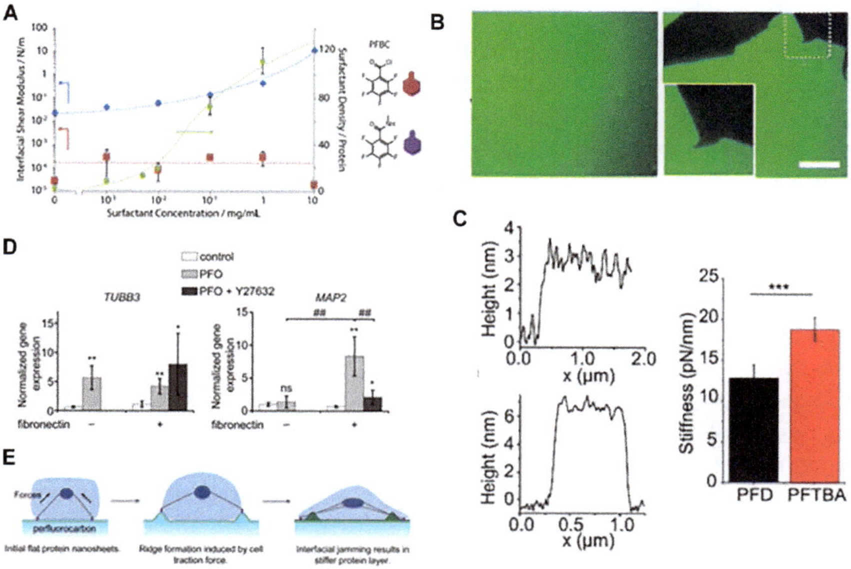

Figure 6.6 Liquid–liquid interfacial culture: (A) Effect of a reactive surfactant on the interfacial shear modulus at the water-perfluorocarbon interface. Reproduced from ref. 106, https://doi.org/10.1021/acs.nanolett.7b05339, under the terms of the CC BY 4.0 license https://creativecommons.org/licenses/by/4.0/. (B) Robust solid-like brittle nature of BSA nanofilms formed at the interface. Reproduced from ref. 108 with permission from John Wiley & Sons, Copyright © 2019 WILEY-VCH Verlag GmbH & Co. KGaA, Weinheim. (C) AFM characterisation of the thickness (left) and stiffness (right) of protein nanofilms formed on two different perfluorocarbons, which allowed good (PFTBA) or bad (PFD) spreading. Reproduced from ref. 108 with permission from John Wiley & Sons, Copyright © 2019 WILEY-VCH Verlag GmbH & Co. KGaA, Weinheim. (D) Enhanced neural differentiation makers in MSCs cultured on perfluorooctane (PFO) coated with fibronectin. Reproduced from ref. 109 with permission from John Wiley & Sons, Copyright © 2019 WILEY-VCH Verlag GmbH & Co. KGaA, Weinheim. (E) Speculated mechanisms of enhanced fibronectin fibril formation *via* cellular traction forces on the adaptive fluid interface. Reproduced from ref. 109 with permission from John Wiley & Sons, Copyright © 2019 WILEY-VCH Verlag GmbH & Co. KGaA, Weinheim.

moduli were sufficiently high for both compared to the known physiological ranges of hydrogels. This result indicates that cellular mechanosensing mechanisms are different between normal hydrogels and liquid–liquid interfaces. They have also identified the adaptive nature of protein nanofilms from the observation of fibronectin fibril formation within one day, which was not detected on normal glass substrates at that time scale.[109] It should be noted that MSCs cultured on the fluid interfaces differentiate into neuronal cells without the addition of any differentiation induction factors (Figure 6.6D). This seems to be possible because of the enhanced focal adhesion maturation on fibrillar fibronectin (Figure 6.6E).

One of the biggest advantages of a liquid–liquid interfacial culture is that it can be applied to oil-in-water emulsions by utilising the deformable nature of liquids. This reminds us of microcarrier culture systems[100,110] but replacing solid carriers with perfluorocarbon liquids. By taking advantage of microfluidic systems for liquid handling, they can be used as a trypsin-free culturing system with a large surface area, conductive for stem cell expansion.

6.3.3 Force-actuating Materials

So far, we have discussed dynamic interfacial materials that are compliant with cellular traction forces. Recently, Almquist *et al.* proposed a unique method to utilise the adaptive nature of interfaces for the local delivery of bioactive molecules (Figure 6.7A).[111] Their idea was based on the elegant activation system of TGF-β in nature.[112] TGF-β is a secretory protein that activates the EMT signalling pathway. In the resting state, TGF-β is trapped in a protein called large latent complex (LLC) bound to the ECM, keeping TGF-β in the inactive state. However, upon binding of cellular integrin molecules to the RGD sequence existing in LLC, cellular traction forces induce a conformational change in LLC, reducing the binding affinity between TGF-β. This results in the local release of TGF-β molecules from the ECM, and the cells that pull the LLC

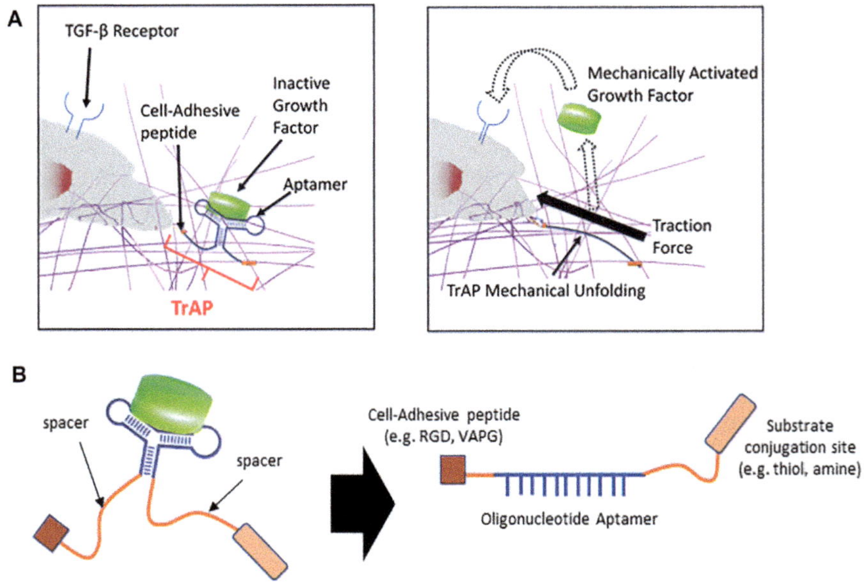

Figure 6.7 Dynamic interfaces with autonomous secretion of proteins. (A) TrAP: a biomimetic approach to release growth factors from their corresponding aptamers bearing cell-adhesive peptides. (B) Schematic drawing of TrAP and its folded structures. TrAP is composed of an oligonucleotide aptamer bearing a cell-adhesive peptide at one end and a substrate-adhesive unit at the other end. Redrawn based on ref. 111, https://doi.org/10.1002/adma.201806380, under the terms of the CC BY 4.0 license https://creativecommons.org/licenses/by/4.0/.

are selectively exposed to TGF-β release. Overall, this system can be considered as an adaptive interface (LLC) that can release a drug (TGF-β) in response to mechanical stimuli (cellular traction forces).

The authors replaced TGF-β and LLC with platelet-derived growth factor (PDGF) and a stem-loop type of aptamer against PDGF, respectively, and conjugated an RGD peptide to the aptamer (Figure 6.7B). The capability of stem-loop DNA to customise the force thresholds for denaturing was discussed in Chapter 3.[113] This smart interface can release PDGF in response to cellular traction forces. Furthermore, the authors have demonstrated selective activation of TGF-β signalling in primary human dermal fibroblasts (HMEC-1) among other cell types, depending on the expression levels of cell-adhesive receptors. Moreover, the use of aptamers adds further versatility to the present strategy in utilising the system to release various kinds of molecules in response to the application of cellular forces.

6.4 Conclusions and Future Perspectives

A wide variety of extracellular biochemical and mechanical cues dictate cellular activity and fate. In this chapter, we discussed two kinds of dynamic interfaces: one that has a switchable element in response to external stimuli and another that has a mobile nature adaptable to cellular traction forces. One of the greatest advantages of switchable interfaces is their ability to alter the cellular mechanical equilibrium from one state to another with controlled kinetics. Earlier work has been successful in resolving cell migration and stem cell lineage determination using two-state chemistry (Sections 6.2.1 and 6.2.2). However, there is still a broad space for their innovation and application to mimic physiological and pathological processes, such as the transition from healthy to disease states, whose dynamics are highly variable between individuals. The recapitulation of such personalised diseased states will provide a platform for screening in precision medicine.[114]

Furthermore, a wide variety of oscillatory mechanical cues are found in living tissues, and cells have been shown to be more sensitive to force loading rates than to force *per se*.[115] These findings encouraged materials scientists to develop dynamic materials that are durable to fast repeatable switching (Section 6.2.3) and force dissipation (Sections 6.3.1 and 6.3.2) at various spatial and temporal scales, further promoting our understanding of mechanobiology and mechanoarchitectonics.

Lastly, our deep understanding of cellular activities and fates through materials-based mechanobiological approaches will reveal the future path of a new-generation biomaterials, such as smart autonomous actuating devices (Section 6.3.3) for medical applications. These facts further emphasise the significance of the marriage between interfacial science and mechanobiology.

Acknowledgements

This work was supported in part by the Japan Society for the Promotion of Science, Kakenhi (No. 20K20645).

References

1. A. J. Engler, S. Sen, H. L. Sweeney and D. E. Discher, *Cell*, 2006, **126**, 677–689.
2. J. Seong, A. Tajik, J. Sun, J.-L. Guan, M. J. Humphries, S. E. Craig, A. Shekaran, A. J. García, S. Lu, M. Z. Lin, N. Wang and Y. Wang, *Proc. Natl. Acad. Sci.*, 2013, **110**, 19372–19377.
3. P. Roca-Cusachs, N. C. Gauthier, A. del Rio and M. P. Sheetz, *Proc. Natl. Acad. Sci.*, 2009, **106**, 16245–16250.
4. H. B. Schiller, M.-R. Hermann, J. Polleux, T. Vignaud, S. Zanivan, C. C. Friedel, Z. Sun, A. Raducanu, K.-E. Gottschalk, M. Théry, M. Mann and R. Fässler, *Nat. Cell Biol.*, 2013, **15**, 625–636.
5. C. Diaz, S. Neubauer, F. Rechenmacher, H. Kessler and D. Missirlis, *J. Cell Sci.*, 2020, **133**, jcs232702.
6. J. Z. Kechagia, J. Ivaska and P. Roca-Cusachs, *Nat. Rev. Mol. Cell Biol.*, 2019, **20**, 457–473.
7. P. Ringer, G. Colo, R. Fässler and C. Grashoff, *Matrix Biol.*, 2017, **64**, 6–16.
8. B. Cheng, W. Wan, G. Huang, Y. Li, G. M. Genin, M. R. K. Mofrad, T. J. Lu, F. Xu and M. Lin, *Sci. Adv.*, 2020, **6**, eaax1909.
9. S. E. Winograd-Katz, R. Fässler, B. Geiger and K. R. Legate, *Nat. Rev. Mol. Cell Biol.*, 2014, **15**, 273–288.
10. E. R. Horton, J. D. Humphries, J. James, M. C. Jones, J. A. Askari and M. J. Humphries, *J. Cell Sci.*, 2016, **129**, 4159–4163.
11. C. M. Nelson, R. P. Jean, J. L. Tan, W. F. Liu, N. J. Sniadecki, A. A. Spector and C. S. Chen, *Proc. Natl. Acad. Sci. U. S. A.*, 2005, **102**, 11594–11599.
12. N. Gjorevski, E. Boghaert and C. M. Nelson, *Cancer Microenviron.*, 2012, **5**, 29–38.
13. A. Cipitria and M. Salmeron-Sanchez, *Adv. Healthcare Mater.*, 2017, **6**, 1700052.
14. S. Saha, T. L. Nagy and O. D. Weiner, *Philos. Trans. R. Soc., B*, 2018, **373**, 20170145.
15. J. Winkler, A. Abisoye-Ogunniyan, K. J. Metcalf and Z. Werb, *Nat. Commun.*, 2020, **11**, 5120.
16. V. Vogel, *Annu. Rev. Physiol.*, 2018, **80**, 353–387.
17. J. Nakanishi, *Chem. – Asian J.*, 2014, **9**, 406–417.
18. P. Bugga and M. Mrksich, *Curr. Opin. Colloid Interface Sci.*, 2018, **38**, 80–87.
19. K. Uto, J. H. Tsui, C. A. DeForest and D.-H. Kim, *Prog. Polym. Sci.*, 2017, **65**, 53–82.
20. W. Li, Z. Yan, J. Ren and X. Qu, *Chem. Soc. Rev.*, 2018, **47**, 8639–8684.
21. Y. Ma, X. Tian, L. Liu, J. Pan and G. Pan, *Acc. Chem. Res.*, 2019, **52**, 1611–1622.
22. S. Tang, B. M. Richardson and K. S. Anseth, *Prog. Mater. Sci.*, 2021, **120**, 100738.
23. M. Cantini, H. Donnelly, M. J. Dalby and M. Salmeron-Sanchez, *Adv. Healthcare Mater.*, 2020, **9**, 1901259.

24. O. Chaudhuri, J. Cooper-White, P. A. Janmey, D. J. Mooney and V. B. Shenoy, *Nature*, 2020, **584**, 535–546.
25. N. Yamada, T. Okano, H. Sakai, F. Karikusa, Y. Sawasaki and Y. Sakurai, *Macromol. Rapid Commun.*, 1990, **11**, 571–576.
26. K. M. Park, A. Shrinidhi, J. Murray and K. Kim, *Isr. J. Chem.*, 2018, **58**, 461–465.
27. C. De Pascalis and S. Etienne-Manneville, *Mol. Biol. Cell*, 2017, **28**, 1833–1846.
28. X. Trepat and E. Sahai, *Nat. Phys.*, 2018, **14**, 671–682.
29. X. Y. Jiang, D. A. Bruzewicz, A. P. Wong, M. Piel and G. M. Whitesides, *Proc. Natl. Acad. Sci. U. S. A.*, 2005, **102**, 975–978.
30. R. A. Desai, L. Gao, S. Raghavan, W. F. Liu and C. S. Chen, *J. Cell Sci.*, 2009, **122**, 905–911.
31. M. N. Yousaf, B. T. Houseman and M. Mrksich, *Proc. Natl. Acad. Sci. U. S. A.*, 2001, **98**, 5992–5996.
32. E. W. L. Chan, S. Park and M. N. Yousaf, *Angew. Chem., Int. Ed.*, 2008, **47**, 6267–6271.
33. X. Y. Jiang, R. Ferrigno, M. Mrksich and G. M. Whitesides, *J. Am. Chem. Soc.*, 2003, **125**, 2366–2367.
34. I. Choi, Y.-K. Kim, D.-H. Min, S. Lee and W.-S. Yeo, *J. Am. Chem. Soc.*, 2011, **133**, 16718–16721.
35. S. Park, N. P. Westcott, W. Luo, D. Dutta and M. N. Yousaf, *Bioconjugate Chem.*, 2014, **25**, 543–551.
36. W. S. Dillmore, M. N. Yousaf and M. Mrksich, *Langmuir*, 2004, **20**, 7223–7231.
37. E. R. Ruskowitz and C. A. DeForest, *Nat. Rev. Mater.*, 2018, **3**, 17087.
38. H. P. Lee and A. K. Gaharwar, *Adv. Sci.*, 2020, **7**, 2000863.
39. T. L. Rapp and C. A. DeForest, *Adv. Healthcare Mater.*, 2020, **9**, 1901553.
40. J. Nakanishi, *Chem. Rec.*, 2017, **17**, 611–621.
41. C. G. Rolli, H. Nakayama, K. Yamaguchi, J. P. Spatz, R. Kemkemer and J. Nakanishi, *Biomaterials*, 2012, **33**, 2409–2418.
42. M. Kamimura, O. Scheideler, Y. Shimizu, S. Yamamoto, K. Yamaguchi and J. Nakanishi, *Phys. Chem. Chem. Phys.*, 2015, **17**, 14159–14167.
43. M. Kamimura, M. Sugawara, S. Yamamoto, K. Yamaguchi and J. Nakanishi, *Biomater. Sci.*, 2016, **4**, 933–937.
44. Y. Shimizu, M. Kamimura, S. Yamamoto, S. A. Abdellatef, K. Yamaguchi and J. Nakanishi, *Anal. Sci.*, 2016, **32**, 1183–1188.
45. S. A. Abdellatef and J. Nakanishi, *Biomaterials*, 2018, **169**, 72–84.
46. S. Petersen, J. M. Alonso, A. Specht, P. Duodu, M. Goeldner and A. del Campo, *Angew. Chem. Int. Ed.*, 2008, **47**, 3192–3195.
47. S. Weis, T. T. Lee, A. del Campo and A. J. García, *Acta Biomater.*, 2013, **9**, 8059–8066.
48. T. T. Lee, J. R. García, J. I. Paez, A. Singh, E. A. Phelps, S. Weis, Z. Shafiq, A. Shekaran, A. del Campo and A. J. García, *Nat. Mater.*, 2015, **14**, 352–360.
49. H. Kang, K. Zhang, H. J. Jung, B. Yang, X. Chen, Q. Pan, R. Li, X. Xu, G. Li, V. P. Dravid and L. Bian, *Adv. Mater.*, 2018, **30**, 1803591.

50. C. Khatua, S. Min, H. J. Jung, J. E. Shin, N. Li, I. Jun, H.-W. Liu, G. Bae, H. Choi, M. J. Ko, Y. S. Jeon, Y. J. Kim, J. Lee, M. Ko, G. Shim, H. Shin, S. Lee, S. Chung, Y. K. Kim, J.-J. Song, V. P. Dravid and H. Kang, *Nano Lett.*, 2020, **20**, 4188–4196.
51. P. Han, J. E. Frith, G. A. Gomez, A. S. Yap, G. M. O'Neill and J. J. Cooper-White, *ACS Nano*, 2019, **13**, 11129–11143.
52. Y. Shimizu, H. Boehm, K. Yamaguchi, J. P. Spatz and J. Nakanishi, *PLoS One*, 2014, **9**, e91875.
53. G. Roman, M. Martin and P. S. Joachim, *Nanotechnology*, 2003, **14**, 1153.
54. M. Arnold, E. A. Cavalcanti-Adam, R. Glass, J. Blummel, W. Eck, M. Kantlehner, H. Kessler and J. P. Spatz, *ChemPhysChem*, 2004, **5**, 383–388.
55. Y. Wei, X. Mo, P. Zhang, Y. Li, J. Liao, Y. Li, J. Zhang, C. Ning, S. Wang, X. Deng and L. Jiang, *ACS Nano*, 2017, **11**, 5915–5924.
56. K. X. Zhang, R. J. Deng, Y. P. Sun, L. Zhang and J. H. Li, *Chem. Sci.*, 2017, **8**, 7098–7105.
57. D. Karna, M. Stilgenbauer, S. Jonchhe, K. Ankai, I. Kawamata, Y. Cui, Y.-R. Zheng, Y. Suzuki and H. Mao, *Bioconjugate Chem.*, 2021, **32**, 311–317.
58. S. Min, M. J. Ko, H. J. Jung, W. Kim, S.-B. Han, Y. Kim, G. Bae, S. Lee, R. Thangam, H. Choi, N. Li, J. E. Shin, Y. S. Jeon, H. S. Park, Y. J. Kim, U. K. Sukumar, J.-J. Song, S.-K. Park, S.-H. Yu, Y. C. Kang, K.-B. Lee, Q. Wei, D.-H. Kim, S. M. Han, R. Paulmurugan, Y. K. Kim and H. Kang, *Adv. Mater.*, 2021, **33**, 2008353.
59. B. Trappmann, J. E. Gautrot, J. T. Connelly, D. G. T. Strange, Y. Li, M. L. Oyen, M. A. Cohen Stuart, H. Boehm, B. Li, V. Vogel, J. P. Spatz, F. M. Watt and W. T. S. Huck, *Nat. Mater.*, 2012, **11**, 642–649.
60. D. S. H. Wong, J. Li, X. Yan, B. Wang, R. Li, L. Zhang and L. Bian, *Nano Lett.*, 2017, **17**, 1685–1695.
61. S. H. D. Wong, X. Xu, X. Chen, Y. Xin, L. Xu, C. H. N. Lai, J. Oh, W. K. R. Wong, X. Wang, S. Han, W. You, X. Shuai, N. Wong, Y. Tan, L. Duan and L. Bian, *Nano Lett.*, 2021, **21**, 3225–3236.
62. S. Yadav, H. T. Ta and N.-T. Nguyen, *VIEW*, 2021, **2**, 20200080.
63. N. Y. K. Li, H. K. Heris and L. Mongeau, *J. Cytol. Mol. Biol.*, 2013, **1**, 001.
64. C. F. Dionello, D. Sá-Caputo, H. V. Pereira, C. R. Sousa-Gonçalves, A. I. Maiworm, D. S. Morel, E. Moreira-Marconi, L. L. Paineiras-Domingos, D. Bemben and M. Bernardo-Filho, *J. Musculoskelet. Neuronal Interact.*, 2016, **16**, 193–203.
65. K. Kapat, Q. T. H. Shubhra, M. Zhou and S. Leeuwenburgh, *Adv. Funct. Mater.*, 2020, **30**, 1909045.
66. H. Nikukar, S. Reid, P. M. Tsimbouri, M. O. Riehle, A. S. G. Curtis and M. J. Dalby, *ACS Nano*, 2013, **7**, 2758–2767.
67. S. N. Robertson, P. Campsie, P. G. Childs, F. Madsen, H. Donnelly, F. L. Henriquez, W. G. Mackay, M. Salmerón-Sánchez, M. P. Tsimbouri, C. Williams, M. J. Dalby and S. Reid, *Philos. Trans. R. Soc., A*, 2018, **376**, 20170290.

68. T. Hodgkinson, P. M. Tsimbouri, V. Llopis-Hernandez, P. Campsie, D. Scurr, P. G. Childs, D. Phillips, S. Donnelly, J. A. Wells, F. J. O'Brien, M. Salmeron-Sanchez, K. Burgess, M. Alexander, M. Vassalli, R. O. C. Oreffo, S. Reid, D. J. France and M. J. Dalby, *Sci. Adv.*, 2021, **7**, eabb7921.
69. A. Sutton, T. Shirman, J. V. I. Timonen, G. T. England, P. Kim, M. Kolle, T. Ferrante, L. D. Zarzar, E. Strong and J. Aizenberg, *Nat. Commun.*, 2017, **8**, 14700.
70. L. F. Kadem, K. G. Suana, M. Holz, W. Wang, H. Westerhaus, R. Herges and C. Selhuber-Unkel, *Angew. Chem., Int. Ed.*, 2017, **56**, 225–229.
71. Z. Liu, Y. Liu, Y. Chang, H. R. Seyf, A. Henry, A. L. Mattheyses, K. Yehl, Y. Zhang, Z. Huang and K. Salaita, *Nat. Methods*, 2016, **13**, 143–146.
72. A. N. Ramey-Ward, H. Su and K. Salaita, *ACS Appl. Mater. Interfaces*, 2020, **12**, 35903–35917.
73. Q. Zhou, J. Chen, Y. Luan, P. A. Vainikka, S. Thallmair, S. J. Marrink, B. L. Feringa and P. Van Rijn, *Sci. Adv.*, 2020, **6**, eaay2756.
74. O. Chaudhuri, L. Gu, M. Darnell, D. Klumpers, S. A. Bencherif, J. C. Weaver, N. Huebsch and D. J. Mooney, *Nat. Commun.*, 2015, **6**, 7365.
75. A. R. Cameron, J. E. Frith and J. J. Cooper-White, *Biomaterials*, 2011, **32**, 5979–5993.
76. A. R. Cameron, J. E. Frith, G. A. Gomez, A. S. Yap and J. J. Cooper-White, *Biomaterials*, 2014, **35**, 1857–1868.
77. J.-H. Seo, S. Kakinoki, Y. Inoue, T. Yamaoka, K. Ishihara and N. Yui, *J. Am. Chem. Soc.*, 2013, **135**, 5513–5516.
78. S. Kakinoki, J.-H. Seo, Y. Inoue, K. Ishihara, N. Yui and T. Yamaoka, *Acta Biomater.*, 2015, **13**, 42–51.
79. J.-H. Seo, S. Kakinoki, T. Yamaoka and N. Yui, *Adv. Healthcare Mater.*, 2015, **4**, 215–222.
80. J.-H. Seo, M. Hirata, S. Kakinoki, T. Yamaoka and N. Yui, *RSC Adv.*, 2016, **6**, 35668–35676.
81. L. Yu, Y. Hou, W. Xie, J. L. C. Camacho, C. Cheng, A. Holle, J. Young, B. Trappmann, W. Zhao, M. F. Melzig, E. A. Cavalcanti-Adam, C. Zhao, J. P. Spatz, Q. Wei and R. Haag, *Adv. Mater.*, 2020, **32**, 2002566.
82. L. Yu, Y. Hou, W. Xie, J. L. Cuellar-Camacho, Q. Wei and R. Haag, *Adv. Mater.*, 2020, **32**, 2006986.
83. R. Glazier and K. Salaita, *Biochim. Biophys. Acta*, 2017, **1859**, 1465–1482.
84. K. Morigaki, *Anal. Sci.*, 2021, **32**, 683–689.
85. M. Bennett, M. Cantini, J. Reboud, J. M. Cooper, P. Roca-Cusachs and M. Salmeron-Sanchez, *Proc. Natl. Acad. Sci.*, 2018, **115**, 1192–1197.
86. K. Uto, S. S. Mano, T. Aoyagi and M. Ebara, *ACS Biomater. Sci. Eng.*, 2016, **2**, 446–453.
87. S. S. Mano, K. Uto, T. Aoyagi and M. Ebara, *AIMS Mater. Sci.*, 2016, **3**, 66–82.
88. S. S. Mano, K. Uto and M. Ebara, *Int. J. Mol. Sci.*, 2020, **21**, 1757.
89. M. Najmina, K. Uto and M. Ebara, *Polym. J.*, 2020, **52**, 985–995.
90. A. C. Chang, K. Uto, K. Homma and J. Nakanishi, *Biomaterials*, 2021, **274**, 120861.

91. E. Ban, J. M. Franklin, S. Nam, L. R. Smith, H. Wang, R. G. Wells, O. Chaudhuri, J. T. Liphardt and V. B. Shenoy, *Biophys. J.*, 2018, **114**, 450–461.
92. H. Wang, A. S. Abhilash, C. S. Chen, R. G. Wells and V. B. Shenoy, *Biophys. J.*, 2014, **107**, 2592–2603.
93. B. M. Baker, B. Trappmann, W. Y. Wang, M. S. Sakar, I. L. Kim, V. B. Shenoy, J. A. Burdick and C. S. Chen, *Nat. Mater.*, 2015, **14**, 1262–1268.
94. E. E. Charrier, K. Pogoda, R. G. Wells and P. A. Janmey, *Nat. Commun.*, 2018, **9**, 449.
95. M. Shri, H. Agrawal, P. Rani, D. Singh and S. K. Onteru, *Sci. Rep.*, 2017, **7**, 1203.
96. A. K. Epstein, T.-S. Wong, R. A. Belisle, E. M. Boggs and J. Aizenberg, *Proc. Natl. Acad. Sci.*, 2012, **109**, 13182–13187.
97. M. D. Rosenberg, in *Cellular Control Mechanisms and Cancer*, eds. P. Emmelot and O. Mohbock, Elsevier, Amsterdam, 1964, pp. 146–164.
98. C. R. Keese and I. Giaever, *Proc. Natl. Acad. Sci.*, 1983, **80**, 5622–5626.
99. I. Giaever and C. R. Keese, *Proc. Natl. Acad. Sci.*, 1983, **80**, 219–222.
100. C. Keese and I. Giaever, *Science*, 1983, **219**, 1448–1449.
101. J. Ando, S. M. Albelda and E. M. Levine, *In Vitro Cell. Dev. Biol.: Anim.*, 1991, **27**, 525–532.
102. Y. Shiba, T. Ohshima and M. Sato, *Kagaku Kogaku Ronbunshu*, 1998, **24**, 343–345.
103. M. Pilarek, I. Grabowska, M. A. Ciemerych, K. Dąbkowska and K. W. Szewczyk, *Biotechnol. Lett.*, 2013, **35**, 1387–1394.
104. C. R. Keese and I. Giaever, *Exp. Cell Res.*, 1991, **195**, 528–532.
105. D. Kong, K. D. Q. Nguyen, W. Megone, L. Peng and J. E. Gautrot, *Faraday Discuss.*, 2017, **204**, 367–381.
106. D. Kong, W. Megone, K. D. Q. Nguyen, S. Di Cio, M. Ramstedt and J. E. Gautrot, *Nano Lett.*, 2018, **18**, 1946–1951.
107. K. Minami, T. Mori, W. Nakanishi, N. Shigi, J. Nakanishi, J. P. Hill, M. Komiyama and K. Ariga, *ACS Appl. Mater. Interfaces*, 2017, **9**, 30553–30560.
108. X. Jia, K. Minami, K. Uto, A. C. Chang, J. P. Hill, T. Ueki, J. Nakanishi and K. Ariga, *Small*, 2019, **15**, 1804640.
109. X. Jia, K. Minami, K. Uto, A. C. Chang, J. P. Hill, J. Nakanishi and K. Ariga, *Adv. Mater.*, 2020, **32**, 1905942.
110. A.-C. Tsai, R. Jeske, X. Chen, X. Yuan and Y. Li, *Front. Bioeng. Biotechnol.*, 2020, **8**, 640.
111. A. Stejskalová, N. Oliva, F. J. England and B. D. Almquist, *Adv. Mater.*, 2019, **31**, 1806380.
112. B. Hinz, *Matrix Biol.*, 2015, **47**, 54–65.
113. Y. Zhang, C. Ge, C. Zhu and K. Salaita, *Nat. Commun.*, 2014, **5**, 5167.
114. B. A. Aguado, J. C. Grim, A. M. Rosales, J. J. Watson-Capps and K. S. Anseth, *Sci. Transl. Med.*, 2018, **10**, eaam8645.
115. A. Elosegui-Artola, X. Trepat and P. Roca-Cusachs, *Trends Cell Biol.*, 2018, **28**, 356–367.

CHAPTER 7

Micro- and Nanopatterned Substrates for Studies on the Mechanobiology of Cell–Matrix Adhesions

JOEL CHRISTIAN AND ELISABETTA ADA CAVALCANTI-ADAM*

Max Planck Institute for Medical Research, Department of Cellular Biophysics, Jahnstraße 29, D-69120 Heidelberg, Germany
*Email: eacavalcanti@mr.mpg.de

7.1 Chemomechanical Coupling at the Cell–Extracellular Matrix Interface

In tissues, the extracellular space comprises several factors that exert both chemical and mechanical effects on cells. These factors, represented by extracellular matrix (ECM) proteins and carbohydrates, biochemical factors embedded in the matrix and neighboring cells, are targets for the design of materials aimed at engineering the cell microenvironment.[1] The ECM presents chemical and physical properties that are distinct for each tissue, due not only to variation in composition but also to differences in the density and organization of ECM proteins. Hence, different topographies at the micro- and nanoscales, as well as different spacing and geometry of adhesion sites, are present at the interface with cells. Additionally, the ECM may show variations in stiffness and viscoelasticity; these traits are observed not only for different tissues but also for the same tissue during disease.[2]

Biomaterials Science Series No. 12
Material-based Mechanobiology
Edited by Jun Nakanishi and Koichiro Uto

Published by the Royal Society of Chemistry, www.rsc.org

Mechanosensing and mechanotransduction processes in cells are related to the distinct chemical and mechanical properties of the ECM. The distribution of chemical and mechanical cues within the ECM is fundamental in the regulation of cell shape, adhesion, and migration and directs several cell functions.[3,4]

A challenging aspect in developing materials that allow us to recapitulate all these properties of the ECM is to maintain a control of adhesion to the substrate at the cellular and molecular levels, while mimicking the complexity of ECM mechanical function (Figure 7.1). Micropatterning of substrates allows precise control over cell adhesion and shape. Adhesion proteins of the ECM are patterned into islands of different sizes and geometries, surrounded by a nonadhesive background.[5] Nanopatterning of extracellular ligands has been developed with the scope of controlling receptor binding sites and receptor lateral clustering.[6–8] Recently, it has been elucidated that the nanoscale spacing of integrin ligands regulates cell forces and coupling to the cytoskeleton.[9,10] Both micro- and nanopatterning can be applied to hydrogels with different stiffness by tailoring the chemistry used for surface functionalization and ligand immobilization.

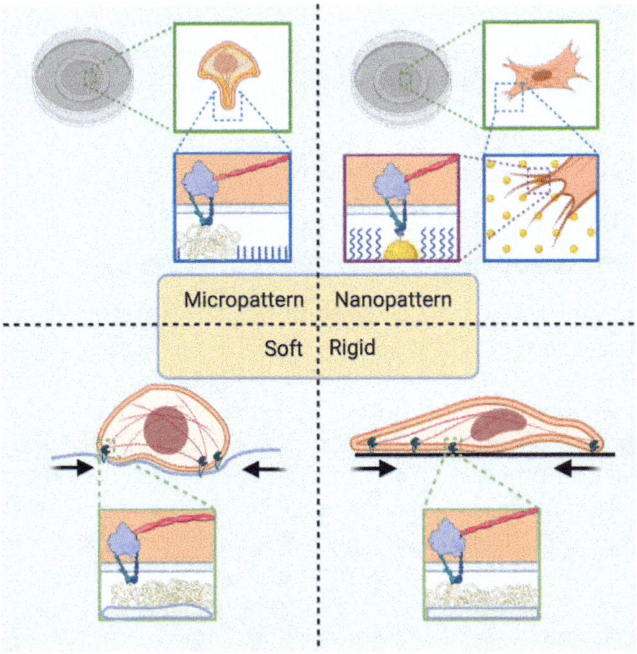

Figure 7.1 Surface patterning of extracellular matrix molecules at the micro- and nanoscales on rigid and soft materials for mechanobiological studies. At the micron scale, patterning allows the control of cell shape and organization and their relation to cell-generated forces. At the nanoscale, patterning is applied to control the adhesion and clustering of cell surface receptors and molecular forces generated at receptor–ligand binding sites.

7.2 Spatial Control of Cell–Extracellular Matrix Adhesion by Surface Patterning

The multiscale structure present in solid tissues has inspired the design of surface materials to control the different signals provided by the extracellular environment at different-length scales.[11] Structural hierarchies strongly influence the mechanical properties of tissues:[12] Besides calcified tissues, where the microarchitecture of matrix collagen fibers strongly regulates the mechanical load distribution, the structure of other tissues has been further related to mechanical functions. For cardiac tissue, inspired by its ECM mechanical properties, patterns at the micro- and nanoscales of biochemical cues have been applied to restore tissue function.[13] Interestingly, when focusing on the nanoscale level, protein structures are highly conserved across different tissues and appear to be altered in diseases, where either mutation or disruption in architecture affect the spatial organization of cell adhesion sites. This aspect constitutes the primary building block that regulates adhesive interactions and mechanics between the ECM and cells.

7.2.1 Micropatterning of Matrix Proteins to Direct Cell Shape and Architecture

The control over tissue and cell architecture *in vitro* can be achieved by surface micropatterning of ECM proteins. Adhesive micropatterning to control cell spreading was first introduced more than a half century ago.[14] Until today, micropatterning methods of matrix proteins have been used extensively in cell biology labs to study cellular physiology ranging from early signaling, actin organization, to collective cell migration.[15–17]

There are several methods for the micropatterning of matrix proteins, and the most popular is microcontact printing.[5] Matrix proteins are deposited on the pre-etched polydimethylsiloxane (PDMS) master and are subsequently stamped onto the coverslip. An alternative method is based on the use of photolithography on polyethylene glycol– (PEG-) coated glass surfaces. Micropatterning is achieved by placing a photomask between a deep UV light source and the PEG-coated substrate. The local exposure to the deep UV will cleave the cell-repellent PEG layer, allowing matrix proteins to be adsorbed on the substrate. Recently, photolithography-based micropatterning of matrix proteins has been achieved without the need for a photomask. This allows micropatterning on a broader range of substrate, such as glass, plastic, and transmission electron microscopy (TEM) grids.[18–20]

Micropatterning can be applied to individual cells or multicellular groups. An important application is the control of cell shape to study the adaptation of cell adhesion and the cytoskeleton to geometrical cues of the extracellular environment.[5] Micropatterns designed to vary the size, shape, and distribution of focal adhesions, which are cell–ECM adhesions mediated by integrins and linked to the actin cytoskeleton, have a strong influence on spreading and migration, as well as on growth and survival.[21,22] Micropatterning of single cells allows to us establish the relation between cell shape and traction forces. Further, actin stress fiber geometries in patterned cells have been investigated to determine how they contribute to the elastic energy dissipation of the stress fiber.[23] To relate micron-sized geometrical features of the ECM and cell migration, micropatterned stripes of different widths and on different types of surfaces (rigid or elastic) have been applied by using microcontact printing of ECM proteins or by UV-photopatterning. Migration tracks as narrow as 2 μm mimic the fibrillar nature of ECM proteins; in recent work from the Ballard Lab,[24] the use of such an approach allowed the elucidation of the early steps of symmetry breaking and contribution of adhesion dynamics for cell migration in the absence of cytoskeleton polarity. For multicellular aggregates, micropatterning allows the study of cell–cell interactions as a function of adhesion size and shape to the underlying ECM. Circular micropatterns of varying size unraveled the importance of arrangement of neighboring cells in the coordination of collective motility.[25] The emergence of coordinated movements in epithelial collectives has been observed after establishment of single cell polarity in patterned microtracks.[17]

7.2.2 Nanopatterning of Matrix Ligands for the Control of Receptor Clustering

Due to the complexity and dynamic architecture of the extracellular environment, it is difficult to obtain precise information regarding the spatial distribution of single binding sites between cells and ECM proteins. Rather than using full-length ECM proteins, studies centered on molecular and cellular responses to adhesive cues have been also performed on material surfaces coated with peptides derived from ECM proteins. A prominent example is the RGD (arginine–glycine–aspartic acid) sequence.[26] The RGD motifs present in ECM proteins are recognized by different types of integrins, and to further achieve selectivity in responses regulated by adhesion to the ECM, peptidomimetic ligands have been designed.[27]

A key aspect when designing the presentation of ligands to cells is their density and distribution, keeping in mind that many cell surface receptors cluster at the membrane and further regulate the recruitment of intracellular proteins involved in the formation of adhesion complexes.[28] The structural information about fibrillar ECM proteins has been of help in engineering the nanoscale presentation of adhesive ligands for cell adhesion: for example, the GFOGER peptide sequence, which is an integrin-binding motif, is predicted to be at a regular distance of 68 nm along the fibril surface of the D-staggered array of collagen type I.[29] To achieve precise control over the spatial presentation of ECM ligands and the resulting cell surface receptor clustering at the nanoscale, surface nanopatterning strategies have been developed over the last two decades.[6,30,31] These have been further expanded not only to measure integrin-mediated mechanotransduction responses to the extracellular environment but also to present multiple extracellular cues to recapitulate the local crosstalk of different types of receptors.

Michael Sheetz and colleagues[32] investigated nanoscale geometries of fibers, consisting of nanolines and nanodot arrays prepared by electron-beam lithography in order to support integrin-mediated adhesion to RGD ligands. An interesting finding is that the formation of a stable integrin-mediated cluster does not take place on fibers <40 nm wide, even if the ligand surface density is sufficient. Further, the lateral interactions between single integrins are crucial for the maintenance of stable adhesions, evidencing the importance of unliganded receptors in building bridges between nanoclusters of bound integrins.

A versatile tool for studying cell membrane receptor clustering and crosstalk during adhesion is based on block copolymer micellar nanolithography (BCMN).[6] Here, hexagonally packed arrays of metal nanoparticles decorated with extracellular ligands and small molecules are patterned on different types of surfaces, at a spacing ranging from 30 to 200 nm. For patterning substrates with gold nanoparticles, the self-assembly of polystyrene-*block*-poly[2-vinylpyridine$(HAuCl_4)_{0.5}$] diblock copolymer micelles is used. An important aspect to consider when engineering the binding sites of single receptors is the nanoparticle size, which should not exceed the lateral size of the receptor. To control integrin lateral clustering and the resulting forces generated at adhesion sites, each particle has a diameter of approximately 8 nm; the interparticle spacing can be varied, ranging from 30 to >200 nm, depending on the molecular weight of the block copolymer and the concentration of the polymer solution. Furthermore, the interparticle spacing is dependent on the micellar film deposited on the surface: This procedure is done by either dip coating or spin coating, and the retracting or rotating speed used for creating the film on the surface can be varied. For the interactions with cells, ligand type, density, and spacing at the nanometer scale influence cell adhesion and spreading dynamics.[10,33] The fabrication of micro-nanopatterned substrate offers an additional way to control these parameters. Micro-nanopatterned substrate can be achieved by combining nanopatterning using the BCMN technique and electron beam lithography or direct laser writing.[34,35]

The versatility of BCMN lies not only on the varying spacing that can be easily created but also on the combination of different metal nanoparticles on the same surface to create binary functionalities.[36,37] Additionally, thiolated integrin-adhesive peptides or peptidomimetic compounds can be immobilized to address specific functions of integrin types.[33] To further recapitulate the complexity of interactions between integrins and other types of receptors involved in cell adhesion, in recent work. integrin selective ligands and growth factors were co-immobilized on nanopatterned substrates (Figure 7.2).[38]

Using this mode of presentation, which is based on the spatial proximity of ligands immobilized on the surface, it is possible to investigate the specific interaction of different integrin types and growth factor receptors. Interestingly, surface co-presentation of integrin ligands and growth factors, namely bone morphogenetic protein-2 (BMP-2) enhances focal adhesion formation in cells. Cells adhering to integrin ligands in the presence of BMP-2 added to the culture media could still assemble focal adhesions but to a lesser extent. The spatial proximity between the two classes of receptors revealed that the activation of α5β1 integrins for adhesion is favored by the

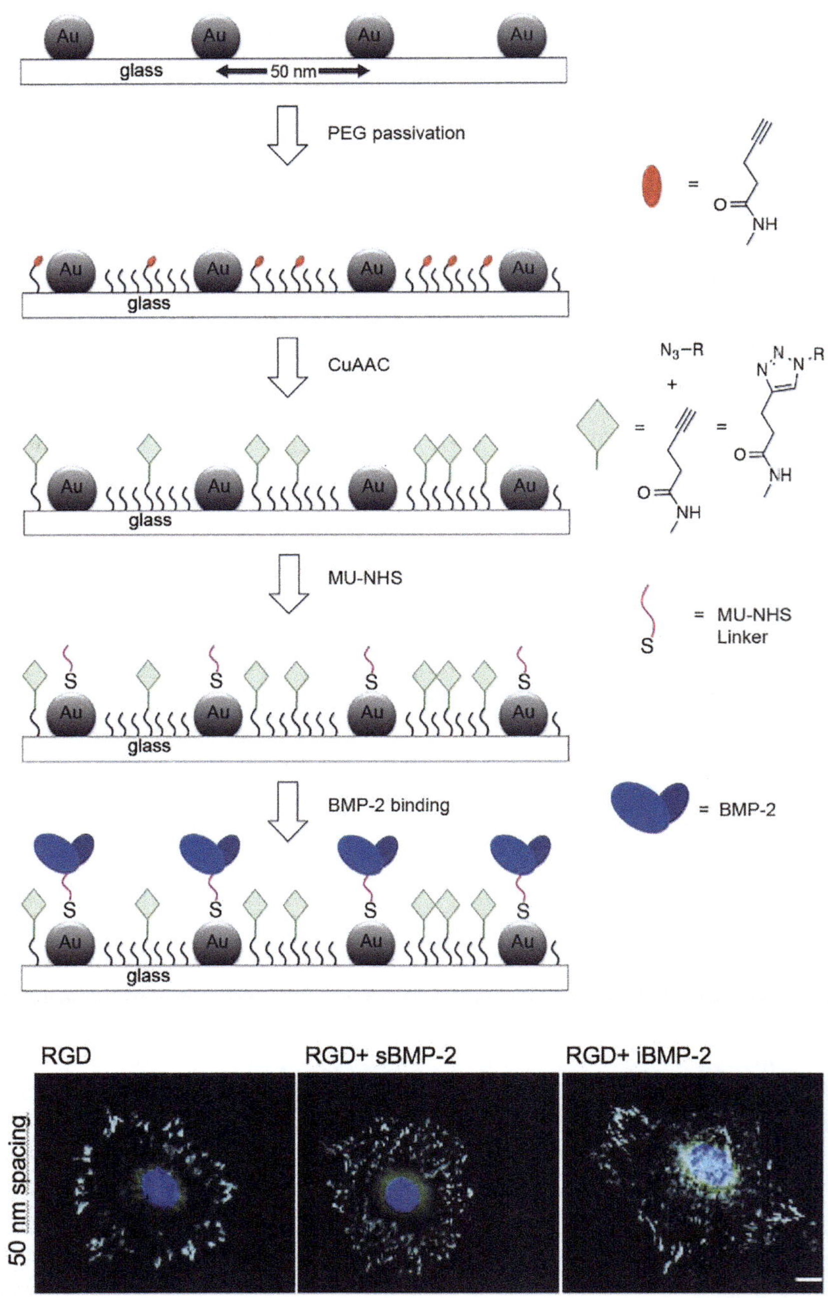

presence of surface immobilized BMP-2, a finding that complements existing information on the selective nature of interactions between integrin types and growth factor receptors.[39] Furthermore, it is possible to engineer a 1:1 adhesion site on single nanoparticles by using chimeric peptides for integrin and growth factor receptor binding.[40] Here, the proximity of αvβ3 integrins and BMP-receptor type I promotes spreading dynamics and BMP-dependent signaling. Taken together, these recent works highlight the importance of nanoscale regulation of binding sites not only for integrins but also for other receptors, in regulating adhesion assemblies and function.

7.3 Surface Patterning for Cell-generated Force Measurements

It has been widely known that cells require not only genetic and biochemical factors but also physical features to survive and grow. The combination of these parameters is shown by high variability in cell spreading and polarization, especially on two-dimensional substrates. In order to measure force, it is crucial to reduce differences in cell behavior by introducing a controlled environment.[41] Micro- and nanopatterning approaches represent a possibility to control biological systems by providing a defined boundary condition that results in controlled cell shape and structure organization, which in turn influence cell signaling and function. Surface patterning has been applied for the measurement of traction forces in single cells or cell collectives, as well as for the analysis of molecular forces generated at single adhesion sites to the ECM.[10,42–44] A crucial aspect is the development of surface chemistry approaches that allow the application of micropatterning to soft materials to perform traction force microscopy, a commonly used materials-based mechanobiology tool. For the measurement of molecular forces, nanopatterning of ECM ligands combined with molecular tension probes has proven to be a very useful tool in determining the relationship between receptor and protein spatial arrangement and force generation.

Figure 7.2 Surface nanopatterning of growth factors (BMP-2) immobilized on gold nanoparticles. Nanopatterns are prepared by block copolymer micellar nanolithography. A PEG layer carrying alkyne functional groups is deposited on the surface between the nanoparticles and subsequently azide-functionalized RGD ligands are bound to it by click chemistry. Single growth factor molecules are linked at their amine resides to the particles using a 11-Mercaptoundecanoyl-*N*-hydroxysuccinimide ester (MU-NHS) heterobifunctional linker. Reproduced from ref. 38, https://doi.org/10.1016/j.biomaterials.2020.120484, under the terms of the CC BY 4.0 license https://creativecommons.org/licenses/by/4.0/. The assembly of focal adhesions is enhanced in the presence of immobilized BMP-2 (iBMP-2), in comparison to BMP-2 added to the culture media (sBMP-2) or when only RGD is present on the surface (cyan, vinculin; scale bar 10 μm).

7.3.1 Micro- and Nanopatterning of Soft Materials to Probe Cell Rigidity Sensing and Measure Cellular Traction Forces

Surface adhesive patterning is getting more accessible to a broader range of substrates, owing to the development of a variety of material science technologies. Viscoelastic materials like hydrogels allow studies of the influence of mechanical rigidity on cellular physiology and functions.[45–47] Micropatterning methods such as microcontact printing and photolithography have been successfully used to generate patterned cells on the surface of polyacrylamide hydrogels.[48,49] Despite its popularity, microcontact printing requires a PDMS stamp that is generally fabricated by the etching process. Photolithography, however, requires a deep UV source and a photomask that is usually grafted by electron or laser beams. These fabrication techniques are commonly less accessible by cell biology laboratories. Another drawback is that microcontact printing and photolithography *via* photomask yield a nonflexible micropatterning strategy. For each type of pattern, a new PDMS stamp or photomask has to be produced. As an alternative, a micropatterning technique that allows flexibility in changing patterns is achieved with "maskless" photolithography.

Instead of using deep UV illumination, the polymer brushes are cleaved upon a low dose of near-UV illumination with the presence of a water-soluble photoinitiator, allowing the patterning of pattern arrays within seconds.[20] The maskless patterning technique is achieved by the illumination of arbitrary grayscale patterns of UV light generated by a wide-field digital micromirror device–based projection system coupled to an epifluorescence microscope. In the end, micropatterned hydrogels can be produced by combining maskless photolithography and direct stamping on a functionalized polyacrylamide hydrogel (Figure 7.3). The hydrogel functionalization can be achieved by incorporating either reactive aldehyde groups or NHS moiety, leading to a covalent bond between the protein and the surface of polyacrylamide hydrogel.[49,50]

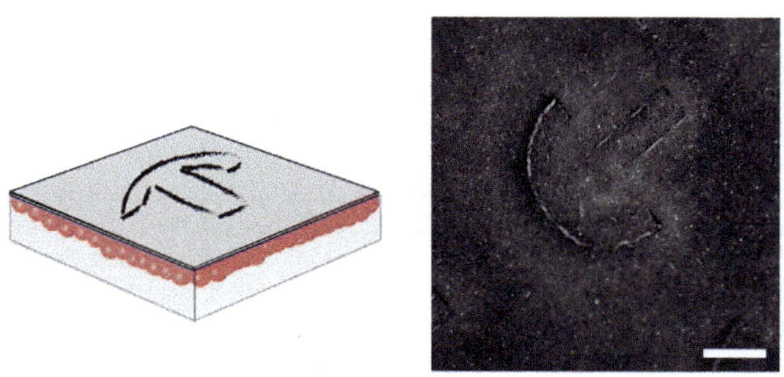

Figure 7.3 Left: Cartoon showing a crossbow patterned ECM protein stamped on a polyacrylamide gel with fluorescent beads confined near the surface. Right: Fluorescent micrograph of a crossbow pattern of fibronectin on the surface of a polyacrylamide hydrogel (scale bar = 20 μm).

Single-cell micropatterning on hydrogels has been applied to cells expressing β1 and/or αv integrin types for rigidity sensing and force generation mediated by integrins.[51] Areas of high tension at the cell–ECM interface are enriched in αv integrins, which further activate myosin II. Interestingly, fibroblasts lacking αv integrins present large focal adhesions and low myosin II and traction forces. The adjustment of fibroblast contractility to varying rigidity of fibronectin-coated substrates relies on the cooperation between β1 and αv integrins, where β1 integrins regulate force generation at adhesion sites, and αv integrins are responsible for the adaptation of focal adhesions to forces.[51,52]

Traction force microscopy (TFM) is a leading approach to measure cell contractility forces by tracking the displacement of submicron fluorescent beads embedded in protein-functionalized elastic substrates.[53] Forces are reconstructed from the substrate deformation at adhesion sites, and they are in the nanonewton (nN) range.[54] Nanopatterned hydrogels are produced by transferring gold nanoarrays prepared by BCMN on the surface of functionalized polyacrylamide hydrogels.[10,55] First, gold nanoparticles with a diameter of approximately 8 nm are deposited into a quasi-hexagonal pattern with a tunable interparticle distance on a glass substrate[6] (Figure 7.4). The transfer of the nanoparticles to the surface of a polyacrylamide hydrogel is achieved by incorporation of *N*, *N*′-bis-(acryloyl)cystamine that links gold nanoparticles to the crosslinked polyacrylamide. By independently varying hydrogel rigidity (1.5–150 kPa) and interparticle spacing of gold nanoparticles on polyacrylamide hydrogels (50–100 nm), it has been possible to address the relationship between force loading and integrin ligand spacing at cell–ECM adhesion sites.[10] Surprisingly, for low-rigidity substrates, increasing ligand spacing promotes focal adhesion growth, whereas on rigid substrates, large spacing induces the collapse of focal adhesions. TFM measurements and analysis of actin flow speed support the existence of a "molecular clutch" at focal adhesions:[46] Here, recruitment of further integrin bonds at the ECM interface takes place in response to force loading to redistribute the overall force and reduce the single loading per clutch. Thus cells are able to precisely adjust to the geometrical and physical cues of the ECM and regulate the force distribution at adhesion sites. As a result, mechanosensing and mechanotransduction signaling and generation of cellular traction forces appear to be regulated by the tuning of rigidity and spacing sensing.[56] An integrin ligand spacing of >200 nm hinders adhesion and spreading on stiff substrates but further promotes ion channel activation and YAP/TAZ signaling for the osteogenic differentiation of mesenchymal stem cells on soft hydrogels. Using cellular traction force microscopy in combination with micro- and nanopatterning strategies it has been possible to unravel the mechanical regulation of cell–ECM adhesion as a function of ligand spacing.

It should be mentioned that a drawback of conventional TFM, which is to date still difficult to overcome, is represented by the limited temporal and spatial resolution of TFM measurements. These issues are due to (1) random and high-density placement of the fluorescent beads within the deformable

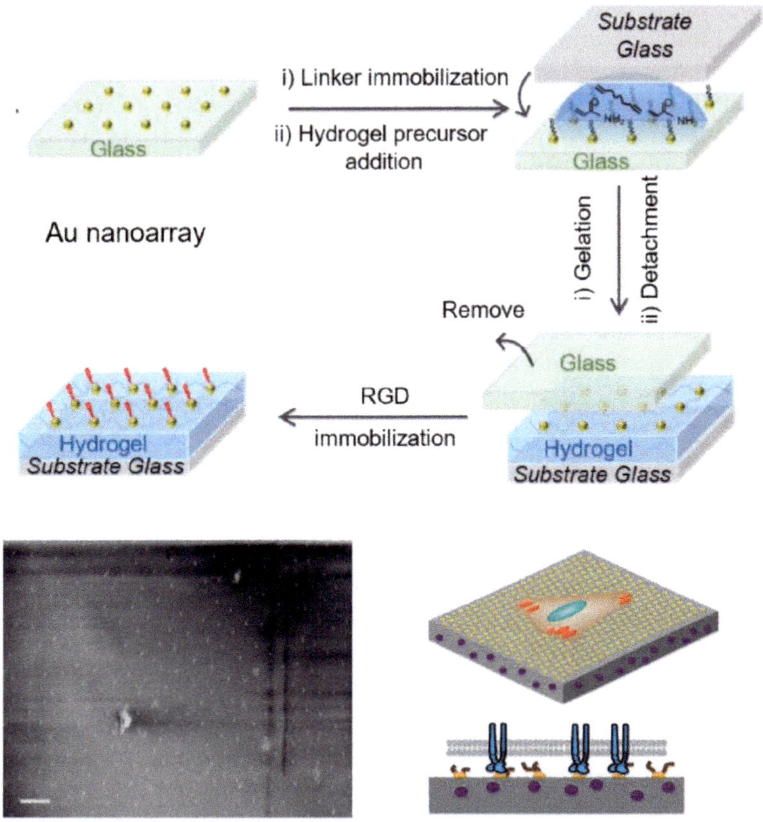

Figure 7.4 Schematic of the procedure for the nanopatterning of hydrogels. Gold nanoarrays on glass are transferred to the hydrogel surface during polymerization by using a *N*,*N*′-bis-(acryloyl)cystamine linker. Thiolated integrin ligands such as RGD are then immobilized on the gold nanoparticles for mechanobiology studies on receptor clustering. Bottom left: Cryo-SEM micrograph showing the gold nanoparticles on the hydrogel surface. Reproduced from ref. 56 with permission from Elsevier, Copyright 2021; lower part is reproduced from ref. 10 with permission from Springer Nature, Copyright 2017.

materials and (2) reference fields obtained only after material relaxation. To overcome these limitations, patterning of fluorescent quantum dots (100 nm lateral size and <20 nm thickness) by electrohydrodynamic printing with spatial positioning accuracy of 20 nm has made it possible to achieve a reference-free TFM platform combined with super-resolution microscopy.[57,58] Here, quantum dots are printed in monocrystalline nanodisc arrays by using a electrohydrodynamic nanodrip printing, based on the periodic ejection and acceleration of ink droplets by an electric field.[59] Further development of these types of approaches is necessary to make the technology more accessible to biology labs in the near future not only in terms of equipment needed but also for the software analysis routine.

7.3.2 Nanopatterning and Measurements of Molecular Forces at Cell–Matrix Adhesion Sites

Surface nanopatterning for cell adhesion studies offers the advantage of creating a defined presentation of extracellular ligands; this in turn allows the control of the spatial organization of cell surface receptor binding sites. A great advancement in the field of mechanobiology has been represented by the combination of nanopatterning with extracellular molecular tension sensors (MTS).[9,60,61] For molecular force measurements and the detection of piconewton (pN) forces generated by cell surface receptors, methods based on single-molecule force spectroscopy such as atomic force microscopy or optical and magnetic tweezers, measure forces for the dissociation of ligand–receptor bonds.[62] The dynamics of different adhesion sites in living cells cannot be measured with these methods, since they allow measurements of forces generated by one molecule or adhesion site at a time. Moreover, they are applied at the dorsal side of cells, whereas adhesion to material surfaces takes place at the ventral side. This aspect is of particular relevance for cells with basoapical polarization (*e.g.* polarized epithelial cells), which is related to differences in cytoskeletal organization and force distribution. To this end, Salaita and colleagues developed molecular tension fluorescence microscopy (MTMF), based on the use of polymer-, nucleic acid–, or peptide-based extendible springs as extracellular tension sensors for measuring receptor–ligand binding forces.[9,63,64] Such springs are designed to carry a fluorescent donor and acceptor/quencher, *e.g.* FRET pair of chromophores, to detect the changes in extension. These changes are due to receptor-generated tension upon binding to a ligand immobilized on one end of the spring, while the other end is linked to the substrate.[65]

MTMF has been applied to elucidate how integrin lateral clustering regulates the generation of molecular forces upon coupling of focal adhesions to the actomyosin machinery.[9] Here the MTS are immobilized on gold nanoarrays to control the lateral clustering of integrins at 50 and 100 nm spacing. The gold nanoparticles act as anchor and quencher; thus the mechanism for energy transfer is nanometal surface energy transfer, which offers a more linear fluorescence-distance response than the one measured with FRET probes. With this method, it has been possible to determine that integrin lateral clustering is necessary for developing forces above a threshold of 2 pN at single bonds. These forces are generated to harness actin cytoskeleton tractions, as stress fibers form and focal adhesions mature at a spacing of 50 nm. For larger spacing, the molecular forces remain low, and as soon as actin fibers polymerize and are coupled to early focal adhesions, the tension drops until adhesions disassemble.

Such a system has been also applied to study forces generated during endocytosis of single viral particles[60] (Figure 7.5). When interacting with cells, viral particles cause deformation of the cell membrane; some viruses present on their surfaces' integrin-binding sites, which trigger receptor binding and clathrin-mediated endocytosis.

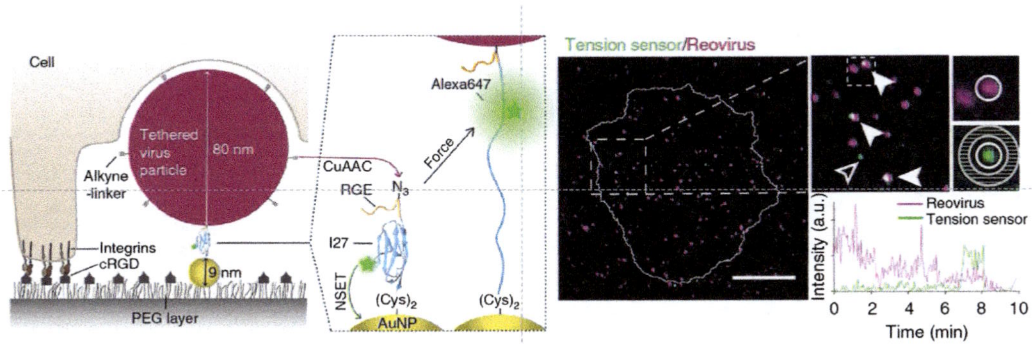

Figure 7.5 Left: Schematic of the experimental approach to combine MTS consisting of Titin I27 domain and surface nanopatterning of immobilized viruses for molecular force measurements at the ventral side of cells. Right: Imaging and temporal detection of fluorescent signal (green) generated when cells bind to reoviruses (pink). Reproduced from ref. 60, https://doi.org/10.1038/s41467-019-13877-w, under the terms of the CC BY 4.0 license https://creativecommons.org/licenses/by/4.0/.

MTMF combined with surface nanopatterning served here as a platform to immobilize mammalian reoviruses and track forces generated at the ventral side of cells adhering to the surfaces. With this mechanobiology approach based on nanopatterning, it was possible to determine that forces as high as 100 pN are generated at the ventral side of cells in the presence of viral particles. In comparison to other studies performed with AFM on the apical side, forces appear to be in the same range, although the force generation mechanism is different, considering the contribution of substrate adhesion forces for the uptake process.

7.4 Outlook

The spatial patterning of ECM ligands and extracellular molecules has allowed researchers to gain a better understanding of the cellular and molecular processes regulating cell adhesion to the ECM. Recent advances in materials-based mechanobiology that combine micro- and nanopatterning with cellular and molecular force measurements proved the central role of receptor clustering in the regulation of mechanosensing and mechanotransduction. Future research directions should serve to further elucidate how different types of cell surface receptors and their bond lifetimes act in stabilizing and reinforcing bonds to the ECM at the nanoscale. Further, it remains unknown how receptor-ligand binding forces are regulated as a function of receptor crosstalk at the nanoscale and of mechanical signaling on varying substrate stiffness.

A current challenge is represented by the application of surface micronanopatterning for TFM and MTMF measurements to relate the control of cell shape to the molecular forces at integrin adhesion sites to the ECM. The fabrication of micronanopatterned substrate requires advanced technologies such as photolithography, electron beam lithography, and laser machining that are often not really accessible to a cell biology lab.[34] Combining TFM and MTMF on nanopatterned materials represents a further technical challenge that will require further work on the design of the MTS, preparation of the hydrogels for TFM, and suitability of the samples to perform super-resolution microscopy techniques.

Abbreviations

BCMN	block copolymer micellar nanolithography
BMP-2	bone morphogenetic protein 2
ECM	extracellular matrix
FRET	fluorescence resonance energy transfer
MTMF	molecular tension fluorescence microscopy
MTS	molecular tension sensor
PDMS	polydimethylsiloxane
PEG	polyethylene glycol
TFM	traction force microscopy
UV	ultraviolet

Acknowledgements

The authors thank group members from Max Planck Institute of Medical Research for helpful discussions and acknowledge the financial support of the Max Planck Society and the German Science Foundation (DFG SFB1129 Project nr. 240245660 P15 to E.A.C.-A.).

References

1. A. Cipitria and M. Salmeron-Sanchez, *Adv Healthcare Mater*, 2017, **6**, 170052.
2. P. M. Gilbert and V. M. Weaver, *Semin. Cell Dev. Biol.*, 2017, **67**, 141–152.
3. J. Z. Kechagia, J. Ivaska and P. Roca-Cusachs, *Nat. Rev. Mol. Cell Biol.*, 2019, **20**, 457–473.
4. K. L. Mui, C. S. Chen and R. K. Assoian, *J. Cell Sci.*, 2016, **129**, 1093–1100.
5. M. Thery, *J. Cell Sci.*, 2010, **123**, 4201–4213.
6. M. Arnold, E. A. Cavalcanti-Adam, R. Glass, J. Blummel, W. Eck, M. Kantlehner, H. Kessler and J. P. Spatz, *Chem. Phys. Chem.*, 2004, **5**, 383–388.
7. E. A. Cavalcanti-Adam, T. Volberg, A. Micoulet, H. Kessler, B. Geiger and J. P. Spatz, *Biophys. J.*, 2007, **92**, 2964–2974.
8. R. Changede and M. Sheetz, *BioEssays*, 2017, **39**, 1–12.
9. Y. Liu, R. Medda, Z. Liu, K. Galior, K. Yehl, J. P. Spatz, E. A. Cavalcanti-Adam and K. Salaita, *Nano Lett.*, 2014, **14**, 5539–5546.
10. R. Oria, T. Wiegand, J. Escribano, A. Elosegui-Artola, J. J. Uriarte, C. Moreno-Pulido, I. Platzman, P. Delcanale, L. Albertazzi, D. Navajas, X. Trepat, J. M. Garcia-Aznar, E. A. Cavalcanti-Adam and P. Roca-Cusachs, *Nature*, 2017, **552**, 219–224.
11. K. Liu and J. Lei, *ACS Nano*, 2011, **5**, 6786–6790.
12. P. Fratzl and R. Weinkamer, *Prog. Mater. Sci.*, 2007, **52**, 1263–1334.
13. B. Zhang, Y. Xiao, A. Hsieh, N. Thavandiran and M. Radisic, *Nanotechnology*, 2011, **22**, 494003.
14. S. B. Carter, *Exp. Cell Res.*, 1967, **48**, 4.
15. V. Fitzpatrick, L. Fourel, O. Destaing, F. Gilde, C. Albiges-Rizo, C. Picart and T. Boudou, *Sci. Rep.*, 2017, **7**, 41479.
16. X. Wang, X. Hu, I. Dulinska-Molak, N. Kawazoe, Y. Yang and G. Chen, *Sci. Rep.*, 2016, **6**, 28708.
17. S. Jain, V. M. L. Cachoux, G. Narayana, S. de Beco, J. D'Alessandro, V. Cellerin, T. Chen, M. L. Heuze, P. Marcq, R. M. Mege, A. J. Kabla, C. T. Lim and B. Ladoux, *Nat. Phys.*, 2020, **16**, 802–809.
18. V. Nier, S. Jain, C. T. Lim, S. Ishihara, B. Ladoux and P. Marcq, *Biophys. J.*, 2016, **110**, 1625–1635.
19. M. Toro-Nahuelpan, I. Zagoriy, F. Senger, L. Blanchoin, M. Thery and J. Mahamid, *Nat. Methods*, 2020, **17**, 50–54.
20. P. O. Strale, A. Azioune, G. Bugnicourt, Y. Lecomte, M. Chahid and V. Studer, *Adv. Mater.*, 2016, **28**, 2024–2029.

21. C. S. Chen, M. Mrksich, S. Huang, G. M. Whitesides and D. E. Ingber, *Science*, 1997, **276**, 1425–1428.
22. D. Lehnert, B. Wehrle-Haller, C. David, U. Weiland, C. Ballestrem, B. A. Imhof and M. Bastmeyer, *J. Cell Sci.*, 2004, **117**, 41–52.
23. E. Kassianidou, C. A. Brand, U. S. Schwarz and S. Kumar, *Proc. Natl. Acad. Sci. U. S. A.*, 2017, **114**, 2622–2627.
24. K. Hennig, I. Wang, P. Moreau, L. Valon, S. DeBeco, M. Coppey, Y. A. Miroshnikova, C. Albiges-Rizo, C. Favard, R. Voituriez and M. Balland, *Sci. Adv.*, 2020, **6**, 12.
25. F. J. Segerer, F. Thuroff, A. Piera Alberola, E. Frey and J. O. Radler, *Phys. Rev. Lett.*, 2015, **114**, 228102.
26. K. M. Yamada, *J. Biol. Chem.*, 1991, **266**, 4.
27. T. G. Kapp, F. Rechenmacher, S. Neubauer, O. V. Maltsev, E. A. Cavalcanti-Adam, R. Zarka, U. Reuning, J. Notni, H. J. Wester, C. Mas-Moruno, J. Spatz, B. Geiger and H. Kessler, *Sci. Rep.*, 2017, **7**, 39805.
28. B. B. Geiger and A. Bershadsky, *Cell*, 2002, **110**, 4.
29. M. D. Shoulders and R. T. Raines, *Annu. Rev. Biochem.*, 2009, **78**, 929–958.
30. M. Schvartzman, M. Palma, J. Sable, J. Abramson, X. Hu, M. P. Sheetz and S. J. Wind, *Nano Lett.*, 2011, **11**, 1306–1312.
31. J. Malmstrom, J. Lovmand, S. Kristensen, M. Sundh, M. Duch and D. S. Sutherland, *Nano Lett.*, 2011, **11**, 2264–2271.
32. R. Changede, H. Cai, S. J. Wind and M. P. Sheetz, *Nat. Mater.*, 2019, **18**, 1366–1375.
33. V. Schaufler, H. Czichos-Medda, V. Hirschfeld-Warnecken, S. Neubauer, F. Rechenmacher, R. Medda, H. Kessler, B. Geiger, J. P. Spatz and E. A. Cavalcanti-Adam, *Cell Adhes. Migr*, 2016, **10**, 505–515.
34. M. Arnold, M. Schwieder, J. Blummel, E. A. Cavalcanti-Adam, M. Lopez-Garcia, H. Kessler, B. Geiger and J. P. Spatz, *Soft Matter*, 2009, **5**, 72–77.
35. L. Yang, J. Wei, Z. Ma, P. Song, J. Ma, Y. Zhao, Z. Huang, M. Zhang, F. Yang and X. Wang, *Nanomaterials (Basel)*, 2019, 9, 1789.
36. J. Guasch, B. Conings, S. Neubauer, F. Rechenmacher, K. Ende, C. G. Rolli, C. Kappel, V. Schaufler, A. Micoulet, H. Kessler, H. G. Boyen, E. A. Cavalcanti-Adam and J. P. Spatz, *Adv. Mater.*, 2015, **27**, 3737–3747.
37. J. R. Polleux, M. Rasp, I. Louban, N. Plath, A. Fedlhoff and J. P. Spatz, *ACS Nano*, 2011, **5**, 10.
38. F. Posa, E. H. Baha-Schwab, Q. Wei, A. Di Benedetto, S. Neubauer, F. Reichart, H. Kessler, J. P. Spatz, C. Albiges-Rizo, G. Mori and E. A. Cavalcanti-Adam, *Biomaterials*, 2021, **267**, 120484.
39. L. Fourel, A. Valat, E. Faurobert, R. Guillot, I. Bourrin-Reynard, K. Ren, L. Lafanechere, E. Planus, C. Picart and C. Albiges-Rizo, *J. Cell Biol.*, 2016, **212**, 693–706.
40. L. Oliver-Cervello, H. Martin-Gomez, L. Reyes, F. Noureddine, E. Ada Cavalcanti-Adam, M. P. Ginebra and C. Mas-Moruno, *Adv Healthcare Mater*, 2020, e2001757.

41. P. J. Albert and U. S. Schwarz, *Cell Adhes. Migr.*, 2016, **10**, 516–528.
42. S. R. Vedula, A. Ravasio, E. Anon, T. Chen, G. Peyret, M. Ashraf and B. Ladoux, *Methods Cell Biol.*, 2014, **120**, 235–252.
43. V. Maruthamuthu, B. Sabass, U. S. Schwarz and M. L. Gardel, *Proc. Natl. Acad. Sci. U. S. A.*, 2011, **108**, 4708–4713.
44. T. Wiegand, M. Fratini, F. Frey, K. Yserentant, Y. Liu, E. Weber, K. Galior, J. Ohmes, F. Braun, D. P. Herten, S. Boulant, U. S. Schwarz, K. Salaita, E. A. Cavalcanti-Adam and J. P. Spatz, *Nat. Commun.*, 2020, **11**, 32.
45. R. Sunyer, V. Conte, J. Escribano, A. Elosegui-Artola, A. Labernadie, L. Valon, D. Navajas, J. M. Garcia-Aznar, J. J. Munoz, P. Roca-Cusachs and X. Trepat, *Science*, 2016, **353**, 1157–1161.
46. A. Elosegui-Artola, R. Oria, Y. Chen, A. Kosmalska, C. Perez-Gonzalez, N. Castro, C. Zhu, X. Trepat and P. Roca-Cusachs, *Nat. Cell Biol.*, 2016, **18**, 540–548.
47. A. J. Engler, S. Sen, H. L. Sweeney and D. E. Discher, *Cell*, 2006, **126**, 677–689.
48. Q. Tseng, I. Wang, E. Duchemin-Pelletier, A. Azioune, N. Carpi, J. Gao, O. Filhol, M. Piel, M. Thery and M. Balland, *Lab Chip*, 2011, **11**, 2231–2240.
49. B. Sarker, C. Walter and A. Pathak, *ACS Biomater. Sci. Eng.*, 2018, **4**, 2340–2349.
50. C. E. Kandow, P. C. Georges, P. A. Janmey and K. A. Beningo, *Cell Mech.*, 2007, 29–46.
51. H. B. Schiller, M. R. Hermann, J. Polleux, T. Vignaud, S. Zanivan, C. C. Friedel, Z. Sun, A. Raducanu, K. E. Gottschalk, M. Thery, M. Mann and R. Fassler, *Nat. Cell Biol.*, 2013, **15**, 625–636.
52. M. Bharadwaj, N. Strohmeyer, G. P. Colo, J. Helenius, N. Beerenwinkel, H. B. Schiller, R. Fassler and D. J. Muller, *Nat. Commun.*, 2017, **8**, 14348.
53. W. J. Polacheck and C. S. Chen, *Nat. Methods*, 2016, **13**, 415–423.
54. P. Roca-Cusachs, V. Conte and X. Trepat, *Nat. Cell Biol.*, 2017, **19**, 742–751.
55. D. Aydin, I. Louban, N. Perschmann, J. Blummel, T. Lohmuller, E. A. Cavalcanti-Adam, T. L. Haas, H. Walczak, H. Kessler, R. Fiammengo and J. P. Spatz, *Langmuir*, 2010, **26**, 15472–15480.
56. M. Zhang, Q. Sun, Y. Liu, Z. Chu, L. Yu, Y. Hou, H. Kang, Q. Wei, W. Zhao, J. P. Spatz, C. Zhao and E. A. Cavalcanti-Adam, *Biomaterials*, 2020, **268**, 120543.
57. M. Bergert, T. Lendenmann, M. Zundel, A. E. Ehret, D. Panozzo, P. Richner, D. K. Kim, S. J. Kress, D. J. Norris, O. Sorkine-Hornung, E. Mazza, D. Poulikakos and A. Ferrari, *Nat. Commun.*, 2016, **7**, 12814.
58. A. Ferrari, *Biophys. Rev.*, 2019, **11**, 679–681.
59. P. Galliker, J. Schneider, H. Eghlidi, S. Kress, V. Sandoghdar and D. Poulikakos, *Nat. Commun.*, 2012, **3**, 890.

60. T. Wiegand, M. Fratini, F. Frey, K. Yserentant, Y. Liu, E. Weber, K. Galior, J. Ohmes, F. Braun, D.-P. Herten, S. Boulant, U. S. Schwarz, K. Salaita, E. A. Cavalcanti-Adam and J. P. Spatz, *Nat. Commun.*, 2020, **11**, 32.
61. F. Pennarola and E. A. Cavalcanti-Adam, *Methods Mol. Biol.*, 2021, **2217**, 183–195.
62. J. Helenius, C. P. Heisenberg, H. E. Gaub and D. J. Muller, *J. Cell Sci.*, 2008, **121**, 1785–1791.
63. Y. Zhang, C. Ge, C. Zhu and K. Salaita, *Nat. Commun.*, 2014, **5**, 5167.
64. J. M. Brockman, A. T. Blanchard, V. M. Pui-Yan, W. D. Derricotte, Y. Zhang, M. E. Fay, W. A. Lam, F. A. Evangelista, A. L. Mattheyses and K. Salaita, *Nat. Methods*, 2018, **15**, 115–118.
65. C. Jurchenko and K. S. Salaita, *Mol. Cell. Biol.*, 2015, **35**, 2570–2582.

CHAPTER 8

Role of Topographic Cues in Engineering the Muscle Niche

JESSE MACADANGDANG,[a] CHRISTIAN MANDRYCKY,[b] CHANGHO CHUN,[b] NICHOLAS A. GEISSE,[a] DAVID L. MACK[c,d] AND ALEC S. T. SMITH*[d,e]

[a] Curi Bio Inc., 3000 Western Avenue, Seattle, WA 98121, USA; [b] Department of Bioengineering, University of Washington, Seattle, WA 98195, USA; [c] Department of Rehabilitation Medicine, University of Washington, Seattle, WA 98195, USA; [d] Institute for Stem Cell and Regenerative Medicine, University of Washington, Seattle, WA 98109, USA; [e] Department of Physiology and Biophysics, University of Washington, Seattle, WA 98195, USA

*Email: astsmith@uw.edu

8.1 Introduction

The microscopic world that cells inhabit is a complex milieu of biophysical structures, chemicals, and forces. Each of these external stimuli helps to shape the behaviour of resident cells to form the amazingly specialized organ systems of the living organism. Advances in the development of molecular biology tools, microscopy techniques, and methods for gene editing have enabled scientists to begin to unravel the multifaceted interactions that make up a cell's microenvironment. Exactly how cells respond to the various biochemical signals present within these microenvironments, including growth factors, cytokines, and insoluble factors tethered to the extracellular

Biomaterials Science Series No. 12
Material-based Mechanobiology
Edited by Jun Nakanishi and Koichiro Uto

Published by the Royal Society of Chemistry, www.rsc.org

matrix (ECM), has become an important area of study in recent years.[1] While perturbation of chemical factors is straightforward and easily modulated over time, it has become increasingly understood that the mechanical and physical properties of the microenvironment play an equally important role in determining cell function.[1] ECM topography, elasticity (the capacity of a surface to return to its normal shape after being stretched or compressed), and stiffness (the capacity of a surface to resist deformation in response to stretching or compression) help guide cellular phenotypes in a profound manner and can be manipulated within *in vitro* systems to enhance the predictive power of such platforms.[2] In this chapter, we will focus on the ways scientists design and utilize engineered surfaces to replicate the micron- and submicron-scale physical ECM microenvironment of skeletal and cardiac muscle to study the innate biology of myogenic development, function, and disease. We will also discuss the future of such technologies and their potential application in the development of next-generation pre-clinical assays.

8.1.1 Historical Background

In terms of generating controlled physical microenvironments to guide cellular function, efforts have been made as far back as the first half of the 20th century. In 1914, R.G. Harrison cultured embryonic frog cells on a spider web attached to a cover slip.[3] He found that seeded cells grew and elongated along the direction of the web fibres. In 1934, P. Weiss reported on the important role of mechanical structures and developed the principle of topographical guidance by culturing aligned neural fibres on coverslips covered in an oriented, rod-like fibrin.[4] While these seminal works demonstrated the potential for ECM topography to influence cell behaviour and polarization, it was many more years before the mechanisms underlying contact guidance could be elucidated due to a lack of sophisticated fabrication techniques.

Substrate patterning in biomedical engineering is the process by which surfaces are fabricated to have topographical features on the micron or submicron scale to influence cell behaviour. The introduction of fabrication techniques from other engineering disciplines to biomedical applications opened the door to the production of more controlled, cost-effective structures with which to study the intracellular signalling processes responsible for these cellular behaviours. Photolithography has been a well established, commercially available technique in the semiconductor industry since the 1970s. It was not until 1988, however, that Kleinfeld *et al.* first introduced the technique to the biomedical community by demonstrating guided neurite outgrowth and grid formations.[5,6] Electrospinning was first patented in 1934 by A. Formhals,[7,8] but it was not until the early 21st century that biodegradable poly(lactic acid) or poly(lactic-*co*-glycolic acid) fibres were first utilized in a biomedical application.[9,10] The introduction of soft lithography in the 1990s by G. Whitesides[11] has led to an explosion of projects and investigations utilizing polydimethylsiloxane (PDMS). Together, these techniques have allowed micro- and nanoscale topographic features to be reproducibly fabricated in a low-cost, low-expertise manner to enable more profound questions relating to structure–function relationships in cellular biology to be explored.

8.1.2 Structure–Function Relationship in Muscle Biology

Heart and skeletal muscle are complex systems. They incorporate electrical, biochemical, and physical stimuli to achieve force production to propel blood through the circulatory system or to move the body. This force production must be done in a coordinated and efficient fashion to minimize energy consumption. To do so, skeletal and cardiac muscles are aligned in a specific orientation so the force production vectors of each sarcomere (the contractile unit of muscle) are working optimally and in unison. Our efforts to understand and mimic these systems must span from the macroscale all the way to the nanoscale (*i.e.* from synchronized muscle contraction to the molecular machinery of the sarcomere).

Skeletal muscle generates force through the contraction of striated multinucleated myotubes that span the length of the muscle. These myotubes have a distinct directionality and are arranged in parallel, allowing for force propagation along a common axis.[12] In mature skeletal muscle tissue, myotubes mature into myofibres, which are arranged in bundles. These fibre bundles are surrounded by a thick ECM layer termed the perimysium.[13] The perimysium consists of collagen fibres that are organized into discrete populations that can extend both along and across the muscle fibres. Transverse collagen fibres interconnect muscle fibres at discrete points termed perimysial plates. Tissue-engineered skeletal muscle must mimic the organization and physiological function of skeletal muscle. Under standard culture conditions, differentiated myotubes lack any coordinated direction and are randomly organized, resulting in less functional tissue and aberrant fibre morphologies.[12] Thus substrate topographies mimicking native ECM guidance cues can be employed to direct the formation of organized, functional myotubes and skeletal muscle *in vitro*.

Diffusion tensor magnetic resonance imaging of the heart has revealed a slightly more complex helical structure of cardiac tissue compared to skeletal muscle.[14] From the inner to outer wall of the ventricles, cardiac muscle fibres rotate approximately 90 degrees.[15] This results in a twisted, wringing motion of the muscle that facilitates efficient expulsion of blood from the heart chambers. Individual cardiomyocytes, however, are organized into arrays of cells aligned in parallel to facilitate directionally controlled contraction similar to skeletal muscle.[16] Cardiac ECM is made up of fibrillar collagens, basement membrane proteins, and proteoglycans, all of which are important for organ function.[17] Ultrastructural analysis of this ECM reveals that fibrillar collagens in direct contact with cardiomyocytes align to be highly anisotropic.[16] Cellular alignment is strongly correlated with the direction of the ECM fibres, again highlighting the ability of the ECM structure to influence cell behaviour. With the heart, fibres have been shown to have feature sizes on the micron and nanometre scale in width, while extending for much longer distances in length with a high level of fidelity. Thus fabrication techniques intending to imitate the native ECM components must be robust and reproducible over fairly large areas.

8.1.3 Importance of Substrate Patterning for Muscle Tissue Engineering and Its Applications

Cardiovascular disease remains the leading cause of death for both men and women worldwide, with a rapidly growing impact on developing nations.[18] The socioeconomic burden of cardiovascular disease is forecasted to exceed $1 trillion per year by 2030.[19] This devastating impact is due, in large part, to the heart's inability to regenerate, with only 1–2% cardiomyocyte turnover per year.[20] Cardiac tissue engineering aims to alleviate the burdens of cardiovascular disease by fabricating a functional piece of human heart muscle to either (1) repair or replace damaged myocardium after disease or (2) to serve as an *in vitro* model of the heart to screen drugs, study disease mechanisms, and better understand heart development.[21–24]

Although skeletal muscle has far greater regenerative capacity than cardiac muscle, its abilities are not limitless.[25] As such, skeletal muscle tissue engineering also holds promise as a means to generate replacement human tissue for patients that have suffered volumetric muscle loss due to severe injury.[26–28] Similar to cardiac tissues, the *in vitro* modelling of human skeletal muscle is also being pursued as a predictive platform with which to evaluate novel therapeutics, study disease onset and progression, and study myogenesis during development and in response to injury.[29–38]

While regenerative medicine will someday revolutionize patient treatment, from a translational perspective, *in vitro* muscle tissue engineering applications are a much more attainable goal in the near future. The advent of stem cell technology has been a significant leap forward for the muscle tissue engineering field because it has exponentially increased the availability and utility of human cardiomyocytes[39–41] and skeletal muscle myoblasts.[42–45] Unlike animal models, human pluripotent stem cell–derived cardiomyocytes and skeletal muscle, derived from either embryonic stem cell (ESC) or induced pluripotent stem cell (iPSC) sources, theoretically express the full array of ion channels, sarcomeric proteins, and metabolic machinery found in patients and should therefore more closely recapitulate the human phenotype *in vitro*.

Both stem cell–derived cardiomyocytes and skeletal muscle hold significant promise as models with which to enhance our understanding of the aetiology of human muscular diseases. However, the cells obtained from such differentiation methods tend to adopt foetal or early neonatal phenotypes, which limits their utility for the study of late onset diseases.[46] Furthermore, the lack of physiological guidance cues in cultured environments tends to result in the formation of disorganized cells that do not mimic the organization, structure, or alignment of skeletal muscle or cardiac tissue *in vivo*.[16]

Despite recent advances in the derivation of cardiomyocytes and skeletal muscle from stem cells, the lack of cell and tissue maturity remains a significant hurdle to the adoption of these cells in preclinical applications. To that end, methods to engineer more accurate and biomimetic muscle niches *in vitro* to improve cellular development and maturation have become a focal point of much biomedical research.

The central tenet of tissue engineering is that by generating an environment that more closely recapitulates the *in vivo* niche, it is possible to develop *in vitro* culture systems that promote the physiological and functional maturation of cells in a more biomimetic manner. While native muscle tissues consist of arrays of cells aligned in parallel to facilitate directionally controlled macroscopic contraction, conventional *in vitro* culture platforms lack any form of anisotropic signal, leading to random alignment of developing cells. Therefore, in order to promote correct physiological function and development, muscle engineering strategies must promote uniaxial alignment of the cultured cells to mimic native tissue structure and enable the subsequent maturation of skeletal and cardiac myocytes towards an adult phenotype.[47]

To improve the performance of cultured stem cell–derived cardiomyocytes and skeletal muscle, investigators make use of the cells' ability to sense and respond to microscale topographic features. Cells are connected to the ECM through integrins. Integrins bind to intracellular filaments through a plethora of linker proteins and aid in mechanically coupling intracellular and extracellular structures.[48,49] Biomimetic grooved surfaces have been shown to improve cell adhesion, proliferation, and migration, as well as myogenic differentiation and structural development.[50–53] These anisotropic grooves and ridges help maintain the cell polarity and alignment critical for organ development by providing contact guidance cues that influence the organization of microtubules, focal adhesions, and actin filaments in parallel with the underlying topography.[16] In addition to improving cellular alignment, topographic cues that match the dimensions of native ECM fibres have been shown to promote substantial improvements in maturation, leading to a more adult-like phenotype.[16,52] This observation lends credence to the theory that accurate recapitulation of the native structural niche promotes the development of seeded muscle cells towards a more mature phenotype. The scale of ECM fibres in relation to muscle cells and the comparable size of engineered surfaces used to mimic these structures are illustrated in Figure 8.1.

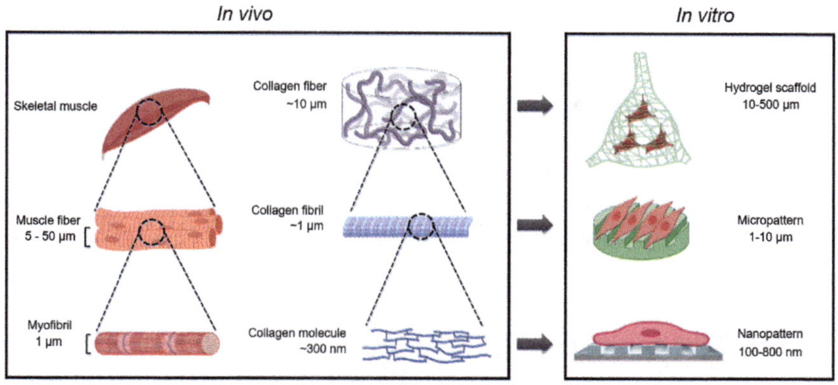

Figure 8.1 Comparing the scale of muscle cells, ECM structures, and engineered *in vitro* mimics. Muscle cells exist in the micron scale, with diameters typically in the tens of μm. ECM fibres such as collagen have a multiscale structure in which individual molecules combine to form fibrils, which are in turn combined into fibrous networks. Engineering strategies for modelling each of these scales exist, and cells interact differently with each, based on feature size. At the sub-μm scale, topographic features represent individual ECM molecules or small fibres. Cells can therefore bind to multiple features simultaneously, and organization is achieved through regulating how and where focal adhesions can form on the substrate. At larger scales, hydrogels model the fibrous nature of the ECM to encase cells within their structure. Cells in these environments tend to adhere to one or two fibres, and orientation is typically achieved through macroscale organization of fibre orientation.

8.2 Fabricating Substrate Topographies for Use in Cell Culture

In general, the fabrication methods used to generate substrate topographies can be divided into two categories: template-assisted and template-free methods. When seeking to model the mechanical properties of the ECM for *in vitro* applications, it is important to remember that the topography of the matrix does not act on cells in isolation. Other factors, such as substrate stiffness, likewise play critical roles in modulating cellular phenotypes. Specifically, stiffnesses that more closely match those of native tissues have been shown to be superior for promoting muscle cell maturation.[54–57] Therefore, while this chapter focuses on substrate topographies, methods for their fabrication, and their role in regulating cellular development, it must be noted that topography is only one aspect of cells' mechanical niche and so should not be considered in isolation when designing novel culture platforms and substrates.

8.2.1 Template-assisted Fabrication

Template-assisted fabrication methods require a mould that can be (1) transparent or opaque depending on the utilization of light, (2) rigid or flexible depending on the pressure requirements, and (3) compact or permeable with respect to the usage of solvents. These template-assisted fabrication techniques can be further categorized with respect to the energy sources required for physical modification.

The most established template-assisted fabrication techniques for biomedical research are based on lithography processes. Historically, micro- and nanoscale lithography techniques have been widely applied to semiconductor device fabrication. However, their high-dimensional fidelity and resolution have made them attractive for use in the tissue engineering and mechanobiology research fields to facilitate the fabrication of complex biomimetic structures. The most widely used lithography techniques for biomedical applications are based on photolithography.

Photolithography uses specific wavelengths of light to transfer desired patterns from a photomask to photoresist. Photoresists can either be degraded or crosslinked *via* exposure to specific wavelengths of light. The advantages of photolithography are its resolution (as low as 10 nm feature sizes depending on the wavelength of the light source employed), its capacity for application in a noncontact manner to prevent contamination, and its flexibility in terms of its compatibility with a wide range of polymer solutions. However, the process requires an environmentally controlled clean room space, free from any airborne impurities. This generates high initial investment cost and has clear limitations with regard to the direct use of biomaterials. Several research labs have recently attempted to develop combinatorial photolithographic techniques as more cost-effective and facile processes to broaden their application in biomedical experiments. For example, Yin *et al.* developed hybrid nanopatterning by integrating inverse photolithography with conventional hot embossing to create 130–150 nm grooves with significant cost reduction and improved pattern transfer reproducibility.[58] Besides using photolithography for 2D nanopatterning, Lee *et al.* developed a technique to create 3D microscale networks of PDMS using a single moulding process. Pyrolysis, in combination with photolithography, resulted in 90% shrinkage of the polymer mould to produce nano- and microsized hollow channels in a 3D environment.[59] Liu *et al.* demonstrated the successful application of photolithography in combination with an etching process to facilitate the use of a protein itself as an efficient photoresist. They used silk fibroin as a "biophotoresist" that enabled direct creation of micropatterned protein structures in response to UV exposure for the purpose of providing spatial guidance cues to cultured neural stem cells.[60]

Although photolithography is already one of the best established fabrication tools for generating patterned surfaces, studies such as those just discussed highlight how conventional methods are currently undergoing extensive transformations in order to increase its versatility. Such improvements will not only broaden the range of features that can be patterned but also stand to improve surface fidelity as well as fabrication cost and throughput. One such effort in this area has led to the establishment of a useful lithography subcategory; soft lithography. This name is derived from the use of lithographic techniques in conjunction with elastomeric, organic biomaterials. Although traditional photolithography carries advantages for substrate patterning, in terms of resolution and pattern fidelity, its exacting fabrication processes, involving substrate exposure to high temperatures, ion etching solvents, and high-energy light sources, often cause biomaterial degradation. Soft lithography techniques, such as microcontact printing, utilize elastomeric stamps, moulds, and photomasks to replicate micro- and nanostructures on culture surfaces.[61] The elastomeric stamp, usually made of inert PDMS, is submerged in the material solution and then placed in direct contact with the target substrate to generate a 2D pattern with the "material ink". This simple process enables the rapid deposition of geometrically defined patterns, even on cell-repellent flat surfaces. Adhesive biomaterials, such as collagen, fibronectin, or immunoglobulin antibodies, can be deposited by manipulating the hydrophilicity of the substrate and/or PDMS stamp.

Since James *et al.* first made a millimetre-scale polylysine pattern with microcontact printing in 1998,[62] various improvements have been made to the technique in order to diversify its utility in biomedical research applications. For example, Khadpekar *et al.* devised low-cost lithography-free microcontact printing by adapting polystyrene beads and injection needles with collagen I and demonstrated the use of these surfaces for patterning C2C12 myoblasts.[63] Bhujbal *et al.* recently developed microcontract printing for single-cell arrays by introducing 1–10 mm Matrigel-coated spots on glass surfaces pretreated with polyethylene glycol.[64] Borowiec *et al.* reported on a 3D microcontact printing technique and its use for one-step construction of 3D structures inside microchannels by combining microthermoforming with conventional microcontact printing. This technique was able to transfer collagen, laminin, fibronectin onto the 3D geometries, all of which promoted enhanced cell adhesion and growth.[65]

8.2.2 Template-free Fabrication

Template-free methods do not require the use of moulds but instead apply materials to substrates directly, drawing on the surface using fine tips loaded with the desired material. The most representative and well established techniques in this category are e-beam/laser lithography, electrospinning, and 3D printing. E-beam lithography is specialized for creating patterns with feature sizes below 10 nm and uses a focused electron beam under precise current control for intensity and positioning.[66] Alternatively, a laser can be used instead of an electron beam to make patterns by applying the point exposure of a femtosecond laser to produce complex 3D structures with various heights at a submicron scale.[67] Bruk *et al.* tweaked conventional "wet" e-beam lithography to develop "dry" e-beam lithography based on the chain depolymerization of polymethyl methacrylate to provide its monomer source under dry conditions, which in turn solidifies to a desired pattern upon exposure to the electron beam. The method requires two to three orders of magnitude lower electron beam energy than that of conventional e-beam lithography, with up to 1 nm and 100 nm of vertical and lateral resolution, respectively. Given these differences, dry e-beam lithography represents a potentially more suitable option for creating high-throughput and cost-effective topographic patterns than the conventional wet method.[68] E-beam lithography techniques have been diversified *via* the use of combinatorial photoresists to enable the production of a broad range of surface stiffnesses coupled with fine control of surface topography. Chen and colleagues summarized recent fabrication results with various combinations of photoresist to guide the proper selection of material to be used for substrate patterning in terms of the purpose of conducting the downstream biomedical experiment.[69]

Electrospinning is a facile and scalable method for fabricating stacks of nanosized polymer fibres, providing a web-like scaffold that effectively mimics ECM structures. The process uses a highly viscous polymer solution inside a syringe fitted with a conductive tip that ejects polymer fibres upon voltage application. The polymer initially forms a protrusion, called a Taylor cone, at the end of the tip. This cone narrows in diameter during flight to make nano-sized fibres that collect on a distal metal plate carrying the opposite charge.[70,71] The beauty of the electrospinning technique is its ability to tune fibre size, porosity, biocompatible material composition, and fibre alignment to give great flexibility in terms of recapitulating the diverse structures of native ECM scaffolds. However, electrospun structures show poor stability in aqueous solution and are susceptible to swelling and collapse – a significant limitation when applied to cell culture. Also, the porous environments of 2D electrospun mats are not favourable for active cell infiltration and longevity. To address these issues, expanded forms of electrospinning have emerged that have diversified scaffold materials and spinning techniques to produce more stable 3D scaffolds while maintaining their biomimetic features. For example, gas foaming technology has enabled the stable stacking of 2D electrospun mats using a hydrolysis reaction of sodium borohydride that generates hydrogen gas bubbles to lift up each layer.[72] Song *et al.* applied an origami approach to build layer-by-layer human foetal osteoblast constructs while electrospinning collagen blended with polycaprolactone.[73] Nanofibre microspheres developed by Chen *et al.* expanded the application of electrospun fibres to *in vivo* injection through the use of self-assembling gelatin/PLA spheres that are small enough for syringe injection with minimal invasiveness. Chondrocytes cultured inside of these nanofibre spheres exhibited enhanced cartilage regeneration ability in a rabbit-based model of articular cartilage injury.[74] Given the high expandability and scalability of electrospinning techniques, these methods hold great potential for future biomedical applications in tissue engineering and fundamental mechanobiology studies for the development of clinically applicable regenerative therapies.

Since 3D printing enables significant customization and flexibility, while simultaneously providing simplicity to iterative design methods, the technique is well suited to trial-and-error-based optimization of extracellular 3D microstructures for use in biomedical research. The available list of printing materials was recently expanded beyond conventional thermoplastics to include photocrosslinkable polymers, extracellular proteins, PDMS, and SiO_2 for transparent object printing.[75,76] One of the main goals of 3D bioprinting within the biomedical research fields is to create scaffold structures capable of modulating endogenous differentiation, migration, and regeneration by recapitulating native extracellular environments in normal and diseased conditions. As such, different printable inks bearing variable topographic features for promoting cellular development *in situ* have been explored.

Since generating structures with homogeneous features (*e.g.* size, stiffness, surface roughness, transparency, *etc.*) fails to create a representative extracellular environment, 3D bioprinting is advancing to integrate time and/or stimulus factors to dynamically alter characteristics of printed constructs in response to various stimuli or in a time-dependent manner.[77] Temperature change is the most frequently used stimulus with which to deform printed constructs to alter topographic feature size, regularity, and/or orientation. Biocompatible thermoresponsive materials, such as poly(*N*-isopropylacrylamide), are often used to fold, shrink, or swell established structures as their temperature crosses a specific threshold.[78,79] The typical process includes blending of thermoresponsive polymer solutions with growth factors and/or cells to create a printable bioink that then undergoes deformation upon temperature change postprinting.

Four-dimensional bioprinting utilizes the self-folding, swelling, and releasing nature of specific polymer scaffolds exposed to a physiological aqueous solution for tissue engineering applications. Such water-responsive scaffolds can be fully customized to recreate anatomically relevant microscale geometries with excellent biocompatibility and minimal cell damage. Another stimulus used for the postprinted transformation of printed scaffolds is magnetism. Magnet-responsive materials can be mixed with biocompatible base polymer solutions to control the orientation of printed anisotropic particles under controlled application of a magnetic field.[77,80] These postprinting time and/or stimulus-dependent manipulations of 3D constructs are a useful technique to better recapitulate dynamic and complex 3D tissue environments. The development of functional microvascular tissues, multicellular 3D organoids, and matrices for use in drug delivery studies can all be advanced *via* the timely degradation of printed 3D topographies. The methods described in this section are illustrated in Figure 8.2.

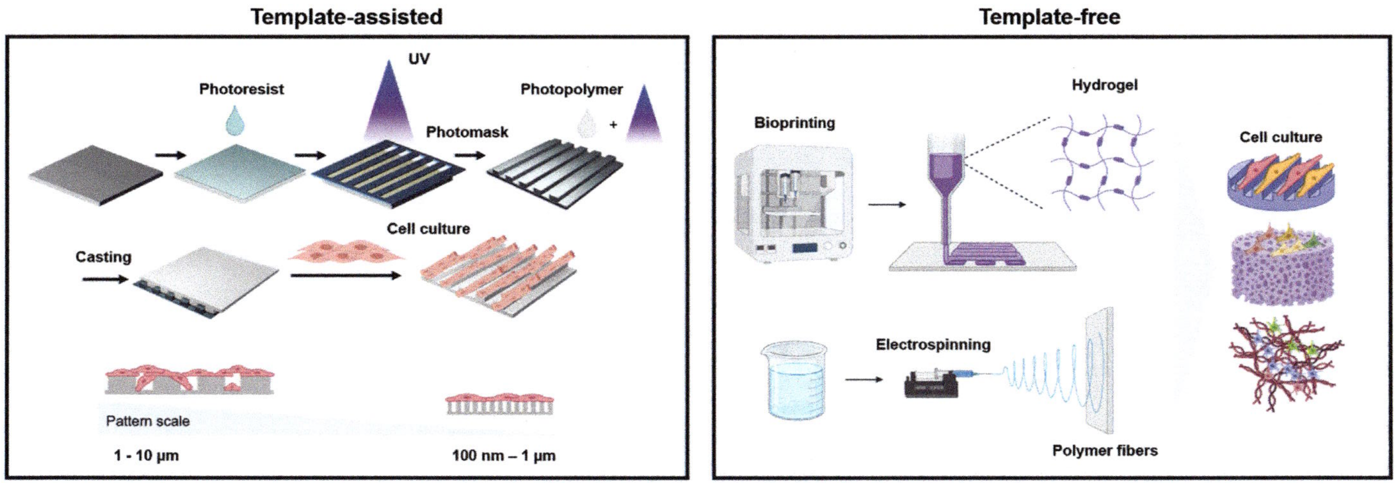

Figure 8.2 Common fabrication methods used to engineer surface topographies for guiding cell development. Template-assisted methods rely on a mould to guide the deposition of topographic patterns in the nano- to microscale. How cells interact with these substrates depends on the dimensions of the features. Submicron scale features tend to promote cell monolayer formation on top of the pattern and guide alignment *via* regulation of focal adhesion complex shape and orientation. Larger patterns promote alignment *via* confinement of cells to a specific orientation within the confines of the larger features. Cells can either sit on the bottom of these features, on top, or a combination depending again on feature size. Template-free methods are more common for synthesizing 3D scaffolds and involve deposition of fibres with specific dimensions to provide scaffold support and guidance signals. Cells can then be seeded on top or within these substrates depending on porosity.

8.3 Use of Patterned Surfaces to Enhance Muscle Cells in Culture

Over the last two decades, a voluminous literature has been established detailing the use of topographic patterns to organize cultured cardiac cells and promote their structural maturation for downstream use. We do not intend to discuss this literature in detail here as these studies have been reviewed elsewhere.[81–84] Instead, this section will focus on several specific case studies that highlight the ways in which topographic cues can be manipulated to improve the predictive power of *in vitro* cardiac assays. We will then provide an overview of the ways in which topographic features have so far been used to study skeletal muscle development, disease phenotypes, and response to compound exposure. Lastly, we will briefly discuss efforts to employ similar strategies in the patterning of smooth muscle cells as well as potential applications for such endeavours.

8.3.1 Patterned Cardiomyocyte Case Studies

8.3.1.1 Using Topographic Cues to Stratify Disease Phenotypes in a Stem Cell-derived Cardiac Model of Becker's Muscular Dystrophy

Human iPSC-derived cardiomyocytes offer unprecedented opportunities to study inherited heart conditions *in vitro*, but are phenotypically immature, limiting their ability to effectively model adult-onset diseases. Cardiomyopathy is becoming the leading cause of death in patients with Duchenne muscular dystrophy (DMD), but the pathogenesis of this disease phenotype is not fully understood. Once cardiomyopathy has developed in patients, there is no known cure outside of heart transplantation, a procedure rarely performed in DMD patients due to weak respiratory muscle function, respiratory insufficiency, and short supplies of donor hearts.[85] Therefore, it is imperative to discover new therapies that can slow the progression of diseases (such as DMD) that cause cardiomyopathy in order to improve the quality of life for these patients until a more definitive, long-term cure is found.[86]

Macadangdang *et al.* tested whether biomimetic microenvironments would further stratify the disease phenotype in a dystrophin-mutant iPSC model of Becker's muscular dystrophy by adding more *in vivo*-like ECM cues that would facilitate the development of physiologically representative cell–ECM interactions and activate the mutated dystrophin-glycoprotein complex (DGC).[87] Such a platform could be used to identify simple, high-throughput end points with which to screen novel therapeutics for their capacity to ameliorate the cardiac phenotype *in vitro*.

When normal and mutant cardiomyocytes were cultured on traditional flat substrates, there was very little difference between their structural and functional phenotypes. Cell shape, cytoskeleton, and contractile organization were indistinguishable between the two cell types. When cells were grown on biomimetic topographic surfaces, however, a striking difference in the structure of the mutant dystrophic cardiomyocytes became apparent. In these experiments, anisotropic topographies consisting of 800 nm wide and 600 nm high grooves and ridges were used to mimic the size and orientation of extracellular matrix fibres present in the native heart. Mutant cells cultured on such substrates displayed a much lower level of cellular and tissue alignment compared to healthy controls, resulting in a lower level of alignment (see Figure 8.3).

This lower level of alignment on the nanotopography was found to be due in part to a lower level of actin turnover in the dystrophin-mutant cardiomyocytes. The loss of dystrophin is known to disrupt the DGC, but this disruption is not lethal and does not interfere with cardiac differentiation. Therefore, compensatory mechanisms may help to offset the loss of dystrophin. Other groups have postulated that utrophin may play an important role in compensating for the loss of dystrophin due to its homology with dystrophin.[88] Other proteins, such as integrins and focal adhesion proteins, that help to maintain connection with the ECM might also be differentially expressed in dystrophic cardiomyocytes.[89] These compensatory mechanisms, however, may be less dynamic than the DGC and thus inhibit actin cytoskeletal remodelling. Thus, when dystrophic mutant cardiomyocytes establish their connections with the ECM, they may be less able to respond to changes to the microenvironment, such as dissociation and replating. The

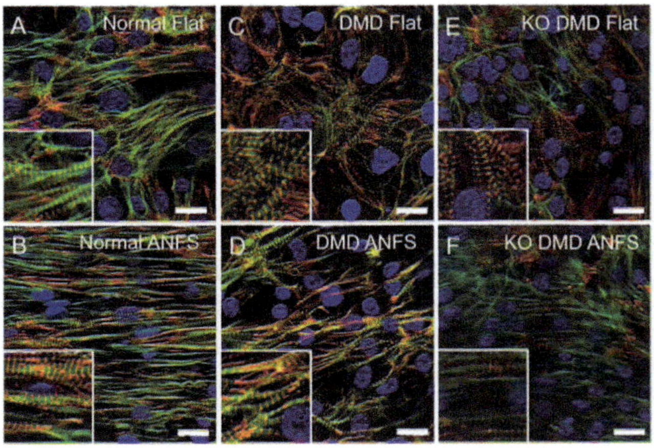

Figure 8.3 Immunofluorescent confocal images of sarcomere structure in normal (A and B), patient-derived dystrophic (DMD; C and D), and CRISPR-edited dystrophin knockout (KO DMD; E and F) human iPSC-derived cardiomyocytes on flat or anisotropic nanofabricated substrates (ANFS) after two weeks of culture. Cells have highly aligned cytoskeletal and sarcomere structures on ANFS compared to flat substrates, but DMD cardiomyocytes show less alignment than normal controls. Insets show higher magnification details of striated sarcomeres. Green = F-actin, red = α-actinin, blue = DAPI (scale bar = 20 μm). Reproduced from ref. 87 with permission from Springer Nature, Copyright 2015.

assay system developed in this study, utilizing biomimetic nanotopography, represents a useful tool to further test this theory.

Engineered topographic surfaces enable clearer distinction between dystrophic mutant and normal structural and functional phenotypes in cultured stem cell–derived cardiomyocytes. Specifically, the differences in the observed responses of diseased and normal cardiomyocytes to topographic cues are suggestive of a structural cardiomyopathy-like phenotype in Becker's muscular dystrophy cells. Moreover, the collected data accentuate that topography can be used to stratify the structural and functional differences between normal and dystrophic cardiomyocytes in a way not possible on conventional flat substrates. The developed platform also offers a simple, reliable, and reproducible method for screening phenotypic changes in these cells in response to novel therapeutic treatment. Such technology could be invaluable to the development of a new test bed for drug discovery, enabling the provision of better care to patients suffering from this debilitating condition in the near future.

8.3.1.2 Long-term Culture on Anisotropic Topographic Substrates Promotes Myofibril Development and T-tubule Formation in Stem Cell–derived Cardiomyocytes

Structurally and functionally mature cardiac tissues are lacking for use in cardiac tissue engineering applications. Various methods, such as microcontact printing, electrospinning, and microtopography have previously been employed to control the structure of engineered cardiac tissues on the microscale.[90–92] While these techniques have proven to be successful in gaining cellular alignment, it is likely that the structure and function of *in vivo* cardiac tissue is governed by smaller nanotopographical cues comparable to those of the native ECM. Through contact guidance cues, anisotropic substrates with 800 nm grooves and ridges promote both tissue anisotropy and single-cell elongation in the direction of the topographic features.[16,52] This anisotropic behaviour mimics *in vivo* morphologies and promotes structural development of cultured cardiomyocytes.

Such surfaces have multiple practical and technical advantages over microscale methods such as microcontact printing or electrospinning. Capillary force lithography, the method used to create these organized topographies, is relatively cheap and more cost-effective at scale compared to electrospinning.[93] These patterned surfaces can also be fabricated with much more controllable and tuneable topographic features compared to electrospinning. Microcontact printing, on the other hand, relies on cells adhering to specific lanes to gain structural anisotropy. In order to create a functional monolayer, cells must then either proliferate or migrate out of these lanes to form tight junctions. This is much more difficult for cells with little to no proliferative capacity such as stem cell–derived cardiomyocytes. Microtopography is also limited by the inability to create tightly coupled cell monolayers. Due to the scale of the cells compared to the microtopography, cells reside on top of or inside the microridges. This limits the amount of cell–cell interaction and decreases cell–cell coupling. Conversely, nanotopographic feature sizes are at least an order of magnitude smaller than the cells themselves. This allows for more biomimetic, tightly coupled cell monolayers to be created.

Taking advantage of this structural maturation of stem cell–derived cardiomyocytes, Pioner *et al.* developed novel methods with which to isolate mature cardiac myofibrils from healthy and diseased cells cultured on bioinspired 800 nm topographies.[94,95] Immature stem cell–derived cardiomyocytes are round-shaped and demonstrate sporadic signs of myofibrillogenesis with sparse expression of sarcomeric elements and assembly. Obtaining mechanical measurements from these premyofibrils is quite challenging due to their short length and fragility. Therefore, to achieve myofibril lengths of a more suitable size (>50 μm) and stability on the experimental platform, cardiomyocytes were cultured on 800 nm topographic substrates for 60 days. The resulting cells were elongated and had many dimensions similar to those found in cultured adult cardiomyocytes. Phase contrast imaging and immunohistochemistry demonstrated that later-stage mature cardiomyocytes were characterized by aligned myofibrils (in the direction of the underlying topography) and clearly defined Z-bands spanning the width of the cells.

Without this elongated morphology and robust sarcomerogenesis, stem cell–derived cardiomyocytes cultured on traditional flat cell culture surfaces are too short and often very fragile. The patterned surfaces promote robust in-register sarcomere formation that results in myofibres capable of being handled and loaded into a custom-built apparatus required to measure single myofibre mechanical properties.[95]

Interestingly, Pioner *et al.* also found that, beyond the structural development of the sarcomeres, 800 nm topographies promoted the early-stage formation of transverse (T)-tubule-like structures. By staining the cells for caveolin-3, a scaffolding protein found in the sarcolemma, which is associated with the formation of caveolae and T-tubules during the development of striated muscle,[96] patterned cells were found to exhibit finger-like membranous invaginations tightly associated with the Z-bands of the sarcomere, likely representing the formation of T-tubules.[95] This finding provides evidence that transverse-axial tubule system formation may occur together with myofibrillogenesis and suggests a strict interplay between tubules and Z-bands during cardiac development on 800 nm patterned topographic surfaces.

8.3.2 Replicating Skeletal Muscle Structure *In Vitro* Using Topographic Patterns for Disease Modelling and Drug Screening Applications

As is the case for cardiac muscle, skeletal muscle is arranged in a highly ordered and organized structure. Individual skeletal muscle fibres are anisotropic and can be many centimetres long in mature muscle found *in vivo*.[97] Volumetric muscle loss and diseases that lead to deficiencies in skeletal muscle size or muscle performance are also widespread and have few, if any, effective clinical treatments. These include muscular dystrophies such as Duchenne muscular dystrophy, some types of congenital contractures that are thought to arise from skeletal muscle dysfunction, as well as a variety of other myopathies.[98,99] A deeper understanding of these diseases as well as possible new avenues for treating them may come from skeletal muscle cells and tissues grown in the lab. Replicating the structure and function of skeletal muscle is an ongoing effort, but considerable progress has come with the use of topographically patterned surfaces to promote the formation of mature skeletal muscle *in vitro*.

Engineered topographical features at a variety of length scales have been used in skeletal muscle engineering. Early investigations with lithographically patterned cell-scale rectangular islands of type I collagen found that the confinement of cells during myotube differentiation yielded more functionally mature and aligned myotubes than circularly patterned controls.[100] Confining cells to a small rectangular area, however, inherently limits the number of cells that can undergo fusion and the size the resulting myotubes can ultimately achieve. Rather than confining cells, the entire substrate can be regularly patterned to promote the orientation of cells across the entire culture surface. Line patterns at cellular and subcellular scales have been used to successfully align myoblasts and generate more mature and aligned myotubes[12,101] (see Figure 8.4).

Both larger linewidth patterns (50 $\mu m \times 50$ μm)[102] as well as smaller linewidth patterns (0.8 $\mu m \times 0.8$ μm)[103] have been used to create confluent sheets of aligned muscle. These patterned sheets can be manipulated and stacked into thicker tissues with layer-by-layer control of cell angle. Interestingly, individual muscle layers fabricated from smaller linewidth patterns maintained their orientation for longer than larger linewidth sheets when they were incorporated into stacked tissues. More recently, this approach has been used to generate contractile tissues. By stacking an aligned skeletal myotube sheet and fibroblast sheet and encapsulating the tissue in a fibrin-Matrigel hydrogel, tissues were formed that displayed twitch and tetanic contractions when stimulated at appropriate frequencies.[104] Incorporating contraction and electrical stimulation will be an essential feature of advanced models of skeletal muscle going forward.

While these approaches have primarily sought pattern cells in a scaffold-free form, other approaches involve the incorporation of patterned features directly onto a cell-supportive scaffold. For example, Hosseini and colleagues generated microgrooved gelatin-methacrylate hydrogels and seeded them with mouse myoblasts.[105] The myotubes that formed on these hydrogels were highly aligned and contracted in response to electrical stimulation. This approach has been extended to micropattern transglutaminase cross-linked gelatin hydrogels, which also support differentiation of aligned myotubes and enable their maintenance in culture for several weeks.[106]

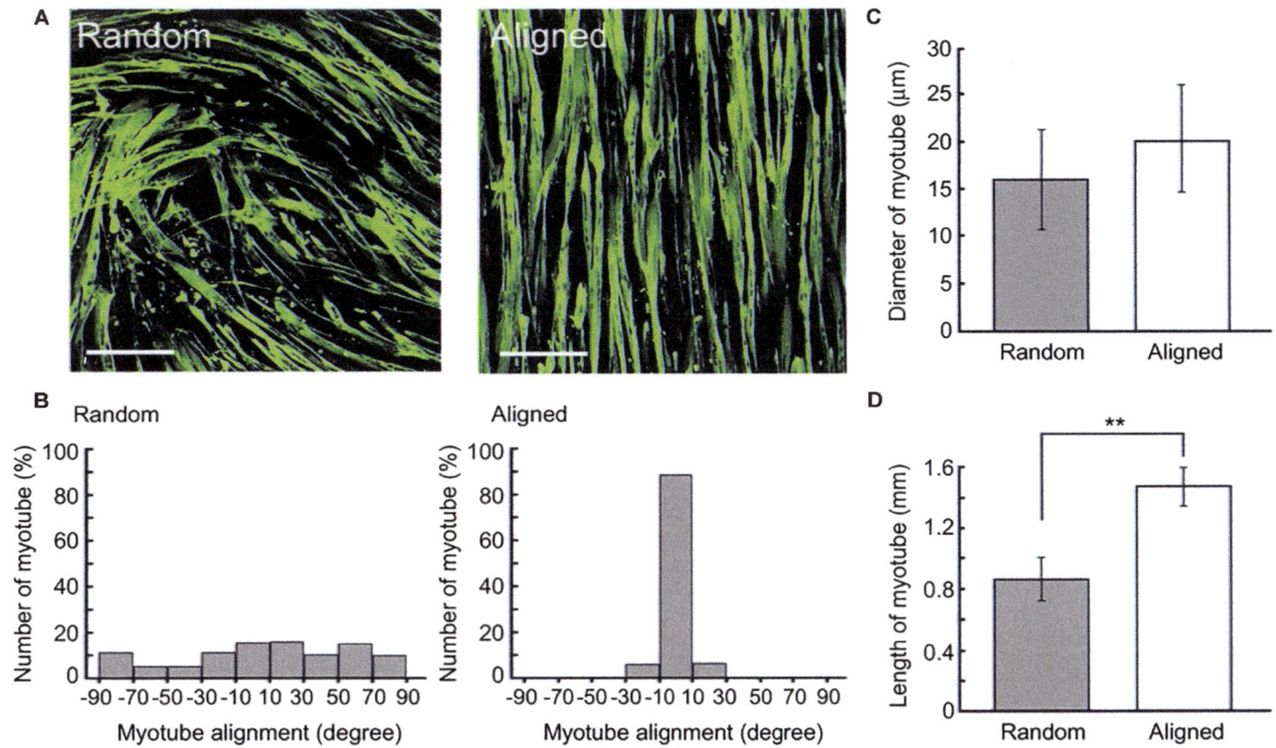

Figure 8.4 Aligned myotube formation on normal cell culture dishes by transferring anisotropic myoblast sheets. (A) Confocal microscopic images of myotubes on tissue culture plastic surfaces after cell sheet transfer from patterned (Aligned) and nonpatterned thermoresponsive surfaces (Random). After transferring, myoblast sheets were incubated in differentiation medium for 5 days, and myosin heavy chains were stained with fluorescein (green) (scale bar = 200 μm). (B) Distribution of myotube alignment in cell sheets transferred from patterned (Aligned) and nonpatterned surfaces (Random) ($n = 5$). (C, D) Diameter and length of myotubes in cell sheets composed of randomly oriented myoblasts (Random) and aligned myoblasts (Aligned) ($^{**}p < 0.01$) ($n = 5$). Reproduced from ref. 102 with permission from Elsevier, Copyright 2013.

These patterned hydrogels can be formed into cantilever designs to perform muscular thin-film assays for investigating the contractile properties of aligned skeletal muscle and the effect of drug treatment.[107]

Micropatterned scaffolds can also be formed from native decellularized tissue, as was recently demonstrated by Cheng and colleagues.[108] They showed that decellularized green onion has a unique topography consisting of repeating ~20 µm wide grooves similar to microfabricated tissues. This cellulose scaffold–supported myotube differentiation into aligned myotubes represented a low-cost and simple alternative approach to generating aligned skeletal muscle.

Electrospinning has been used to generate fibre mesh networks with highly aligned fibres of different dimensions. Human skeletal muscle cells readily attach to electrospun polycaprolactone/collagen nanofibre meshes (~300 nm fibre diameter), which can be fabricated with aligned or randomly oriented fibres.[109] On these surfaces, myotube formation, length, and alignment were significantly improved on unidirectionally aligned mesh networks. Similar results have been achieved using fibres prepared from poly(lactic-*co*-glycolic acid).[110] Further optimization has underscored the importance of fibre diameter as well as fibre orientation in electrospun scaffolds. Narayanan and colleagues found that aligned scaffolds prepared from ~3000 nm diameter fibres improved myoblast growth, differentiation, and expression of muscle proteins myogenin and myosin heavy chain when compared to aligned scaffolds prepared with ~300 nm diameter fibres.[111] Electrospun scaffolds can also be fabricated to achieve more complicated 3D architectures. Apsite and colleagues created a shape-changing electrospun PCL-alginate composite with uniaxially aligned polycaprolactone fibres that could support myoblast proliferation and differentiation.[112] This scaffold was sensitive to calcium ion concentration and, under low calcium conditions, flat scaffold sheets spontaneously rolled to form tubes with myoblasts aligned along the long axis of the tube. Approaches such as this could lead to larger and more complex structures that incorporate vascular conduits to supply the considerable metabolic demand of active skeletal muscle.

Various approaches have shown that skeletal muscle cells readily respond to nano- and microscale topographical cues. Generally, these promote myocyte alignment and improve the fusion of cells into more mature myotubes, which has important implications for future efforts in the areas of disease modelling, drug screening, and regenerative medicine. Much of this work, however, has been performed with cells from the mouse or chick, limiting its applicability to human health. Recent advances in stem cell technology have enabled the efficient generation of stem cell–derived skeletal muscle to address this concern.[43,44] The production of these cells is both more expensive and time-consuming compared to xenogeneic options but represents new opportunities for investigating specific aspects of human disease and physiology.[29,113] It should be noted that while patterns have been optimized in terms of substrate topographic dimensions required to promote cytoskeletal maturation in cardiomyocytes,[52] the range of pattern sizes still being investigated for skeletal muscle cultures suggests that data defining an optimal range for promoting myotube maturation *in vitro* has yet to be provided.

8.3.3 Other Applications: Patterned Smooth Muscle Cells for Vascular Modelling and Tissue Engineering

This chapter has so far focused on cardiac and skeletal muscle patterning using topographic patterns. However, some data have been published demonstrating the value that such features can confer on smooth muscle cell (SMCs) cultures. SMCs lack the highly ordered sarcomere structure found in skeletal and cardiac muscle, but they maintain an anisotropic elongated shape in the direction of contraction that is critical to their function. In particular, vascular SMCs align circumferentially around blood vessels throughout the body regulating vascular tone.[114] Changes in this phenotype and loss of contractility are associated with a variety of vascular pathologies.[115] As a result, methods for generating aligned SMCs have also been highly sought after to develop a deeper understanding of SMC biology. Many of the approaches used for smooth muscle mirror those for the alignment of skeletal and cardiac myocytes. Nanoscale and microscale line patterns promote the elongation of SMCs and inhibit proliferation.[116,117] Culture on micropatterns additionally supports the expression of muscle-specific genes such as miR-145 and α-SMA.[118] Acquiring this specific cell shape can guide human mesenchymal stem cells to differentiate towards a smooth muscle-like phenotype when using micropatterns ranging from 20 to 200 μm linewidths[119] or high aspect ratio (>5 : 1) rectangular islands.[120] These cues can be incorporated into more complex 3D structures, such as self-assembled tubes featuring micropatterned fibronectin lines.[121] SMCs seeded onto these scaffolds align circumferentially, support the growth of an endothelial layer on their surface, and have elevated nitric oxide signalling relative to 2D controls. As with other muscle types, smooth muscle cells are highly sensitive to cellular and subcellular topographical cues. Taking advantage of this sensitivity may lead to new discoveries about fundamental smooth muscle cell biology and the mechanisms of various SMC pathologies.

8.4 Future Directions

8.4.1 Topography Is Only a Piece of the Maturation Puzzle

Current cardiac and skeletal muscle differentiation protocols produce immature cells with structures and functions that resemble foetal tissues. These immature cells are suboptimal for use in modelling adult-onset diseases or the performance of preclinical drug toxicity screening that is predictive for adult patients. As we have described, the use of topographic features provides a means to enhance the structural development of muscle cells, promoting improvements in sarcomere organization and myofibril alignment. However, given the complex interplay of mechanical, molecular, and electrical cues that are temporally regulated to enable normal cardiac and skeletal muscle development *in vivo,* it is incredibly unlikely that provision of any one cue *in vitro* will promote the maturation of cultured cells to adopt a phenotype that is fully representative of the adult tissue.

While it is therefore unlikely that modifying culture substrates to enhance structural development will promote the maturation of stem cell-derived cardiomyocytes and skeletal muscle to adult phenotypes, the simplicity of the techniques we have described means that such methods are easily integrated with other maturation methods. This in turn raises the possibility of developing combinatorial maturation platforms integrating multiple cues to produce engineered tissues that can be used for predictive modelling applications. The available literature is rich with potential additional cues that could be employed in combination with topographic substrates to further enhance muscle development *in vitro.* For example, thyroid hormone is an important regulator of both cardiac and skeletal muscle development[122,123] and has been shown to modulate the development of muscle cells *in vitro.*[123] Similarly, supplementation with other small molecules have been shown to have positive effects on muscle development.[37,124–128] In addition, modulation of micro-RNA expression profiles has been used to drive the development of both skeletal[129] and cardiac cells[130] *in vitro,* promoting a shift to fatty acid metabolism in the latter.[130] Broad field electrical stimulation is another common stimulus that has been shown in numerous models to enhance both engineered cardiac[23,131,132] and skeletal muscle[37,133,134] function. The use of conductive substrates has likewise been shown to confer benefit on cultured muscle cell phenotypes.[135,136]

Many of these examples can be readily integrated into standard multi-well culture plate formats, and so it would be straightforward to investigate them in conjunction with topographic surfaces for their additive benefit to muscle cell maturation. As a proof of concept, our recent work combined topographic substrates with thyroid hormone treatment and Let-7 micro-RNA overexpression to enhance iPSC-derived cardiomyocyte maturation.[137] The application of this combinatorial maturation scheme not only improves cardiomyocyte structure, electrophysiology, and metabolic function but also helps stratify normal and dystrophic disease phenotypes to the point where predictive drug screening can be performed. Results such as these underscore the importance of accurately modelling the muscle niche through provision of multiple cues to enhance all aspects of the cellular phenotype. The simplicity of textured culture surfaces makes provision of structural cues an easy addition to complex cell culture models.

8.4.2 High-throughput Assays and Assay Agnostic Applications

An important consideration for the continued development of substrate topographies in preclinical settings is their integration with extant technologies. The patterning of conventional culture plates is no doubt useful for academic studies into the role of external guidance cues in regulating cellular development. However, leveraging the benefit conferred by these topographic features to cells and applying them in predictive downstream assays necessitates the integration of such cues with high-throughput platforms that enable automated analysis.

Our group and others have demonstrated that the patterning of ion permeable membranes can be employed to enhance cardiac maturation within multielectrode arrays (MEA) and used to enhance the sensitivity of these assays for drug development.[138–140] The use of substrate topographies to organize developing myotubes on top of MEAs to facilitate detection of depolarizing events in cultured muscle has also been demonstrated.[141] Current MEA technologies have been multiplexed into 96-well formats, and higher-throughput modalities are on the horizon. Topographic surface modification therefore offers a simple means to increase reproducibility as well as electrophysiological function within such assays.

While MEAs are focused on assaying electrophysiological function, microcontact printing has been used to promote uniaxial organization of cardiomyocytes on flexible cantilevers for multiplexed analysis of contractile function.[142] Such data illustrate how topographic guidance cues could be employed in combination with cantilever technologies to further enhance sarcomere development within higher-throughput, automated contractile assays. Cantilevers have been utilized for skeletal muscle culture as well as cardiomyocytes,[37,38] and the integration of topographic cantilevers for enhanced contractility is a viable avenue for future experimentation.

Such data highlight that the simple modification of extant assays is a viable pathway to improving maturation and therefore the predictive power of stem cell–derived cardiomyocyte and skeletal muscle-based platforms. In this way, topographic substrates can be thought of as "assay agnostic" in that their use is not specific to a single assay. Instead, they are amenable to integration with a wide range of preclinical screening tools, including fluorometric assays, contraction-based platforms, and metabolic screens. Widespread application of simple technologies for enhancing stem cell–derived muscle performance not only benefits the assay in question but promotes more reproducibly organized and structurally developed cultures across multiple culture surfaces. This in turn promotes consistency of results across assays and experiments performed in multiple laboratories. As such, the integration of topographic guidance cues with organ-on-a-chip as well as more standard screening assays and determination of their additive benefit for enhancing screening paradigms is an exciting avenue for future study.

8.4.3 Fit for Purpose?

In discussing the ability for combinatorial cues to promote muscle cell maturation in culture, it is worth noting that even if an optimal maturation cocktail can be identified, time is always likely to be a limiting factor. Given the months- to years-long time course for human tissue development *in vivo,* achieving similar levels of maturation in the relatively short timeframe required for preclinical screening is a significant challenge.[143] Even if achievable, questions would still likely arise regarding the impact of pushing cells through such a protracted natural maturation scheme in such a short period of time. So, if achievement of a muscle phenotype that accurately recapitulates that of adult tissues is unlikely within typical cell culture timeframes (weeks to months), the questions become what level of maturation is necessary for the generation of predictive data and which cues are sufficient to achieve that level of maturation?

There is unlikely to be a straightforward answer to these questions as the needs will vary based on the end user and the specific assay in question. For example, expression of the correct cohort of ion channels in the cell membrane is likely to be more important for electrophysiological assays than the status of the contractile apparatus of the cell. The goal should not, therefore, be the complete reconstitution of the adult muscle phenotype in the dish but rather the achievement of a cell state that can be relied upon to provide predictive data pertaining to the *in vivo* condition. Now that so much work has been done in characterizing the effect of various biophysical stimuli on muscle maturation *in vitro*, understanding the minimum level of maturation required for establishment of predictive assays is a critical next step.

Previous work has highlighted the significant benefit that topographic features confer on assays related to myofibril development and structural organization of cultured cells.[16,52] As such, one can easily envisage topographic patterns constituting a critical maturation cue in future assays focused on such metrics. Additionally, assaying structural toxicity is becoming increasingly important during preclinical therapy evaluation, particularly in the assessment of a compound's cardiotoxic potential.[144] Given that breakdown of sarcomeric proteins, such as cTnI, is an important indicator of cardiotoxicity,[145] the use of substrates that enhance expression of these proteins in cultured cells is likely to enhance the sensitivity of such assays.[138] Careful assessment of the end points to be evaluated will likely lead to a greater understanding of the maturation cues necessary to ensure that the assay in question provides the most meaningful data possible.

8.4.4 Three-dimensional Applications

While conventional 2D culture formats are incredibly valuable for high-throughput preclinical assays, they confer significant limitations for accurate modelling of tissue development and disease. Cells do not naturally exist in a 2D environment and the provision of substrate cues on one side of the cell, coupled with the rigid and undynamic nature of the culture surface, can be a hindrance when seeking to model how tissues behave in the body and break down in response to disease. As we move towards more widespread adoption of 3D culture platforms for muscle modelling, it is important not to forget the lessons learned from 2D culture in terms of understanding the role of topographic cues in modulating cellular behaviour.

Synthesis of 3D engineered tissues incorporating skeletal and cardiac muscle cells typically involves the seeding of cells within a biopolymer or synthetic matrix to serve as a surrogate ECM and support cellular growth and function. Use of biopolymer matrices based on collagen, fibrin, or similar ECM protein provides seeded cells with topographic cues that closely resemble those in the body. Researchers make use of cells' innate ability to compact matrix material to promote their reorganization within these matrices. Fixing such constructs between two anchors enables cells to generate uniaxial lines of tension throughout the engineered tissue as matrix compaction leads to stress development between the stationary points.[146] This in turn leads to remodelling of the ECM fibres and cells in parallel with the lines of strain that develop within the tissue. The result is a fibrous bundle of muscle cells arranged uniaxially and capable of performing directed work. The kinetics of these contractions can then be measured directly by force transducers[133,147] or indirectly *via* optical tracking[148] or magnetic sensing.[149]

Tissues engineered in this way exhibit structures and cellular morphologies that more closely recapitulate native architectures. However, most technologies in this space rely on relatively simple matrix materials, and, to date, little investigation of the impact of changing topographic features has been performed. While optimal topographic dimensions for promoting muscle maturation have been investigated in 2D,[52] the dynamic environment and complexity of cell–matrix interactions associated with the development of 3D tissues likely necessitates a re-evaluation of the optimal cues required for promoting maturation under these culture conditions. ECM fibres can range from submicron to multiple-micron diameters, and different fibre groups are typically expressed in different tissue locations. In 2015, Fleischer and colleagues investigated the effect of fibrous scaffolds exhibiting dimension scales of distinct fibre populations in the heart matrix on the function of seeded cardiomyocytes.[150] They found that cells maintained on nanoscale-fibre scaffolds exhibited more elastic properties than those maintained in microscale matrices but reduced spreading and elongation. Furthermore, by optimizing blends of different fibre diameters, they were able to produce engineered constructs with superior function than those grown within uniform matrices. These results underscore the importance of continuing to consider the role of topographic features in contributing to engineered muscle function in 3D. It is highly likely that a blend of different topographic feature sizes will be advantageous for engineered skeletal muscle, as has already been demonstrated for cardiac constructs,[150] given the range of fibre diameters present as one transitions from the perimysium, to the endomysium, and to the epimysium.[13]

While the majority of 3D cultures utilize matrix-based platforms, methods for producing matrix-free 3D tissues *via* cell sheet stacking technologies have also been developed.[102] Such technologies facilitate the synthesis of cell-dense tissues without copious amounts of exogenous extracellular material. Given the density of muscle cells present in native tissues, and the importance of this organization for overall tissue function, sheet stacking methods may arguably produce engineered constructs that more closely recapitulate the cell–cell and cell–matrix interactions that occur *in vivo.* Methods for incorporating topographic features into detachable cell sheets have been described and used to facilitate the formation of multilayer, aligned cellular constructs.[102,151] (See Figure 8.4.) In such tissues, the structural organization conferred by the underlying topography persists even after the transfer off of the patterns and the replate of cells on flat surfaces. Evidence suggests that this persistence is due to the organization of ECM matrix fibres synthesized by the cells themselves in parallel with the provided topographic cues. During transfer, these endogenous ECM fibres are transferred along with the cells, thereby providing a persistent guidance cue to help maintain alignment.[103] Such technologies hold potential for engineering organized 3D constructs of variable thicknesses.

As 3D culture becomes more widely adopted, continued optimization of topographical cues applied to tissue engineered muscle is an important avenue for further study and a means to continue investigating the application of combinatorial maturation cues within 3D culture environments. Ultimately, the incorporation of physiologically relevant topographic cues into 3D engineering efforts will lead to more accurate modelling of native tissue architecture. For organs incorporating complex cellular alignments, such as the heart, this will facilitate the development of platforms capable of assaying whole tissue function in a manner that is more representative of *in vivo* endpoints.

8.5 Conclusions

Provision of topographic cues to cultured cells represents one of the simplest ways to enhance the structural organization of cells *in vitro.* For muscle cells, such cues can be relied upon to promote uniaxial alignment and sarcomere development, leading to the establishment of more physiologically representative *in vitro* tissues for subsequent analysis. The simplicity of this technique, which seeks to mirror *in vivo* ECM guidance cues to drive *in vitro* cellular phenotypes, can be readily adapted to a wide range of preclinical assays to enhance sensitivity and improve reproducibility. The continued advancement of *in vitro* assays, for basic biological studies, disease modelling, and drug development necessitates the generation of more accurate and predictive representations of native tissues. Use of patterned surfaces represents one viable path to improving cellular architectures within extant assays as well as novel organ-on-a-chip platforms. Similarly, careful modulation of topographic cues within 3D culture systems has the potential to improve functional outcomes and enable collection of more meaningful data. In both academic and commercial settings, the specific mechanical requirements of the cells being analysed and how best to provide these physical cues within the confines of the assay in question are critical considerations during the development of next-generation *in vitro* assays and engineered tissues.

Disclosures

J. Macadangdang and N. Geisse are employees of Curi Bio Inc., a company that manufacturers topographic cultureware. A. Smith and D. Mack are scientific advisors to Curi Bio Inc.

References

1. T. Dvir, B. P. Timko, D. S. Kohane and R. Langer, *Nat. Nanotechnol.*, 2011, **6**, 13.
2. D.-H. Kim, P. P. Provenzano, C. L. Smith and A. Levchenko, *J. Cell Biol.*, 2012, **197**, 351–360.
3. R. G. Harrison, *J. Exp. Zool.*, 1914, **17**, 521–544.
4. P. Weiss, *J. Exp. Zool.*, 1934, **68**, 393–448.
5. D. Kleinfeld, K. Kahler and P. Hockberger, *J. Neurosci.*, 1988, **8**, 4098–4120.
6. C. A. Goubko and X. Cao, *Mater. Sci. Eng.: C*, 2009, **29**, 1855–1868.
7. A. Formhals and R. Schreiber-Gastell, *Apparatus for Producing Artificial Filaments from Material such as Cellulose Acetate*, Schreiber-Gastell, Richard, 1934.
8. J. Nam, Y. Huang, S. Agarwal and J. Lannutti, *Tissue Eng.*, 2007, **13**, 2249–2257.
9. W. J. Li, C. T. Laurencin, E. J. Caterson, R. S. Tuan and F. K. Ko, *J. Biomed. Mater. Res.*, 2002, **60**, 613–621.
10. J. D. Stitzel, K. J. Pawlowski, G. E. Wnek, D. G. Simpson and G. L. Bowlin, *J. Biomater. Appl.*, 2001, **16**, 22–33.
11. Y. Xia and G. M. Whitesides, *Annu. Rev. Mater. Sci.*, 1998, **28**, 153–184.
12. Y. Zhao, H. Zeng, J. Nam and S. Agarwal, *Biotechnol. Bioeng.*, 2009, **102**, 624–631.
13. A. R. Gillies and R. L. Lieber, *Muscle Nerve*, 2011, **44**, 318–331.
14. D. F. Scollan, A. Holmes, R. Winslow and J. Forder, *American Journal of Physiology-Heart and Circulatory Physiology*, 1998, **275**, H2308–H2318.
15. L. Zhukov and A. H. Barr, *IEEE Visualization, 2003*, 2003, VIS 2003.
16. D.-H. Kim, E. A. Lipke, P. Kim, R. Cheong, S. Thompson, M. Delannoy, K.-Y. Suh, L. Tung and A. Levchenko, *Proc. Natl. Acad. Sci.*, 2010, **107**, 565–570.
17. C. W. Chang, A. J. Dalgliesh, J. E. López and L. G. Griffiths, *Proteomics: Clin. Appl.*, 2016, **10**, 39–50.
18. D. S. Celermajer, C. K. Chow, E. Marijon, N. M. Anstey and K. S. Woo, *J. Am. Coll. Cardiol.*, 2012, **60**, 1207–1216.

19. S. Reddy, F. Riahi, G. Dorling, R. Callahan and H. Patel, *Innovative Approaches to Prevention Tackling the Global Burden of Cardiovascular Disease,* World Innovative Summit for Health, Doha, Qatar, 2016.
20. O. Bergmann, R. D. Bhardwaj, S. Bernard, S. Zdunek, F. Barnabé-Heider, S. Walsh, J. Zupicich, K. Alkass, B. A. Buchholz and H. Druid, *Science*, 2009, **324**, 98–102.
21. M. N. Hirt, A. Hansen and T. Eschenhagen, *Circ. Res.*, 2014, **114**, 354–367.
22. G. Vunjak-Novakovic, N. Tandon, A. Godier, R. Maidhof, A. Marsano, T. P. Martens and M. Radisic, *Tissue Eng., Part B*, 2010, **16**, 169–187.
23. S. S. Nunes, J. W. Miklas, J. Liu, R. Aschar-Sobbi, Y. Xiao, B. Zhang, J. Jiang, S. Masse, M. Gagliardi, A. Hsieh, N. Thavandiran, M. A. Laflamme, K. Nanthakumar, G. Gross, P. H. Backx, G. Keller and M. Radisic, *Nat. Meth.*, 2013, **10**, 781–787.
24. Y. Xiao, B. Zhang, H. Liu, J. W. Miklas, M. Gagliardi, A. Pahnke, N. Thavandiran, Y. Sun, C. Simmons, G. Keller and M. Radisic, *Lab Chip*, 2014, **14**, 869–882.
25. M. B. Baghdadi and S. Tajbakhsh, *Dev. Biol.*, 2018, **433**, 200–209.
26. K. W. VanDusen, B. C. Syverud, M. L. Williams, J. D. Lee and L. M. Larkin, *Tissue Eng., Part A*, 2014, **20**, 2920–2930.
27. Y.-J. Choi, Y.-J. Jun, D. Y. Kim, H.-G. Yi, S.-H. Chae, J. Kang, J. Lee, G. Gao, J.-S. Kong, J. Jang, W. K. Chung, J.-W. Rhie and D.-W. Cho, *Biomaterials*, 2019, **206**, 160–169.
28. B. T. Corona, M. A. Machingal, T. Criswell, M. Vadhavkar, A. C. Dannahower, C. Bergman, W. Zhao and G. J. Christ, *Tissue Eng., Part A*, 2012, **18**, 1213–1228.
29. S. M. Maffioletti, S. Sarcar, A. B. Henderson, I. Mannhardt, L. Pinton, L. A. Moyle, H. Steele-Stallard, O. Cappellari, K. E. Wells and G. Ferrari, *Cell Rep.*, 2018, **23**, 899–908.
30. M. Koning, M. C. Harmsen, M. J. A. van Luyn and P. M. N. Werker, *J. Tissue Eng. Regener. Med.*, 2009, **3**, 407–415.
31. A. Khodabukus and K. Baar, *Tissue Eng., Part A*, 2015, **21**, 1003–1012.
32. T. Broer, A. Khodabukus and N. Bursac, *Expert Opin. Drug Discovery*, 2020, **15**, 643–645.
33. J. Shansky, B. Creswick, P. Lee, X. Wang and H. Vandenburgh, *Tissue Eng.*, 2006, **12**, 1833–1841.
34. H. Vandenburgh, J. Shansky, F. Benesch-Lee, V. Barbata, J. Reid, L. Thorrez, R. Valentini and G. Crawford, *Muscle Nerve*, 2008, **37**, 438–447.
35. H. Vandenburgh, J. Shansky, F. Benesch-Lee, K. Skelly, J. M. Spinazzola, Y. Saponjian and B. S. Tseng, *FASEB J.*, 2009, **23**, 3325–3334.
36. H. Vandenburgh, M. D. Tatto, J. Shansky, J. Lemaire, A. Chang, F. Payumo, P. Lee, A. Goodyear and L. Raven, *Hum. Gene Ther.*, 1996, **7**, 2195–2200.
37. C. W. McAleer, A. S. Smith, S. Najjar, K. Pirozzi, C. J. Long and J. J. Hickman, *J. Appl. Physiol.*, 2014, **117**, 1398–1405.

38. A. S. Smith, C. J. Long, K. Pirozzi, S. Najjar, C. McAleer, H. H. Vandenburgh and J. J. Hickman, *J. Biotechnol.*, 2014, **185**, 15–18.
39. P. W. Zandstra, C. Bauwens, T. Yin, Q. Liu, H. Schiller, R. Zweigerdt, K. B. S. Pasumarthi and L. J. Field, *Tissue Eng.*, 2003, **9**, 767–778.
40. S. Tohyama, F. Hattori, M. Sano, T. Hishiki, Y. Nagahata, T. Matsuura, H. Hashimoto, T. Suzuki, H. Yamashita, Y. Satoh, T. Egashira, T. Seki, N. Muraoka, H. Yamakawa, Y. Ohgino, T. Tanaka, M. Yoichi, S. Yuasa, M. Murata, M. Suematsu and K. Fukuda, *Cell Stem Cell*, 2013, **12**, 127–137.
41. X. Guan, D. L. Mack, C. M. Moreno, J. L. Strande, J. Mathieu, Y. Shi, C. D. Markert, Z. Wang, G. Liu, M. W. Lawlor, E. C. Moorefield, T. N. Jones, J. A. Fugate, M. E. Furth, C. E. Murry, H. Ruohola-Baker, Y. Zhang, L. F. Santana and M. K. Childers, *Stem Cell Res.*, 2014, **12**, 467–480.
42. J. Chal, Z. Al Tanoury, M. Hestin, B. Gobert, S. Aivio, A. Hick, T. Cherrier, A. P. Nesmith, K. K. Parker and O. Pourquie, *Nat. Protoc.*, 2016, **11**, 1833–1850.
43. J. Chal, M. Oginuma, Z. Al Tanoury, B. Gobert, O. Sumara, A. Hick, F. Bousson, Y. Zidouni, C. Mursch, P. Moncuquet, O. Tassy, S. Vincent, A. Miyanari, A. Bera, J.-M. Garnier, G. Guevara, M. Hestin, L. Kennedy, S. Hayashi, B. Drayton, T. Cherrier, B. Gayraud-Morel, E. Gussoni, F. Relaix, S. Tajbakhsh and O. Pourquie, *Nat. Biotechnol.*, 2015, **33**, 962–969.
44. M. R. Hicks, J. Hiserodt, K. Paras, W. Fujiwara, A. Eskin, M. Jan, H. Xi, C. S. Young, D. Evseenko, S. F. Nelson, M. J. Spencer, B. V. Handel and A. D. Pyle, *Nat. Biotechnol*, 2018, **20**, 46–57.
45. H. Xi, W. Fujiwara, K. Gonzalez, M. Jan, S. Liebscher, B. Van Handel, K. Schenke-Layland and A. D. Pyle, *Cell Rep.*, 2017, **18**, 1573–1585.
46. X. Yang, L. Pabon and C. E. Murry, *Circ. Res.*, 2014, **114**, 511–523.
47. J. Kim and R. C. Hayward, *Trends Biotechnol.*, 2012, **30**, 426–439.
48. Z. Kassiri and R. Khokha, *Thromb. Haemostasis*, 2005, **93**, 212–219.
49. M. D. Boppart, D. J. Burkin and S. J. Kaufman, *Am. J. Physiol.*, 2006, **290**, C1660–C1665.
50. P.-Y. Wang, H.-T. Yu and W.-B. Tsai, *Biotechnol. Bioeng.*, 2010, **106**, 285–294.
51. K. Shimizu, H. Fujita and E. Nagamori, *Biotechnol. Bioeng.*, 2009, **103**, 631–638.
52. D. Carson, M. Hnilova, X. Yang, C. L. Nemeth, J. H. Tsui, A. S. Smith, A. Jiao, M. Regnier, C. E. Murry, C. Tamerler and D. H. Kim, *ACS Appl. Mater. Interfaces*, 2016, **8**, 21923–21932.
53. D. H. Kim, Kshitiz, R. R. Smith, P. Kim, E. H. Ahn, H. N. Kim, E. Marbán, K. Y. Suh and A. Levchenko, *Integr. Biol.*, 2012, **4**, 1019–1033.
54. A. G. Rodriguez, S. J. Han, M. Regnier and N. J. Sniadecki, *Biophys. J.*, 2011, **101**, 2455–2464.

55. S. Romanazzo, G. Forte, M. Ebara, K. Uto, S. Pagliari, T. Aoyagi, E. Traversa and A. Taniguchi, *Sci. Technol. Adv. Mater.*, 2012, **13**, 064211.
56. G. Forte, S. Pagliari, M. Ebara, K. Uto, J. K. V. Tam, S. Romanazzo, C. Escobedo-Lucea, E. Romano, P. Di Nardo, E. Traversa and T. Aoyagi, *Tissue Eng., Part A*, 2012, **18**, 1837–1848.
57. S. Calve and H.-G. Simon, *FASEB J.*, 2012, **26**, 2538–2545.
58. Z. Yin, E. Cheng and H. Zou, *Lab Chip*, 2014, **14**, 1614–1621.
59. Y. Lee, Y. Lim and H. Shin, *Nanoscale*, 2016, **8**, 11810–11817.
60. W. Liu, Z. Zhou, S. Zhang, Z. Shi, J. Tabarini, W. Lee, Y. Zhang, S. N. Gilbert Corder, X. Li, F. Dong, L. Cheng, M. Liu, D. L. Kaplan, F. G. Omenetto, G. Zhang, Y. Mao and T. H. Tao, *Adv. Sci.*, 2017, **4**, 1700191.
61. K. T. Tran and T. D. Nguyen, *J. Sci.: Adv. Mater. Devices*, 2017, **2**, 1–14.
62. C. D. James, R. C. Davis, L. Kam, H. G. Craighead, M. Isaacson, J. N. Turner and W. Shain, *Langmuir*, 1998, **14**, 741–744.
63. A. J. Khadpekar, M. Khan, A. Sose and A. Majumder, *Sci. Rep.*, 2019, **9**, 1–8.
64. S. V. Bhujbal, M. Dekov, V. Ottesen, K. Dunker, R. Lale and M. Sletmoen, *Sci. Rep.*, 2020, **10**, 1–13.
65. J. Borowiec, J. Hampl, S. Singh, S. Haefner, K. Friedel, P. Mai, D. Brauer, F. Ruther, L. Liverani and A. R. Boccaccini, *ACS Appl. Mater. Interfaces*, 2018, **10**, 22857–22865.
66. M. Altissimo, *Biomicrofluidics*, 2010, **4**, 026503.
67. A. Selimis, V. Mironov and M. Farsari, *Microelectron. Eng.*, 2015, **132**, 83–89.
68. M. Bruk, E. Zhikharev, A. Rogozhin, D. Streltsov, V. Kalnov, S. Averkin and A. Spirin, *Microelectron. Eng.*, 2016, **155**, 92–96.
69. Y. Chen, *Microelectron. Eng.*, 2015, **135**, 57–72.
70. N. Bhardwaj and S. C. Kundu, *Biotechnol. Adv.*, 2010, **28**, 325–347.
71. S. Chen, J. V. John, A. McCarthy and J. Xie, *J. Mater. Chem. B*, 2020, **8**, 3733–3746.
72. S. Chen, M. A. Carlson, Y. S. Zhang, Y. Hu and J. Xie, *Biomaterials*, 2018, **179**, 46–59.
73. J. Song, G. Zhu, H. Gao, L. Wang, N. Li, X. Shi and Y. Wang, *Bio-Des. Manuf.*, 2018, **1**, 254–264.
74. W. Chen, S. Chen, Y. Morsi, H. El-Hamshary, M. El-Newhy, C. Fan and X. Mo, *ACS Appl. Mater. Interfaces*, 2016, **8**, 24415–24425.
75. G. X. Gu, I. Su, S. Sharma, J. L. Voros, Z. Qin and M. J. Buehler, *J. Biomech. Eng.*, 2016, **138**, 021006.
76. S. V. Murphy and A. Atala, *Nat. Biotechnol.*, 2014, **32**, 773–785.
77. B. Gao, Q. Yang, X. Zhao, G. Jin, Y. Ma and F. Xu, *Trends Biotechnol.*, 2016, **34**, 746–756.
78. S. E. Bakarich, R. Gorkin III, M. I. H. Panhuis and G. M. Spinks, *Macromol. Rapid Commun.*, 2015, **36**, 1211–1217.
79. X. Wang, Y. Sun, C. Peng, H. Luo, R. Wang and D. Zhang, *ACS Appl. Mater. Interfaces*, 2015, **7**, 26131–26136.

80. D. Kokkinis, M. Schaffner and A. R. Studart, *Nat. Commun.*, 2015, **6**, 1–10.
81. M. Tallawi, R. Rai, A. R. Boccaccini and K. E. Aifantis, *Tissue Eng., Part A*, 2015, **21**, 157–165.
82. B. Zhang, Y. Xiao, A. Hsieh, N. Thavandiran and M. Radisic, *Nanotechnology*, 2011, **22**, 494003.
83. S. Marchianò, A. Bertero and C. E. Murry, *Pediatr. Cardiol.*, 2019, **40**, 1367–1387.
84. M. Ahmed, L. Yildirimer, A. Khademhosseini and A. M. Seifalian, *J. Nanosci. Nanotechnol.*, 2012, **12**, 4775–4785.
85. V. L. Roger, A. S. Go, D. M. Lloyd-Jones, E. J. Benjamin, J. D. Berry, W. B. Borden, D. M. Bravata, S. Dai and E. S. Ford, *Circulation*, 2012, **125**, e2–e220.
86. H. S. Yang, N. Ieronimakis, J. H. Tsui, H. N. Kim, K. Y. Suh, M. Reyes and D. H. Kim, *Biomaterials*, 2014, **35**, 1478–1486.
87. J. Macadangdang, X. Guan, A. S. Smith, R. Lucero, S. Czerniecki, M. K. Childers, D. L. Mack and D. H. Kim, *Cell. Mol. Bioeng.*, 2015, **8**, 320–332.
88. B. S. Gordon, D. A. Lowe and M. C. Kostek, *Muscle Nerve*, 2014, **49**, 915–918.
89. S. Sen, M. Tewari, A. Zajac, E. Barton, H. L. Sweeney and D. E. Discher, *Eur. J. Cell Biol.*, 2011, **90**, 249–260.
90. C. Rao, T. Prodromakis, L. Kolker, U. A. Chaudhry, T. Trantidou, A. Sridhar, C. Weekes, P. Camelliti, S. E. Harding and A. Darzi, *Biomaterials*, 2013, **34**, 2399–2411.
91. N. Badie and N. Bursac, *Biophys. J.*, 2009, **96**, 3873–3885.
92. M. R. Badrossamay, H. A. McIlwee, J. A. Goss and K. K. Parker, *Nano Lett.*, 2010, **10**, 2257–2261.
93. J. Macadangdang, H. J. Lee, D. Carson, A. Jiao, J. Fugate, L. Pabon, M. Regnier, C. Murry and D. H. Kim, *J. Visualized Exp.*, 2014, **88**, e50039.
94. J. M. Pioner, X. Guan, J. M. Klaiman, A. W. Racca, L. Pabon, V. Muskheli, J. Macadangdang, C. Ferrantini, M. R. Hoopmann, R. L. Moritz, D. H. Kim, C. Tesi, C. Poggesi, C. E. Murry, M. K. Childers, D. L. Mack and M. Regnier, *Cardiovasc. Res.*, 2020, **116**, 368–382.
95. J. M. Pioner, A. W. Racca, J. M. Klaiman, K. C. Yang, X. Guan, L. Pabon, V. Muskheli, R. Zaunbrecher, J. Macadangdang, M. Y. Jeong, D. L. Mack, M. K. Childers, D. H. Kim, C. Tesi, C. Poggesi, C. E. Murry and M. Regnier, *Stem Cell Reports*, 2016, **6**, 885–896.
96. R. G. Parton, M. Way, N. Zorzi and E. Stang, *J. Cell Biol.*, 1997, **136**, 137–154.
97. J. Feher, in *Quantitative Human Physiology*, ed. J. Feher, Academic Press, Boston, 2nd edn, 2017, pp. 292–304.
98. J. Morgan and T. Partridge, *Dis. Models Mech.*, 2020, **13**, dmm042192.
99. H. Nagy and K. D. Veerapaneni, *StatPearls [Internet]*, 2020.
100. B. Li, M. Lin, Y. Tang, B. Wang and J. H. C. Wang, *J. Biomech.*, 2008, **41**, 3349–3353.

101. N. F. Huang, K. G. Thakar, M. Wong, D. Kim, R. J. Lee and L. Song, *The 26th Annual International Conference of the IEEE Engineering in Medicine and Biology Society*, 2004.
102. H. Takahashi, T. Shimizu, M. Nakayama, M. Yamato and T. Okano, *Biomaterials*, 2013, **34**, 7372–7380.
103. A. Jiao, C. T. Moerk, N. Penland, M. Perla, J. Kim, A. S. T. Smith, C. E. Murry and D.-H. Kim, *J. Biomed. Mater. Res., Part A*, 2018, **106**, 1543–1551.
104. H. Takahashi, T. Shimizu and T. Okano, *Sci. Rep.*, 2018, **8**, 13932.
105. V. Hosseini, S. Ahadian, S. Ostrovidov, G. Camci-Unal, S. Chen, H. Kaji, M. Ramalingam and A. Khademhosseini, *Tissue Eng., Part A*, 2012, **18**, 2453–2465.
106. A. Bettadapur, G. C. Suh, N. A. Geisse, E. R. Wang, C. Hua, H. A. Huber, A. A. Viscio, J. Y. Kim, J. B. Strickland and M. L. McCain, *Sci. Rep.*, 2016, **6**, 28855.
107. M. L. Rexius-Hall, N. N. Khalil, A. M. Andres and M. L. McCain, *FASEB J.*, 2020, **34**, 11562–11576.
108. Y.-W. Cheng, D. J. Shiwarski, R. L. Ball, K. A. Whitehead and A. W. Feinberg, *ACS Biomater. Sci. Eng.*, 2020, **6**, 3046–3054.
109. J. S. Choi, S. J. Lee, G. J. Christ, A. Atala and J. J. Yoo, *Biomaterials*, 2008, **29**, 2899–2906.
110. K. J. Aviss, J. E. Gough and S. Downes, *Eur. Cells Mater.*, 2010, **19**, 193–204.
111. N. Narayanan, C. Jiang, C. Wang, G. Uzunalli, N. Whittern, D. Chen, O. G. Jones, S. Kuang and M. Deng, *Front. Bioeng. Biotechnol.*, 2020, **8**, 203.
112. I. Apsite, J. M. Uribe, A. F. Posada, S. Rosenfeldt, S. Salehi and L. Ionov, *Biofabrication*, 2019, **12**, 015016.
113. L. Rao, Y. Qian, A. Khodabukus, T. Ribar and N. Bursac, *Nat. Commun.*, 2018, **9**, 126.
114. P. B. Canham and K. Mullin, *J. Microsc.*, 1978, **114**, 307–318.
115. A. Frismantiene, M. Philippova, P. Erne and T. J. Resink, *Cell. Signalling*, 2018, **52**, 48–64.
116. R. G. Thakar, F. Ho, N. F. Huang, D. Liepmann and S. Li, *Biochem. Biophys. Res. Commun.*, 2003, **307**, 883–890.
117. E. K. F. Yim, R. M. Reano, S. W. Pang, A. F. Yee, C. S. Chen and K. W. Leong, *Biomaterials*, 2005, **26**, 5405–5413.
118. S. Chang, S. Song, J. Lee, J. Yoon, J. Park, S. Choi, J.-K. Park, K. Choi and C. Choi, *PLoS One*, 2014, **9**, e88089.
119. T. Nakamoto, X. Wang, N. Kawazoe and G. Chen, *Colloids Surf., B*, 2014, **122**, 316–323.
120. C. Y. Tay, Y.-L. Wu, P. Cai, N. S. Tan, S. S. Venkatraman, X. Chen and L. P. Tan, *NPG Asia Mater.*, 2015, **7**, e199.
121. Q. Jin, A. Bhatta, J. V. Pagaduan, X. Chen, H. West-Foyle, J. Liu, A. Hou, D. Berkowitz, S. C. Kuo, F. B. Askin, T. D. Nguyen, D. H. Gracias and L. H. Romer, *Sci. Adv.*, 2020, **6**, eaaz2598.

122. J. Zhou, D. C. Parker, J. P. White, A. Lim, K. M. Huffman, J. P. Ho, P. M. Yen and W. E. Kraus, *Front. Physiol.*, 2019, 10.
123. X. Yang, M. Rodriguez, L. Pabon, K. A. Fischer, H. Reinecke, M. Regnier, N. J. Sniadecki, H. Ruohola-Baker and C. E. Murry, *J. Mol. Cell. Cardiol.*, 2014, **72**, 296–304.
124. N. Sculthorpe, A. M. Solomon, A. C. Sinanan, P. M. Bouloux, F. Grace and M. P. Lewis, *Med. Sci. Sports Exercise*, 2012, **44**, 610–615.
125. M. R. Weist, M. S. Wellington, J. E. Bermudez, T. Y. Kostrominova, C. L. Mendias, E. M. Arruda and L. M. Larkin, *J. Tissue Eng. Regener. Med.*, 2013, **7**, 562–571.
126. W. Bian and N. Bursac, *FASEB J.*, 2012, **26**, 955–965.
127. X. F. Deng, D. G. Rokosh and P. C. Simpson, *Circ. Res.*, 2000, **87**, 781–788.
128. G. Földes, M. Mioulane, J. S. Wright, A. Q. Liu, P. Novak, B. Merkely, J. Gorelik, M. D. Schneider, N. N. Ali and S. E. Harding, *J. Mol. Cell. Cardiol.*, 2011, **50**, 367–376.
129. N. Liu, A. H. Williams, J. M. Maxeiner, S. Bezprozvannaya, J. M. Shelton, J. A. Richardson, R. Bassel-Duby and E. N. Olson, *J. Clin. Invest.*, 2012, **122**, 2054–2065.
130. K. T. Kuppusamy, D. C. Jones, H. Sperber, A. Madan, K. A. Fischer, M. L. Rodriguez, L. Pabon, W. Z. Zhu, N. L. Tulloch, X. Yang, N. J. Sniadecki, M. A. Laflamme, W. L. Ruzzo, C. E. Murry and H. Ruohola-Baker, *Proc. Natl. Acad. Sci. U. S. A.*, 2015, **112**, E2785–2794.
131. K. Ronaldson-Bouchard, S. P. Ma, K. Yeager, T. Chen, L. Song, D. Sirabella, K. Morikawa, D. Teles, M. Yazawa and G. Vunjak-Novakovic, *Nature*, 2018, **556**, 239–243.
132. K. Kujala, A. Ahola, M. Pekkanen-Mattila, L. Ikonen, E. Kerkelä, J. Hyttinen and K. Aalto-Setälä, *Int. J. Biomed. Sci.*, 2012, **8**, 109–120.
133. A. Khodabukus and K. Baar, *Tissue Eng., Part C*, 2011, **18**, 349–357.
134. N. Nikolic, S. S. Bakke, E. T. Kase, I. Rudberg, I. Flo Halle, A. C. Rustan, G. H. Thoresen and V. Aas, *PLoS One*, 2012, **7**, e33203.
135. A. S. T. Smith, H. Yoo, H. Yi, E. H. Ahn, J. H. Lee, G. Shao, E. Nagornyak, M. A. Laflamme, C. E. Murry and D.-H. Kim, *Chem. Commun.*, 2017, **53**, 7412–7415.
136. I. Jun, S. Jeong and H. Shin, *Biomaterials*, 2009, **30**, 2038–2047.
137. J. R. Macadangdang, J. W. Miklas, A. S. T. Smith, E. Choi, W. Leung, Y. Wang, X. Guan, S. Lee, M. R. Salick, M. Regnier, D. Mack, M. K. Childers, H. Ruohola-Baker and D.-H. Kim, *bioRxiv*, 2018, 456301.
138. A. S. T. Smith, E. Choi, K. Gray, J. Macadangdang, E. H. Ahn, E. C. Clark, M. A. Laflamme, J. C. Wu, C. E. Murry, L. Tung and D.-H. Kim, *Nano Lett.*, 2019, **20**, 1561–1570.
139. J. Li, I. Minami, L. Yu, K. Tsuji, M. Nakajima, J. Qiao, M. Suzuki, K. Shimono, N. Nakatsuji, H. Kotera, L. Liu and Y. Chen, *Stem Cells Int*, 2016, **2016**, 2634013.

140. N.-E. Oyunbaatar, Y. Dai, A. Shanmugasundaram, B.-K. Lee, E.-S. Kim and D.-W. Lee, *ACS Sens.*, 2019, **4**, 2623–2630.
141. C. G. Langhammer, M. K. Kutzing, V. Luo, J. D. Zahn and B. L. Firestein, *Ann. Biomed. Eng.*, 2013, **41**, 408–420.
142. A. Agarwal, J. A. Goss, A. Cho, M. L. McCain and K. K. Parker, *Lab Chip*, 2013, **13**, 3599–3608.
143. S. D. Lundy, W.-Z. Zhu, M. Regnier and M. A. Laflamme, *Stem Cells Dev.*, 2013, **22**, 1991–2002.
144. K. R. Doherty, D. R. Talbert, P. B. Trusk, D. M. Moran, S. A. Shell and S. Bacus, *Toxicol. Appl. Pharmacol.*, 2015, **285**, 51–60.
145. M. J. Cross, B. R. Berridge, P. J. M. Clements, L. Cove-Smith, T. L. Force, P. Hoffmann, M. Holbrook, A. R. Lyon, H. R. Mellor, A. A. Norris, M. Pirmohamed, J. D. Tugwood, J. E. Sidaway and B. K. Park, *Br. J. Pharmacol.*, 2015, **172**, 957–974.
146. M. Eastwood, V. Mudera, D. McGrouther and R. Brown, *Cell Motil. Cytoskeleton*, 1998, **40**, 13–21.
147. N. R. Martin, S. L. Passey, D. J. Player, V. Mudera, K. Baar, L. Greensmith and M. P. Lewis, *Tissue Eng., Part A*, 2015, **21**, 2595–2604.
148. O. F. Vila, S. G. M. Uzel, S. P. Ma, D. Williams, J. Pak, R. D. Kamm and G. Vunjak-Novakovic, *Theranostics*, 2019, **9**, 1232–1246.
149. K. S. Bielawski, A. Leonard, S. Bhandari, C. E. Murry and N. J. Sniadecki, *Tissue Eng., Part C*, 2016, **22**, 932–940.
150. S. Fleischer, J. Miller, H. Hurowitz, A. Shapira and T. Dvir, *Nanotechnology*, 2015, **26**, 291002.
151. A. Jiao, N. E. Trosper, H. S. Yang, J. Kim, J. H. Tsui, S. D. Frankel, C. E. Murry and D. H. Kim, *ACS Nano*, 2014, **8**, 4430–4439.

CHAPTER 9

Engineered Substrates with Dynamically Tunable Topography

KOICHIRO UTO

Research Center for Functional Materials, National Institute for Materials Science, 1-1 Namiki, Tsukuba, Ibaraki 3050044, Japan
Email: UTO.Koichiro@nims.go.jp

9.1 Topography

Cells receive physicochemical and biochemical signals from the surrounding microenvironment, or extracellular matrix (ECM), and integrate and interpret these complex stimuli to determine cellular functions and fates, including adhesion, proliferation, migration, and differentiation. In addition to the biochemical properties of the ECM, its physical cues are particularly important, as cells are homeostatic in a microenvironment with tissue-specific viscoelasticity and topography. It has been consistently noted that topography, which is hierarchically arranged from the nanoscale to the macroscale, determines cell fate, affects interactions between adherent cells and their extracellular environment, and directly regulates cell function and fate. For example, the basement membrane is a highly specialized ECM composed of fibrous collagen, proteoglycans, laminin, and fibronectin, which play pivotal roles as biochemical and mechanical cues in regulating cellular functions. In addition, the basement membrane is known to have a

Biomaterials Science Series No. 12
Material-based Mechanobiology
Edited by Jun Nakanishi and Koichiro Uto

Published by the Royal Society of Chemistry, www.rsc.org

complicated, multiscale, three-dimensional topography with fiber structures, reticular pores, and ridges (Figure 9.1A).[1]

Organs are dependent on specific ECM topographies to achieve proper form and function. The heart, for example, is a complex organ with a very diverse arrangement of muscle cells (cardiomyocytes) and nonmuscle cells, which are assembled into a specific three-dimensional structure. The results of *ex vivo* ultrastructural analysis of rat cardiac tissue showed that the myocardial layer adjacent to the ECM fiber layer was not only anisotropically arranged with highly elongated and aligned cells but also that the orientation of the cells was strongly correlated with the orientation of the matrix fibers directly beneath them (Figure 9.1B).[2] The corpus callosum (CC), an axon connecting the cortices of the cerebral hemispheres, is the major white matter fiber bundle in the brain (Figure 9.1C),[3,4] and cells such as neurons and brain tumor cells have been observed to migrate along it.[5,6] The structure of the natural ECMs in various tissues, such as bone, nerves, and skin, and in organs also shows highly oriented fibrous structures with different-length scales.[7]

As previously mentioned, in tissues and organs, topography not only provides diverse adhesion sites for cells and involves mechanical properties but is also closely involved in the regulation of cellular functions such as adhesion, proliferation, migration, and differentiation. Therefore, inspired by the microstructure of natural ECM, the development of artificial substrates that mimic the sophisticated *in vivo* topography by integrating material science and microfabrication techniques is attractive to engineers and biologists in the fields of classical cell biology, tissue engineering, and regenerative medicine. In fact, as described in Chapter 8, topography has been extensively studied to regulate cell functions and reproduce physiological phenomena. However, the physicochemical properties of the extracellular environment *in vivo* gradually change over time, and topography is no exception. While a random, isotropic arrangement of fibrillar ECM components is an indicator of normal or homeostatic parenchyma; a relatively straight organization and anisotropic arrangement of ECM fibrils is known to be characteristic of a pathological microenvironment (*i.e.* desmoplasia),[8] and such linear/anisotropic features are observed in stroma associated with fibrosis and epithelial tumors (Figure 9.1D). Changes in the alignment of structural collagen fibrils associated with tumorigenesis, termed "tumor associated collagen signatures" (TACS),[9] show a unique stromal phenotype and have been suggested to correlate with different stages of advanced breast cancer. TACS-1 is a curly and anisotropic fibril that can be seen in normal resting tissue, whereas TACS-2 and TACS-3 show increasingly thicker, aligned/straight fibers (Figure 9.1D).[10]

When acute liver injury was induced in mice fed diets containing diethoxycarbonyl dihydrocollidine, changes in not only the composition but also the structure of the liver ECM were observed.[11] The ECM structure of the injured liver was more disordered, thinner, and exhibited an increased number of branched reticular collagen fibers compared to the normal ECM structure (Figure 9.1E). Furthermore, Kanazaki *et al.* used scanning electron

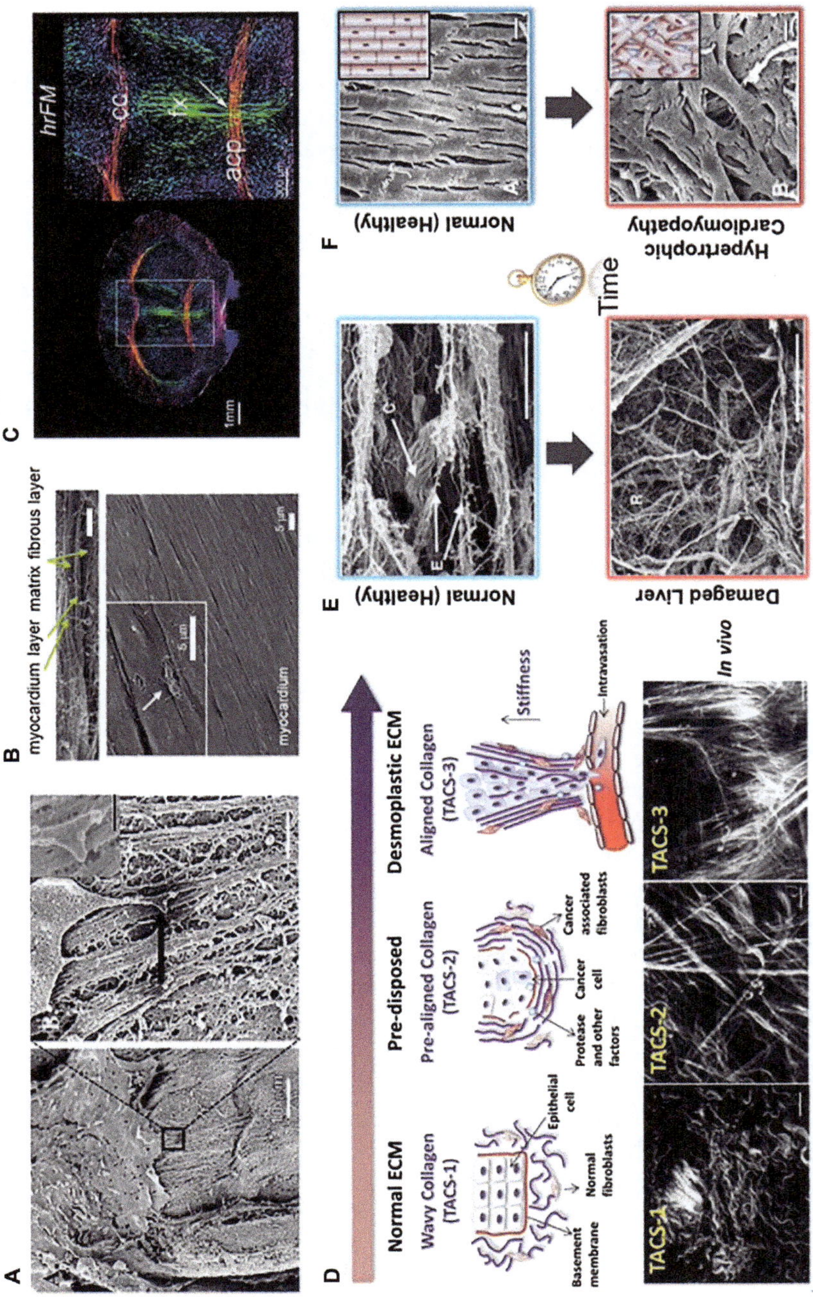
A
B
myocardium layer matrix fibrous layer
myocardium
C
hrFM
D
Normal ECM
Pre-disposed
Desmoplastic ECM
Wavy Collagen (TACS-1)
Pre-aligned Collagen (TACS-2)
Aligned Collagen (TACS-3)
Basement membrane
Epithelial cell
Normal fibroblasts
Protease and other factors
Cancer cell
Cancer associated fibroblasts
Stiffness
Intravasation
TACS-1
TACS-2
TACS-3
In vivo
E
Normal (Healthy)
Damaged Liver
Time
F
Normal (Healthy)
Hypertrophic Cardiomyopathy

microscopy to observe myectomy specimens from patients with hypertrophic cardiomyopathy.[12] Compared to normal heart samples, hypertrophic cardiomyopathy heart samples from which connective tissue has been removed showed cardiomyocytes that were scattered, overlapping, and abnormally branched (Figure 9.1F). When the nonfibrous segments were removed, a thickened interstitial connective tissue called plexiform fibrosis was observed, which also confirmed that the ECM structure was altered. These facts demonstrate that the structure of the ECM is not static but dynamic *in vivo* and that this dynamic nature is closely related to tissue and organ development, regeneration, wound healing, and disease progression.[13,14] Therefore, a dynamic platform with the potential to mimic time-dependent changes in matrix structure similar to those observed *in vivo* could improve understanding of fundamental biological phenomena and mechanobiology

Figure 9.1 Anatomical features of highly organized extracellular matrix (ECM) in various tissues. (A) SEM images showing an aortic endothelial cell with the basement membranes in an intact vessel. The higher magnification allows better visualization of the bumps, ridges, pits, and grooves forming the complex topography of the basement membrane. Reproduced from ref. 1 with permission from Mary Ann Liebert, Inc., Copyright 2009. (B) SEM images of *ex vivo* myocardium of adult rat heart. The side view (top) and top view (bottom) show well aligned myocardium. The inset in the bottom image demonstrates that the structural organization of the myocardium correlates with matrix fibers (arrow) aligned in parallel. Reproduced from ref. 2 with permission from National Academy of Sciences. (C) High-resolution fiber maps (hrFM) generated from the global fiber tracking data of living mouse brain. Various topographical structures of the mouse brain connectional anatomy can be easily recognized using *in vivo* imaging modality: cc, corpus callosum; fx, fornix; acp, anterior commissure posterior part. Reproduced from ref. 3 with permission from National Academy of Sciences. (D) Stromal remodeling cascade initiated by the loss of homeostatic equilibrium leads to production of biomechanically and biochemically altered ECM. In TACS-1 (normal) stage, the collagen fibers are curly and isotropic, indicative of normal tissue. In TACS-2 (predisposed) stage, the collagen fibers are straightened and arranged in parallel around the tumor, indicative of pre-invasive tissue. Eventually, in TACS-3 (aligned) stage, the collagen fibers are thickened and aligned perpendicularly to the tumor boundary, indicative of metastatic stages. Reproduced from ref. 8 with permission from Elsevier, Copyright 2015. (E) Alterations in the microstructure of damaged livers. SEM images show (top) healthy and (bottom) damaged liver: E, elastic fibers; C, collagen fibers; R, reticular fibers. Reproduced from ref. 11, https://doi.org/10.1038/srep27398, under the terms of the CC BY 4.0 license https://creativecommons.org/licenses/by/4.0/. (F) Alterations in the microstructure of cardiac tissue. SEM images show tissue samples from (top) a normal subject and (bottom) a patient with hypertrophic cardiomyopathy. Myocardial disarray, abnormal branching, and bizarre myocardial hypertrophy can be observed in the heart sample from the patient. Reproduced from ref. 12 with permission from Wolters Kluwer Health, Inc., Copyright 2012.

and ultimately lead to the development of tissue engineering and regenerative medicine.

Current efforts to investigate the effects of structure on cells have focused on static structures. Recently, the scientific community has been interested in developing substrates that can dynamically change the structure of cell scaffolds. This chapter focuses on topography as a structure cue and introduces materials that can change their own topography. These can be roughly classified into three material systems: polydimethylsiloxane-(PDMS-) based elastic materials, photochromic/liquid crystal materials, and shape memory materials. These topographically variable substrates can simplify the dynamic changes of topography on cellular behavior because they can induce arbitrary structural changes, unlike *in vivo* ECM. Therefore, research on cell behavior and functional manipulation using topographically variable substrates fabricated from each material will be presented to demonstrate their potential to become attractive tools for future mechanobiology research.

9.2 Silicone Elastomer System for Micro- and Nanofabrication

PDMS-based silicone elastomers are widely used as substrates for cell culture because of their easily tunable elastic modulus, high biocompatibility, and transparency features. Several kits are commercially available for preparing PDMS, including Dow Corning's Sylgard® 184 Silicone Elastomer Kit, which is a two-part kit consisting of a base elastomer and a curing agent. PDMS elastomers have been used as materials for investigating the mechanical properties of culture substrates and the mechanosensing and response of cells to dynamic mechanical stimuli, such as stretching, or for force measurement (see Chapter 4).[15–18] In addition, patterns can be transferred from the template substrate to the PDMS surface by the molding method, in which a liquid PDMS precursor is cured on the template substrate. PDMS substrates with a variety of topographies have been fabricated using this method.[19,20] PDMS itself is a chemically stable material, and the fabricated topography does not change during cell culture. Therefore, it is widely used as a tool to investigate the effects of "static" topography on cell function and fate. This section discusses the design and methodology of PDMS to achieve "dynamic" topographic changes and then outline how it is utilized in cell culture.

9.2.1 Design of Lithography-free Reconfigurable Wrinkles Using PDMS

In addition to this method of fabricating topography by molding, the material surface can be transformed into a wavy structure by creating nonlinear buckling and wrinkling on a soft substrate. This mechanism is analogous to a spring, a stretched structure made of nonstretchable material. Various strategies to induce buckling on soft substrates, such as thermal mismatch,[21] ultraviolet ozone (UVO) surface treatment,[22] mechanical strain,[23,24] and focused ion beam,[25] have been utilized to fabricate substrates with the wavy topography.

Wrinkle structure formation induced by buckling using thermal mismatch on PDMS was first reported by Bowden *et al.* in 1998.[21] In their work, gold-deposited PDMS was fabricated by electron beam evaporation (Figure 9.2A). In this deposition process, the PDMS is heated by the metal source, and its volume expands. Since the coefficient of thermal expansion of PDMS is much larger than that of gold, the gold layer is compressed by the PDMS after cooling, resulting in wrinkles and wavy patterns on the surface, which can be used to design the geometrical structure and create varied topography (Figure 9.2B). From a mechanical point of view, gold-deposited PDMS can be regarded as a bilayer material with a stiff layer on an elastic material. Surface treatment using plasma or ozone is a common approach to form wrinkle patterns, as it allows for the formation of a rigid thin film on the PDMS substrate.[22,26] The mechanism behind this is that carbon atoms are replaced by oxygen atoms on the surface of the PDMS, creating a thin, hard, silica-like layer. In this approach, the thermal mismatch causes compression of the hard layer as the system cools, resulting in wrinkling of the thin film on the surface.

Mechanical strain can also be applied to either a substrate or a two-layer substrate system to provide a strain mismatch between the hard layer and the substrate with precise control at a desired rate.[23] For example, a hard layer can be introduced to the surface of a compliant PDMS under tensile strain. When the substrate is released from the strain, wrinkles appear on the surface if the compressive strain exceeds a critical value.[27] It is also possible to induce surface wrinkling by directly applying compressive strain to a bilayer system of platinum-deposited PDMS.[28] However, this method may induce buckling of the overall system rather than wrinkling of the thin film at the surface, depending on the material properties of the hard layer and the substrate and their thickness ratio.

Interestingly, the complex striations of wrinkles, consisting of convex and concave portions with nearly constant wavelengths, are ordered in the direction of strain and return to their original order with hysteresis when the strain is relaxed. The fact that the wrinkles formed on the PDMS respond reversibly to mechanical stimuli suggests that these substrates can be used as topography-variable materials. This variable surface topography associated with wrinkle formation can be achieved not only by mechanical stimuli such as stretching-compressing but also by physicochemical stimuli such as heat and light.[29] Therefore, the wrinkle formation and dynamic maneuverability exhibited by PDMS with the introduction of a rigid layer makes it one of the promising candidates for materials that can induce dynamic structural changes in cells.

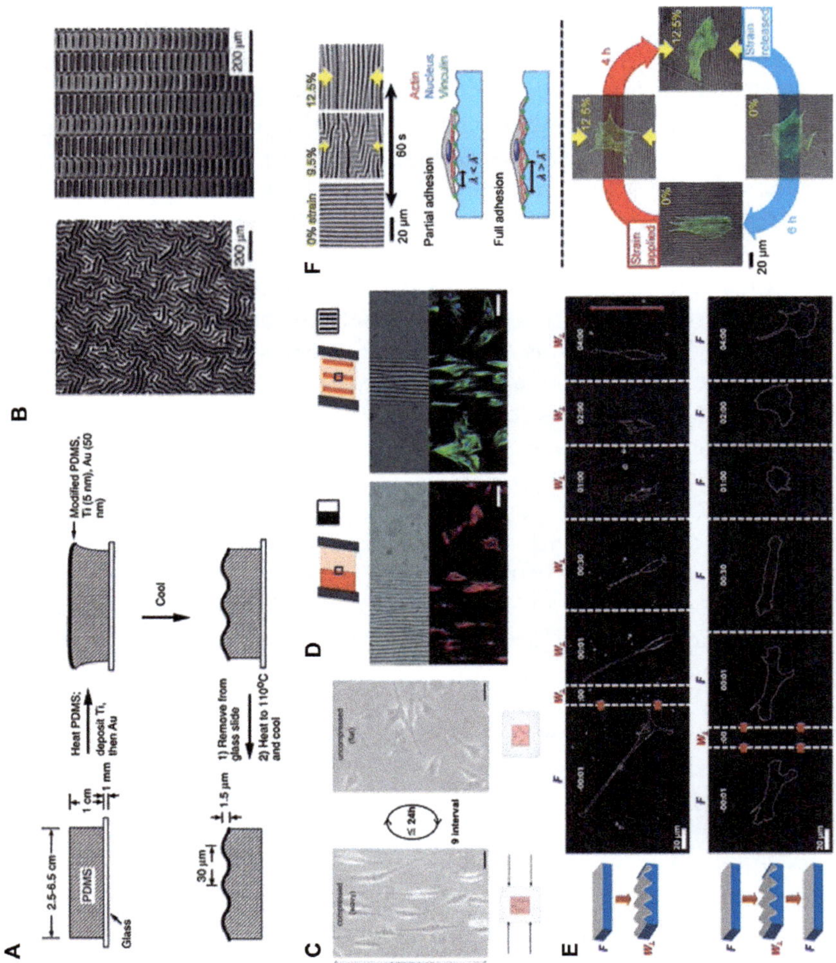
A
B
C
D
E
F
2.5-6.5 cm
PDMS
Glass
1 cm
1 mm
Heat PDMS; deposit Ti, then Au
Modified PDMS; Ti (5 nm), Au (50 nm)
Cool
30 µm
1.5 µm
1) Remove from glass slide
2) Heat to 110°C and cool
200 µm
200 µm
≤ 24h
9 interval
0% strain
9.5%
12.5%
60 s
20 µm
Partial adhesion
Full adhesion
Actin
Nucleus
Vinculin
Strain applied
Strain released
4 h
6 h
12.5%
12.5%
0%
0%
20 µm

9.2.2 PDMS with Reconfigurable Topography to Guide Cell Functions

The development of micro- and nanofabrication techniques has enabled the production of precise material surface features. Replica molding is a soft lithography technique that replicates patterns on relatively soft materials such as PDMS,[30] and the resulting substrates with specific surface topographies are widely used in *in vitro* cell biology studies to manipulate cell adhesion and the resulting cellular responses.[31] While the topography of PDMS obtained by replica molding remains constant during culture, making it useful for studying the effects of shape and its scale on cells, it is difficult to replicate a dynamic *in vitro* topography, such as pattern orientation changes, using this method.

The wavy surface features produced by compression-induced buckling of the thin brittle film on the surface of the PDMS substrate just described can induce reversible topographic changes. Lam *et al.* demonstrated that when the wrinkled PDMS substrate was uncompressed, the microscopic features were also reversibly dispersed, and the wavy topography appeared or

Figure 9.2 Design principles for PDMS with reconfigurable topography and its application to cell culture. (A) Fabrication processes of wrinkling of metal thin film deposited on PDMS substrate *via* thermal strain mismatch. (B) Optical micrographs showing representative patterns of waves that formed when the metals were deposited onto warm (~110 °C) PDMS, and the sample was then cooled to room temperature. Left: Disordered regions covered flat PDMS far from any steps or edge. Right: Rectangular ridge aligned the waves parallel to the direction of the raised portions of the PDMS. Reproduced from ref. 21 with permission from Springer Nature, Copyright 1998. (C) Optical microscopy images of C2C12 myoblasts on the reversible wavy surfaces; cell aligned along the waves and returned to random orientation on the uncompressed, flat surface within 24 h. Reproduced from ref. 32 with permission from Elsevier, Copyright 2008. (D) Spatial control of dynamic surface topography using PDMS substrates. Schematic of the spatial distribution of topography and corresponding masks used during UVO treatment. Bright field images of substrate surface and fluorescence images of hMSCs cultured on the PDMS substrates for 1 day. Reproduced from ref. 33 with permission from John Wiley & Sons, Copyright © 2013 WILEY-VCH Verlag GmbH & Co. KGaA, Weinheim. (E) Morphodynamics of a single fibroblast in response to substrate perturbations induced by dynamic topography change. Single cells are captured before and after topographical switching indicated at zero time point. Top: Flat to perpendicular wavy grating. Bottom: Flat to perpendicular wavy grating back to flat substrate. Reproduced from ref. 34 with permission from Elsevier, Copyright 2015. (F) Dynamic contact guidance of myoblasts by reversible switching of substrate topography. DIC images of wrinkle patterns captured at different strain levels and response of C2C12 myoblast to a dynamic change in the wrinkle orientation. The yellow arrows indicate the direction of axial compression. Reproduced from ref. 38 with permission from American Chemical Society, Copyright 2019.

disappeared by applying or removing compressive strain, respectively.[32] The PDMS wells for cell culture were plasma-treated to form a thin, brittle, oxidized surface layer. Subsequent application of compressive stress induced buckling of the surface, creating a smooth sinusoidal-like wave pattern perpendicular to the direction of compression. The average wavelength and amplitude during compression were 7.3 ± 0.8 μm and 670 ± 120 nm, respectively, and the formed structure was maintained in the medium. This microtopography was reconverted to the original flat surface by decompressing the substrate. These reversible topographical features were used to perform culture experiments of C2C12 myoblasts after laminin coating. On the compressed PDMS substrate, myoblasts were highly oriented along the wavy surface features, but the loss of features by decompression caused orientation disorder and finally a dynamic change to a random orientation state. Similarly, organized orientation of myoblasts was restored upon recompression of the substrate. This phenomenon was consistently observed during a 24-hour period of repeated compression and decompression (Figure 9.2C). This indicates the robustness of the reconfigurable topographical features and the plasticity of the cells. Interestingly, active topographic changes on the culture substrate surface did not interfere with differentiation into myotubes, but once myotubes were formed, reorientation did not occur. This was likely due to an increase in the amount of ECM deposited and a decrease in cell motility as differentiation matured. Although it has been well-known that cells reorient in response to fluid shear stress, multidirectional cyclic stretching, and electric fields, this was the first demonstration to show that cells respond and reorient to dynamic topographic changes.

Guvendiren and Burdick developed a PDMS with a strain-responsive wrinkling pattern that could be tuned for dynamic changes in pattern size and shape.[33] They fabricated a lamellar wrinkling pattern that depended on the degree of strain release by buckling perpendicular to the strain direction of uniaxially stretched UVO treated-PDMS with a rigid layer on the surface. By changing the degree of strain release from 20% to 100% (fully released), the pattern wavelength decreased from 40 μm to 27 μm, and the pattern amplitude increased from 0.4 μm to 5 μm. This lamellar wrinkling pattern disappeared completely when the substrate was stretched to its original strain again. When human mesenchymal stem cells (hMSCs) were cultured on this PDMS substrate, they responded to the switching of the surface pattern by changing their adhesion morphology and orientation, resulting in reversible changes in nuclear area, orientation angle, and aspect ratio. Furthermore, the authors showed that the pattern generation sites could be spatially controlled by selective UVO irradiation of the PDMS substrate through a mask (Figure 9.2D). By combining this patterning method with biaxial stretching, they succeeded in reversibly controlling not only the lamellar wrinkling pattern but also the orientation angle and order (lamella or zig-zag shape) of the pattern, allowing reversible spatial control of cell orientation angle and cell order. This approach is very attractive as a method of replicating dynamic *in vivo* topography because it can induce complex topography changes as well as spatially control the dynamic arrangement of hMSCs *in situ*.

A similar method of strain-induced buckling instabilities has been achieved by forming a rigid, uniform silicon oxide (SiO_2) layer by thermal evaporation on a uniaxially stretched PDMS substrate.[34] Releasing the strain on the SiO_2-PDMS substrate caused buckling, resulting in the formation of wrinkle structures on the surface. Depending on the difference between the initial and applied strain, the substrate could be reversibly changed between three different surface states. Here the net strain ($\Delta\varepsilon$) is defined as the difference between the resulting applied strain (ε_{app}) and the prestrain (ε_{pre}), that is $\Delta\varepsilon = \varepsilon_{app} - \varepsilon_{pre}$. When the prestrain is released, a net compressive strain ($\varepsilon_{app} < \varepsilon_{pre}$) is generated, and a uniaxial wrinkling pattern appears in the direction orthogonal to the axis of the applied strain. A tensile strain ($\Delta\varepsilon = \varepsilon_{app} - \varepsilon_{pre} > 3\%$) produced a wrinkling pattern parallel to the axis of the strain, while setting $\Delta\varepsilon = 0\%$ caused the pattern to disappear. This resulted in a substrate that could transform between a flat substrate ($A = 1.53 \pm 0.55$ nm, $R_{ms} = 0.317 \pm 0.048$ nm), a perpendicular wrinkling pattern ($A_{\perp} = 429.3 \pm 5.8$ nm; $\lambda_{\perp} = 4.95 \pm 0.36$ μm), and parallel wrinkling pattern ($A_{||} = 483.6 \pm 7.8$ nm; $\lambda_{||} = 4.78 \pm 0.16$ μm), where A, R_{ms}, and λ represent the amplitude, roughness, and wavelength, respectively. When the researcher observed cytoskeletal dynamics of 3T3 fibroblasts in response to the surface patterns by optical microscopy, the fibroblasts cultured on "flat-to-wrinkle"-type dynamic substrates showed a robust and rapid change in cellular morphology, with a decrease in circularity after 5 min (Figure 9.2E). By contrast, the steady-state overall morphology did not change significantly for flat-to-wrinkle-to-flat'-type dynamic substrates. It was noteworthy that both flat-to-wrinkle- and flat-to-wrinkle-to-flat-type dynamic substrates induced an initial transient phase of cell rounding immediately after the topographic change of the substrate. In this system, cells may have been accompanied by mechanical stimuli such as stretching and contraction as well as topographic changes, but the magnitude of those stimuli was only about 3%, a range that is apparently smaller than that of systems that provide cyclic stretching stimuli[35] and strains that affect cytoskeletal dynamics.[36,37] These results indicate that substrates that reversibly present topographic structures are useful for elucidating the dynamic aspects of cellular topographic interactions.

Linke *et al.* fabricated PDMS substrates presenting periodic wrinkles using a different approach.[38] The PDMS substrate was treated with ambient air plasma and then spin-coated with polyamic acid dissolved in *N*-methylpyrrolidone (NMP). By curing the sample at 180 °C under axial compression, a thin polyimide layer was formed, and a periodic wrinkle structure was obtained when the axial compression was removed. Interestingly, it was possible to change the direction of the wrinkles by 90° when the PDMS substrate presenting the periodic wrinkles was subjected to axial compression in a direction parallel to the wrinkle direction in air.[39] In other words, the PDMS substrate itself did not reversibly control the appearance and disappearance of topography as reported by Lam *et al.*,[32] but served as a substrate that could reversibly change topography direction (Figure 9.2F). The results obtained by culturing C2C12 myoblasts on the substrates with different wrinkle wavelengths showed that the order parameters of both cells and actin became higher with increasing wrinkle wavelength. However, these order parameters leveled off at $\lambda \geq 3.7$ μm, indicating that the characteristic wavelength (λ^*) existed in the range $2.5 < \lambda^* < 3.7$ μm. The wavelength of formed wrinkles also affected the formation of cell focal adhesion (vinculin). The focal adhesions were concentrated at the top of the wrinkles for the PDMS substrates with 1.7 and 2.7 μm wavelengths, while they were found at both the top and troughs of the wrinkles for the PDMS substrates with 3.7 and 6.3 μm wavelengths. This PDMS substrate enabled them to abruptly change the wrinkle orientation by 90° after the adhesion of C2C12 myoblasts reached a steady state. By observing the dynamic response of myoblasts to this abrupt change in topography direction, the change in cell shape and orientation, coordinated by the nematic ordering of the actin cytoskeleton, was clearly demonstrated.

PDMS substrates with the wrinkle structure just outlined feature completely reversible surface topography changes, which can be varied in terms of pattern appearance and disappearance, spatial control of pattern formation, and pattern direction changes. Such PDMS-based dynamic culture substrates are useful as a new tool for mechanobiological research.

9.3 Surface Relief Grating System

In 1995, the phenomenon of the formation of wavy unevenness in organic materials containing azobenzene when irradiated with spatially patterned light was reported[40,41] and has since attracted strong interest from both the basic science and application aspects as a novel microfabrication technique that differs in principle and characteristics from conventional lithography. Unevenness formation has been found in azobenzene-containing polymer materials, where the photoisomerization of azobenzene between *trans* (rod-like) and *cis* (bent) forms is essential. In studies by Rochon *et al.* and Kim *et al.*, it was reported that interference exposure of an amorphous polymer thin film with azobenzene side chains by a laser with a wavelength that induces photoisomerization of azobenzene results in the formation of unevenness consistent with the interference period; the resulting structure was termed a surface relief grating (SRG). In addition to the fact that the surface structure formed by SRG can be fabricated in a single step of light irradiation, the structure can be erased or reformed because it is formed by the transfer of the material itself.[42] Azobenzene-containing polymers, which have low toxicity to cells, can be used for cell culture to create dynamic substrates with reconfigurable surface topography. In this section, the methodology for realizing "dynamic" topographic changes based on azobenzene-containing polymers is described, followed by an overview of how it can be applied to cell culture.

9.3.1 Photoinduced Surface Relief as Reconfigurable Topography

Azobenzene exhibits photochromism, a reversible color change phenomenon that is based on *cis–trans* photoisomerization upon light irradiation and that is recognized for its large-scale changes in molecular structure compared to those of other photochromic molecules (Figure 9.3A). By coordinating the changes in molecular structure associated with photoisomerization, it was found that light irradiation induced macroscopic motion, a photomechanical effect.[43] Since the discovery of a different photomechanical effect, SRG,[40,41] it was found that various reliefs could be produced not only by interference exposure but also by focused laser,[44] exposure through a photomask,[45] and near-field light;[46] these are collectively termed photoinduced surface relief (PSR). PSR generally maintains a stable formed shape below the glass transition temperature of the polymer, whereas heating to above the glass transition temperature restores the original smooth surface morphology (Figure 9.3B). It has also been found that the surface morphology of the polymer matrix returns to a smooth state under appropriate uniform light irradiation conditions.[47] Continuous-wave argon ion lasers with wavelengths of 488 nm or 514 nm are often used for PSR formation, and the periodicities of the gratings can be adjusted in the range of several hundred nm to several μm, depending on the light interference conditions. A light intensity of a few hundred $mW\,cm^{-2}$ is sufficient for PSR formation, and a relief with an amplitude of several μm is formed. In most cases, the light-irradiated area becomes a concave relief shape, and mass transfer occurs from the bright area to the dark area. Although research has focused on amorphous polymers,[48] the formation of PSR has been found in many azobenzene materials, including liquid crystalline polymers,[49] sol–gel materials,[50,51] amorphous molecular materials,[52] and low-molecular-weight crystals.[53] As previously mentioned, once the surface of the azobenzene-incorporated material is structured, it can be reconstructed by light irradiation, which is in accordance with the general polarization-dependent directionality of the mass migration phenomenon. Upon irradiation with linearly polarized light, structured azobenzene-incorporated materials fluidize at room temperature, their directional flow being parallel to the polarization of the light and reshaping the initial shape of the two-dimensional structure.[54] This process, known as directional photofluidization lithography (DPL), is a light-induced, nonthermal fluidization of azobenzene-incorporated materials by light irradiation. Photofluidization refers to the photo-softening of azobenzene-incorporated materials by repetitive photoisomerization of azobenzene molecules. The coupling of prestructuration and mass transport is particularly important for the fabrication of complex 3D surface textures. The structure of preformed surface texture can be reverted or rewritten many times by means of simple optical steps.[54,55] In addition, complex preformed surface textures can be optically reshaped after fabrication, transforming the initial, regular array shape into complex shapes such as domes, mushrooms, lemons, and canoes.[56] These findings suggest that the azobenzene-incorporated material could be applied as a cell culture substrate with the potential to be used as a reconfigurable substrate for surface topography by PSR.

9.3.2 Reconfigurable Topography Based on PSR for Guiding Cell Functions

As mentioned in Section 9.2.2, PDMS systems can be used to fabricate culture substrates with erasable or reconfigurable surface structures, and research has been developed to apply dynamic topography to cultured cells. Although PDMS is a strain-responsive system with the potential to reversibly control surface topography, it requires the external application of strain and has limitations in patterning versatility. Magin and Anseth developed a polyethylene glycol-(PEG-) based photodegradable hydrogel with tunable microtopography by light irradiation and showed that the cell morphology of two-dimensionally attached hMSCs can be dynamically controlled *in situ*.[57] The use of photochemical reactions allows for high spatiotemporal selectivity and dose control. In addition, photoresponsive systems are

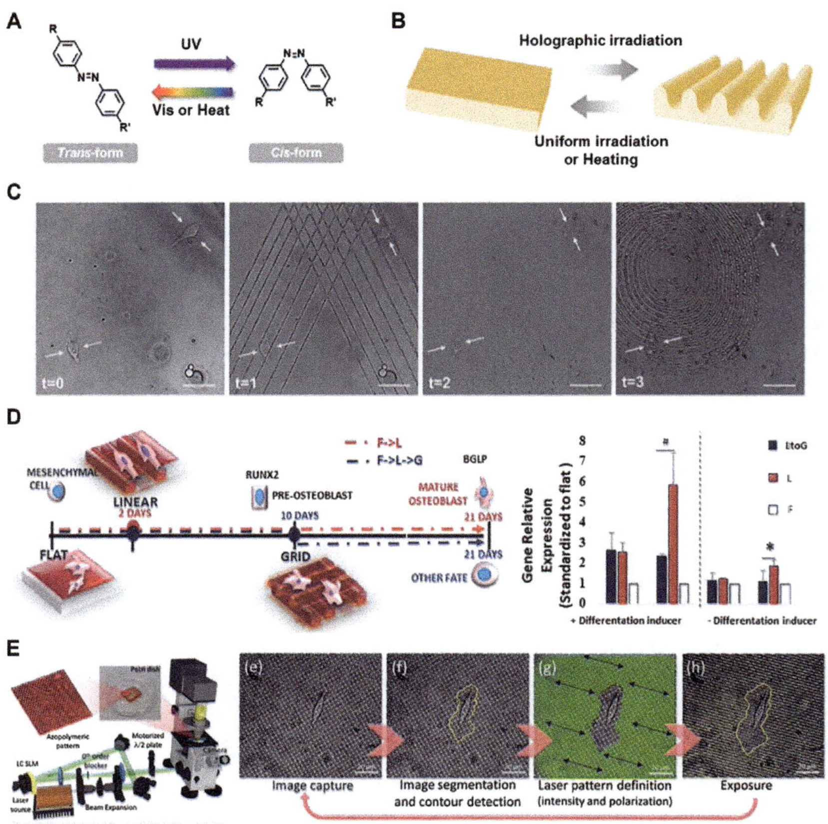

advantageous in the field of biotechnology because they can be applied instantaneously, under specific conditions, and with high precision. However, topographic changes caused by degradation using lithographic techniques limit the variation of surface patterns and make it difficult to realize reversible systems. Furthermore, the effect of degradation products on cells must be considered. Azobenzene has contributed to the most successful incorporation of photoswitches into polymers because it is easy to apply, the reaction is fast and reversible, and no side reactions are associated with isomerization of the azo groups. Baac *et al.* were the first to apply PSR, holographic surface relief grating, of azobenzene-incorporated polymer films with a Lloyd's mirror configuration to cell cultures.[58] They demonstrated that primary human astrocytes responded to the pattern of the substrate by orienting and elongating along the direction of the holographic PSR.

Netti *et al.* intensively studied the focal adhesion formation[59] and migration[60] of MC3T3 cells, an osteoblast-like cell line, to micro- and nanotopography produced by PSR, as well as cellular responses, such as stem cell self-organization and tenogenesis.[61] These studies have focused on static topographic effects, and there has been little knowledge of dynamic systems

Figure 9.3 Design principles for reconfigurable topography based on PSR and its application to cell culture. (A) Molecular photo- and thermal-dynamics of the azobenzene. *Trans*-to-*cis* photoisomerization of azobenzene upon photoirradiation (UV). *Cis*-to-*trans* back-isomerization by visible-light irradiation or thermal energy. (B) Schematic illustration of reversible surface relief grating on the azobenzene-containing polymer films. Exposure to light and temperature leads to random mixture of *cis*- and *trans*-isomers in the film. The film reduces its free energy by the formation of *trans*- and *cis*-rich microdomains. Because the two isomers have different molecular volumes, resulting in undulation and patterns on the surface. (C) On-off controllability of topographic pattern inscription. Two cells on flat pDR1m ($t = 0$), same position after first pattern inscription with Ar laser irradiation during 30 s ($t = 1$), after removing the medium and erasing the pattern by unpolarized and incoherent light irradiation for 60 s ($t = 2$), and after the second pattern inscription with Ar laser irradiation for 30 s ($t = 3$). Reproduced from ref. 63 with permission from John Wiley & Sons, Copyright © 2016 WILEY-VCH Verlag GmbH & Co. KGaA, Weinheim. (D) Schematic representation of topographic modulation of pDR1m on linear (red dashed line) and from linear to grid (blue dashed line). RT-PCR result of osteogenic markers (RUNX2 and BGLAP) on dynamic surface (L-to-G), linear patterned (L), and flat (F) topographies with and without addition of osteogenic differentiation inducer. Reproduced from ref. 64 with permission from John Wiley and Sons, Copyright © 2020 WILEY-VCH Verlag GmbH & Co. KGaA, Weinheim. (E) Schematic representation of laser projection system coupled to an inverted microscope and the sequence of operations illustrating the conceptual workflow of a prototypical use of the designed platform. Reproduced from ref. 67, https://doi.org/10.1002/adhm.202000470, under the terms of the CC BY 4.0 license https://creativecommons.org/licenses/by/4.0/.

with pattern writing and/or erasing in the presence of cells. They used poly-Disperse Red 1-methacrylate (pDR1m) as an azobenzene-incorporated polymer to study dynamic topographic manipulation and cellular behavior.[62] The formed PSR was thermally transformed to a relatively flat state by heating above the glass transition temperature of pDR1m ($T_g = \sim 85$ °C), but this was difficult to apply to cell culture. They first used a single pDR1m substrate to rewrite the topography from flat to pattern to flat and observed that NIH3T3 cells formed focal adhesions corresponding to their surface structure, as well as cell spreading and orientation. However, this was not a truly dynamic culture system because the cells were detached and dried before the rewriting of topography. Eventually, they found that an incoherent, unpolarized beam of light (15 mW intensity, 488 nm filter) integrated into a confocal microscope could be applied to erase the preformed PSR on the pDR1m substrate in an aqueous entity, inducing a topographic change from patterned to flat in the presence of cells. However, incoherent light irradiation in the culture medium was inefficient and resulted in the formation of bubble-like structures. This problem could be solved by removing the medium for about 60 s of the pattern-erasing process, which allowed the system to erase the preformed PSR with better accuracy.[63] In this way, after culturing NIH3T3 cells on a flat pDR1m substrate, they successfully achieved writing-erasing-rewriting of the pattern directly under the cells (Figure 9.3C). The possibility of manipulating cell function and fate using this dynamic platform was later investigated using hMSCs.[64] In this dynamic culture experiment, hMSCs were seeded on a flat pDR1m substrate for 2 days, and parallel gratings of rectangular regions of interest (ROIs) (2.5×495 μm) were then formed at 2.5 μm intervals by irradiation of a 514 nm argon laser in a linear pattern. After 8 days of incubation on the linear topography, the ROIs were rotated 90° and illuminated to transform them into a grid pattern. In a series of dynamic culture experiments on the hMSCs, the expression of RUNX2 and BGLAP, transcription factors related to osteogenesis, was suppressed in a system whose topography was dynamically changed from a linear pattern to a grid pattern, compared to static culture on linear pattern (Figure 9.3D). They speculated that this was due to cyclic stretching of cell bodies and nuclei induced by topographic changes. This highlights the idea that topographic changes can control not only the orientation of cells but also the fate decisions of stem cells.

PSR for azobenzene-incorporated polymers limits the height of the surface structure due to interaction with a coated substrate and can only form submicron-sized structures of less than 1 μm in height.[65] The height direction of the topography is known to affect cell contact guidance and function,[66] limiting its application as a research tool in mechanobiology. To solve this problem, Puliafito *et al.* proposed a method for spatiotemporal control of substrate topography utilizing prepatterned azo-polymeric structures.[67] A pDR1m substrate with pillar structures at different intervals was fabricated by soft lithography using a well-shaped PDMS template (4 μm×4 μm×1.3 μm) set on a glass bottom dish. The topographic changes were induced discontinuously by irradiating the target area with a properly controlled laser distribution in a modified inverted microscope. Indeed, depending on the local polarization direction and intensity of the laser, micropillars extend in a direction parallel to the laser polarization, resulting in an overall anisotropic and nonuniform microtextured surface (Figure 9.3E). To demonstrate the usefulness of such dynamic micropillars, the team performed culture experiments using breast cancer cells (MDA-MB-231) and normal kidney epithelial cells. User-defined topographic changes induced the alignment of MDA-MB-231 cells along the local deformation direction and inhibited the growth of epithelial colonies by biasing the direction of cell division. De Martino *et al.* also fabricated topographic pDR1m substrates with micropillar structures by soft lithography using a PDMS template and applied them to U2OS cells, a human osteosarcoma cell line.[68] When the pDR1m substrate with 1 μm diameter and 1.5 μm spacing of pillar structures was irradiated with a 1.5 mW polarized 532 nm laser for 5 s, the circular pillar structure transformed into an elongated nanobar structure (aspect ratio $\approx$ 0.5). This change in micropillar structure dynamically altered the membrane curvature of U2OS cells cultured on it and activated intracellular curvature-sensing processes, such as promoting local accumulation of actin fibers and actin nucleator Arp2/3 complexes. This work clearly demonstrated that the curvature-dependent reorganization of F-actin occurred about 10 min after the topographic change of the pDR1m substrate by live imaging with Lifeact-GFP-expressing U2OS cells. Based on these results, it can be concluded that the use of prestructured azobenzene-incorporated polymers not only removes the height limitation of topography but also enables the realization of diverse and complex temporal and spatial distributions of topography in combination with live imaging, which can be customized according to biological needs. Therefore, the reconfigurable topography based on PSR described here will be a useful tool for mechanobiology research.

9.4 Shape-memory Polymer System

Artificial materials can be deformed into a shape different from the original shape after molding by applying force. These materials are known as shape-memory materials because their transformed state is temporary, and they can spontaneously recover to their original shape by external stimuli. It has already been found that some alloys, ceramic materials, porous coordination polymers, and organic materials such as polymers and supramolecules exhibit shape-memory effects. In particular, shape-memory polymers (SMPs) are relatively soft and exhibit easy control and manipulation of physical properties as well as high deformability, which is a great advantage in designing biomaterials.

The ability of SMPs to induce preprogrammed shape changes is useful in designing dynamic culture substrates. In 2009, Neuss *et al.* reported for the first time on the effect of shape change in substrates in the presence of cultured cells.[69] They fabricated crosslinked poly(ε-caprolactone) (PCL) and induced a shape change in a 100% strain-fixed substrate by heating it at 54 °C for 10 s. Although the system required a severe temperature change, the contraction of the scaffold associated with the shape change was shown to induce a decrease in cell density due to cell detachment and apoptosis. In addition, the shape memory effect was expressed not only in the bulk but also on the surface of the material[70] and therefore presents a promising culture substrate for realizing dynamic topographic changes. Here, a methodology for realizing "dynamic" topographic changes based on SMPs will be described, followed by an overview of how it is currently used in cell culture.

9.4.1 Design of Programmed Topography Change Based on SMPs

In order to simplify the discussion of the mechanism of shape memory in polymers, an example of a temperature-responsive SMP is adopted (Figure 9.4A). In the case of an amorphous polymer, the transition temperature (T_{trans}) is the glass transition temperature (T_g), while in the case of a (semi)crystalline polymer, the melting temperature (T_m) and crystallization temperature (T_c) are the heating and cooling sides, respectively. When crosslinking is carried out in a certain mold, the shape of the mold is the permanent shape. When the sample is heated above T_m, the crystalline phase melts, allowing for segmental movement of the polymer chains, which can be deformed by a slight external force, similarly to rubber. When the external force is removed from the deformed state, the polymer returns to its permanent shape due to entropic elasticity. On the other hand, when the polymer is cooled below T_c while maintaining the deformation, the segmental motion of the amorphous polymer chains is strongly restricted by crystallization, and the deformed shape can be maintained even after the external force is removed. In this temporary shape, crystallization traps the polymer chains in a low-entropy state. At temperatures lower than T_m, the temporary shape is maintained, but when the crystalline phase melts by heating above T_m, the shape spontaneously recovers to the permanent shape by the driving force of entropic elasticity. In general, networks consisting of amorphous and (semi)crystalline polymers are T_g-driven SMPs and T_m-driven SMPs, respectively.

By performing this temporary shape fixation in a surface-limited manner rather than by bulk deformation of the material, a shape-memory surface that dynamically changes its surface structure can be fabricated (Figure 9.4B). For example, a substrate with a permanent surface structure, indicated by the letter A, is heated above T_{trans}, pressed into a mold with a pattern of interest, indicated by the letter B, and then cooled below T_{trans}. When the mold that was pressed into the substrate is removed, the temporary surface structure is fixed, and the original permanent surface structure is restored by heating the material above T_{trans} again. This process can induce any surface structure change from patterned to flat or from one pattern to another. Even if the same mold is used, different patterns can be fabricated by varying the force with which the mold is pressed against the substrate.[71] This surface patterning based on the shape-memory effect can efficiently immobilize nanoscale to macroscale geometries, depending on the design of the template, allowing for more diverse topography designs than the PDMS substrates and azobenzene-incorporated polymer PSR previously mentioned. Therefore, it is possible to input static or pre-programmed dynamic surface topographic changes to cells if cell culture substrates are fabricated using SMPs, providing a useful tool for mechanobiology research.

9.4.2 SMP Platforms with Programmable Topography for Guiding Cell Functions

The first application of surface shape-memory effects to dynamic cell culture substrates was shown by Davis *et al.*[72] They used an amorphous SMP from UV-curable Norland Optical Adhesive 63 (NOA-63) to achieve surface shape memory during cell culture. NOA-63 is a T_g-driven SMP, its recovery temperature can be adjusted by manipulating the crosslinking conditions,[73] and it exhibits a T_g of 30–37 °C in the wet state. To clarify the feasibility of the surface shape memory effect during cell culture, they embossed an initially flat NOA-63-based SMP substrate by mold with triangular peaks 35 to 40 μm

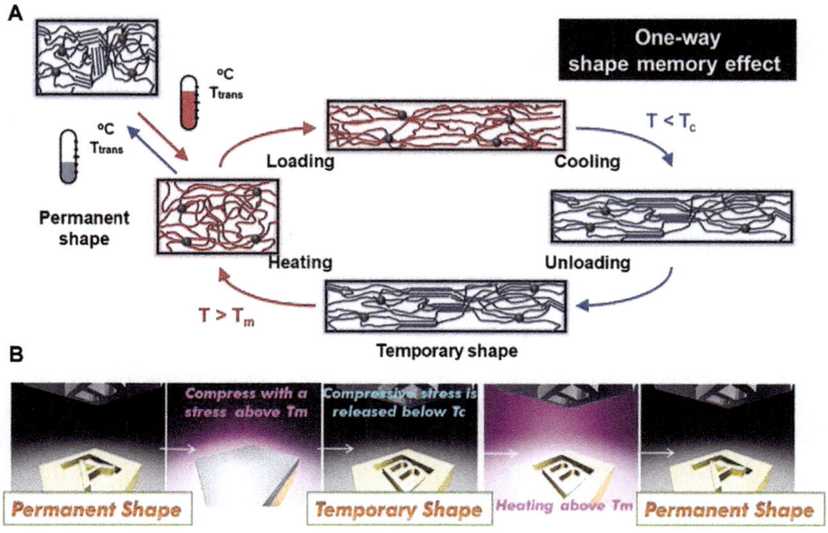

Figure 9.4 (A) Molecular-level mechanism of thermally induced one-way shape memory effect of the crosslinked semicrystalline polymer system. (B) Schematic illustration of fabrication process of shape memory surface pattern that can dynamically change from a temporary pattern to a permanent pattern.

height, 60 μm wide, and 80 μm apart. The temporary topography was stable enough (10 h at 30 °C) to allow for a culture of C3H/10T1/2 mouse embryonic fibroblasts before successfully converting the substrate back to its initial flat surface by shape memory activation at 37 °C. As a result, they found that when the surface topography of the substrate was erased, the alignment of the cells was disrupted, the angular variation increased, and the actin cytoskeleton changed. The cell viability remained above 95% before and after the topography change and temperature increase, indicating that the method is highly cytocompatible.

T_m-driven SMPs have features that allow them to induce shape changes in sharper and narrower temperature ranges than the T_g-driven varieties just described. Semicrystalline PCLs are a common choice for surface SMPs because of their high biocompatibility and the ability to tune T_{trans} to near body temperature. We have already demonstrated that the T_m of PCL can be tuned by controlling the nanoarchitecture of the crosslinked PCL, such as the number of branches and molecular weight.[74] In order to obtain PCL-based substrates for cell culture, shape-memory PCLs were prepared by crosslinking tetrabranched PCL with acrylate end groups in the presence of linear PCL telechelic diacrylate. The T_m of crosslinked PCL decreased in proportion to the increase in tetrabranched PCL content, and a relatively high melting enthalpy (ΔH_m) was maintained for each T_m. A crosslinked PCL with a T_m of about 33 °C was selected for cell culture experiments, and it was found that the temporary topography produced on that material was maintained at 32 °C and disappeared completely within 2 h after incubation at 37 °C.[75] A PCL-based shape-memory culture substrate is capable of erasing, emerging, and changing the orientation of submicron-scale grooves by mild temperature changes to cells. This type of surface topography induces directed adhesion, called contact guidance, causing cells to self-orient along the grooves. For example, NIH3T3 fibroblasts were seeded on fibronectin-coated PCL films with temporary grooved topography (1 μm wide and 300 nm high grooves spaced 2 μm apart) and cultured at 32 °C for 6 h (Figure 9.5A). With culture time, cells adhered to the surface of the substrate and then grew in elongated forms along the grooves. The topography change from grooved to flat resulted in a loss of cell alignment and a morphological change to random adhesion. A similar study using crosslinked PCL as an SMP substrate was reported by Le *et al.*[76] They also demonstrated morphological changes of hMSCs from a highly aligned to a randomly oriented state in response to surface topography changes between a 3 μm and 5 μm channel array and a flat surface. The temperature range for the change from temporary to permanent surface pattern in their group (T_{below}: 28 °C, T_{above}: 37 °C; $\Delta T = 9$ °C) was wider than that of the system reported by Ebara *et al.* (T_{below}: 32 °C; T_{above}: 37 °C; $\Delta T = 5$ °C). In these systems, topographic changes are induced by surface shape-memory effects depending on temperature changes during cell culture; thus sharp transitions indicating that the material acts within a narrow temperature range are highly desirable.

However, if the surface is designed to change the groove direction by 90° degrees, fibroblasts can gradually change their orientation to rotate 90°.[77] Therefore, the orientation and orientation state of the cells can be dynamically controlled by the SMP-based substrates. The change in orientation of adherent cells in response to topographic changes can be achieved by dynamically controlling geometries such as width, height, and spacing, even with the same material composition, creating a dramatic effect on cell adhesion. For example, smaller (500 nm) grooves reduce the time for cells to change orientation by 90° compared to larger (2000 nm) grooves.[78] In one study, cardiac progenitor cells were seeded on PCL films with temporary grooved topography (330 nm wide and 120 nm high grooves spaced 645 nm apart), and the localization of the mechanotransducer Yes-associated protein (YAP)/ transcriptional coactivator with a PDZ-binding motif (TAZ) was followed before and after topographic changes.[79] Before the topographic change, YAP/ TAZ was localized in the nucleus, whereas 90 min after the topographic change from nanogrooved to flat, YAP/TAZ was translocated to the cytoplasm (Figure 9.5B). After a certain period on the flat surface, YAP/TAZ again moved to the nucleus, and mechanical perturbations due to topographic changes and subsequent adaptation phenomena could be observed, demonstrating that cells are mechanically sensitive to dynamic topographic changes.

This approach can be applied not only to single-cell systems in which cell–cell interaction does not exist but also to multicellular systems that more closely resemble tissue structures. By culturing neonatal rat ventricular myocytes on a shape-memory substrate with a submicron-scale groove structure, Mengsteab *et al.* succeeded in realizing the formation of anisotropically oriented cardiomyocyte sheets and associated anisotropic beating with functions similar to those of the *in vivo* environment.[80] Surprisingly, when the direction of the groove was changed by 90°, the cell sheet on the substrate changed orientation while maintaining anisotropic orientation, and the beating direction also changed (Figure 9.5C). The orientation change of the sheet was accompanied by the reorientation of actin fibers and focal adhesion vinculin in response to the orthogonal change in topographic direction. These facts indicate that SMP substrates can dynamically manipulate not only single-cell behavior but also tissue-level cellular behavior, making them attractive as new substrate materials that provide a dynamic environment similar to that of an *in vivo* microenvironment.

Gong *et al.* adopted the method of mixing allyl alcohol with PCL as a plasticizer and introducing the crosslinking structure by free radical reaction in the presence of benzoyl peroxide.[81] The crosslinked PCL with a microgroove pattern of 3 μm width and 5 μm height with 3 μm spacing were prepared by thermal embossing microimprint lithography. The obtained PCLs exhibited a broad transition temperature ranging from 30 °C to 45 °C and a multistep discontinuous shape change when returning from a temporary to a permanent shape. When the crosslinked PCL with the microgroove pattern was stretched to fix its shape, they were able to control the width of the pattern, where temporary substrate with a width of 9 μm at

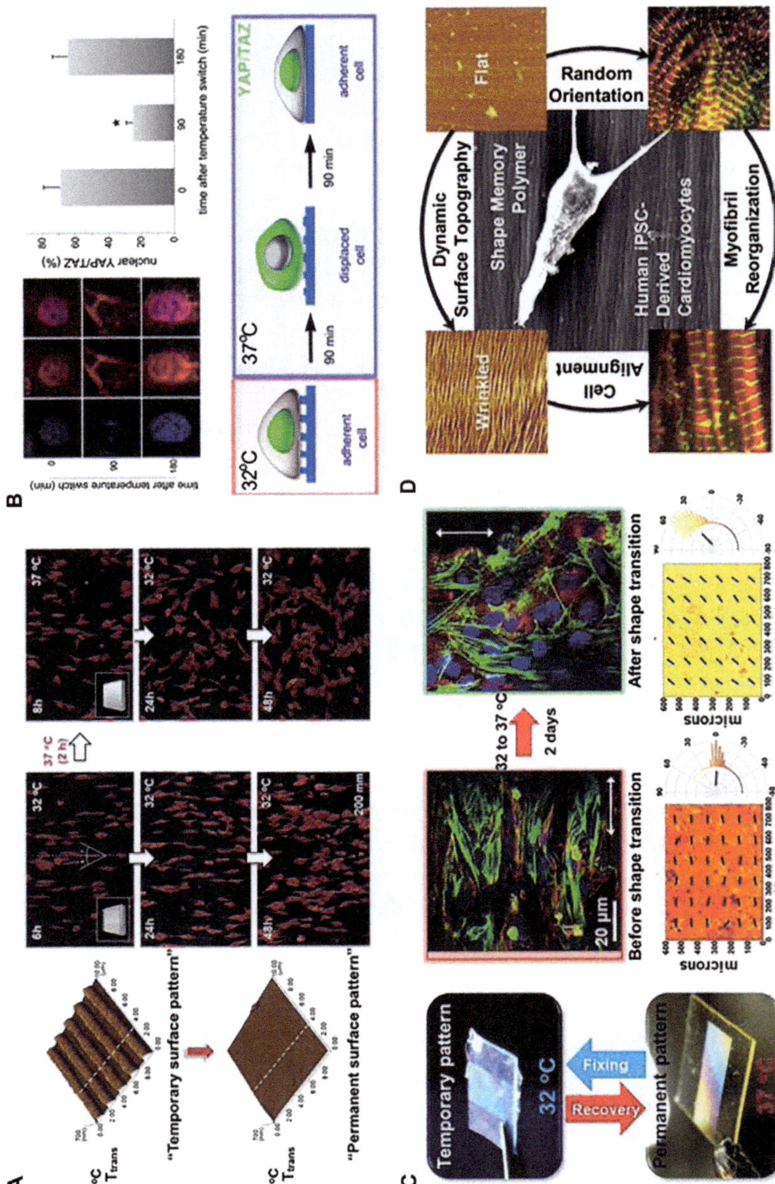
A
"Temporary surface pattern"
"Permanent surface pattern"
37 °C (2 h)
B
time after temperature switch (min)
nuclear YAP/TAZ (%)
32°C
37°C
90 min
adherent cell
displaced cell
YAP/TAZ
C
Temporary pattern
32 °C
Fixing
Recovery
Permanent pattern
37 °C
Before shape transition
After shape transition
32 to 37 °C
2 days
20 µm
microns
D
Dynamic Surface Topography
Shape Memory Polymer
Human iPSC-Derived Cardiomyocytes
Myofibril Reorganization
Random Orientation
Cell Alignment
Flat
Wrinkled

32 °C was changed to 7 μm, 4.5 μm, and 3.1 μm at 35 °C, 38 °C, and 41 °C, respectively, by heat-induced shape recovery. The materials were used as dynamic culture substrates for rat bone marrow mesenchymal stem cells (rBMSCs). The decrease in the width of the stretched microgroove with increasing temperature from 32 °C to 41 °C changed the alignment of patterned cells and cytoskeleton orientation to a more compact morphology. Gene expression and western blot analysis showed that the stretched and elongated cell shape caused by the decrease in microgroove width enhanced gene expression (GATA-4, TNNt2, and β-MHC) related to myogenic differentiation in the absence of any inducer. A similar approach has been reported for the fabrication of crosslinked PCL with micropillars.[82] These micropillars responded to environmental temperature, moving from horizontal (0°) alignment at 32 °C to vertical (90°) alignment at 41 °C while generating a vertical restoring force. The dynamic change in pillar structure induced adhesion and highly directional cell elongation of rBMSCs. Furthermore, it has been reported that SMP substrates with topography can change the culture dimension from a 2D flat shape to a 3D rolled shape and that the selected topography can produce cell structures that mimic vascular structures with an endothelium layer in the inner layer and an oriented vascular smooth muscle cell layer in the outer layer.[83]

This pattern fabrication on SMP substrates requires a patterned template mold, which is pressed against the substrate surface to fix the temporary topography. A method has been proposed that can program the surface topography with a mold-free system using the buckling of the surface rigid layer similar to that used on PDMS. Yang *et al.* fabricated T_g-driven SMPs by

Figure 9.5 (A) Topographic surface images of crosslinked PCL substrate observed by AFM before and after shape memory surface transition by a 37 °C heating (left). Fluorescence microscopy images of NIH3T3 fibroblasts seeded on the substrate with the temporary grooved surface (right). Dynamic transition of cell orientation was demonstrated by shape memory surface activation. Reproduced from ref. 75 with permission from John Wiley & Sons, Copyright © 2012 WILEY-VCH Verlag GmbH & Co. KGaA, Weinheim. (B) Dynamic change of YAP/TAZ localization against the surface topography transition from temporarily nanogrooved to permanently flat surface in cardiac progenitor cells cultured on the crosslinked PCL substrate. Reproduced from ref. 79 with permission from American Chemical Society, Copyright 2014. (C) Photographs of shape memory surface where the direction of the nanogrooves can be dynamically changed orthogonally (left). The results of alignment and contraction analysis of cardiac cell monolayer reorganization induced by orthogonal nanogroove transition of crosslinked PCL substrate (right). Reproduced from ref. 80 with permission from Elsevier, Copyright 2016. (D) Progressive myofibril reorganization and alignment of human iPSC-derived cardiomyocytes on an SMP-based dynamic nanotopographic substrate. The SMP coated with polyelectrolyte multilayers can induce a flat-to-wrinkle surface topographical transition when triggered by a change in incubation temperature. Reproduced from ref. 85 with permission from American Chemical Society, Copyright 2020.

crosslinking two monomers, *tert*-butyl acrylate and butyl acrylate, with tetraethylene glycol dimethacrylate crosslinker.[84] The SMP substrates treated with temporarily fixed uniaxial strain were coated with gold. The resultant surface maintained a flat morphology for a long time at 30 °C, but when the temperature was changed to 37 °C, the spontaneous formation of surface microwrinkles was observed as the substrate shrank. By culturing human adipose-derived stem cells (hASC), which exhibit a flat surface structure, at 30 °C and then changing the temperature to 37 °C, the researchers showed that the cell orientation could be dynamically controlled from a random state to an aligned state, demonstrating for the first time that SMP-based wrinkle formation can be used as a dynamic culture substrate. This SMP substrate with tunable topography from flat to wrinkled was applied to human-induced pluripotent stem cell–derived cardiomyocytes (hiPSC-CMs), and the dynamic topography change not only preferentially promoted the orientation and elongation of hiPSC-CMs but also successfully tracked time-dependent changes in cell morphology and myofibril reorganization (Figure 9.5D).[85] At present, this surface topography transition is one-way, but in the future, materials will be developed to achieve reversible changes similar to those of PDMS.

Fibrous ECMs are often found *in vivo*, and the topographic nature of the fibers plays an important role in cell polarization and migration. It is possible to design SMP fiber substrates in which the constituents themselves act as topography. Tseng *et al.* fabricated a fibrous SMP by electrospinning thermoplastic polyurethane, which consists of a hard segment of polyhedral oligomeric silsesquioxane (POSS) and a biodegradable and amorphous soft segment of polylactide/caprolactone copolymer.[86] When the fibrous SMP was subjected to strain, the fibers became oriented in the direction of the strain, and the aligned state could be fixed as a temporary shape. The temporarily strain-fixed substrate maintained its aligned fiber state at 30 °C but reverted to the original, randomly oriented fiber state when the temperature was raised to 37 °C. When hASCs were cultured on a temporarily strain-fixed fibrous SMP substrate, cell elongation and orientation along the fibers were observed, whereas the original, randomly oriented fiber substrate did not show any cell orientation and promoted random adhesion. In dynamic culture experiments, hASCs successfully changed from an aligned state to a random state by switching fiber orientation due to shape-memory activation, indicating that fibrous SMP substrates can be used as dynamic culture substrates with topography change. For fibrous SMP substrates, it is possible to design substrates that change from random to aligned, as well as those that can change the fiber orientation from aligned to random, as previously described. Wang *et al.* fabricated a fibrous SMP substrate aligned as a permanent shape by electrospinning.[87] The aligned fibers were subjected to uniaxial strain perpendicular to their orientation axis to fix them as a temporary shape, resulting in a fibrous SMP substrate with a relatively random orientation. The temporary shape could be changed from 30 °C to 37 °C in the presence of cells to induce a dynamic topographic change from a random orientation to an aligned state. Due to the thermally induced random-to-aligned or aligned-to-random orientation change of the fibers, both the human fibrosarcoma cell line HT-1080 and multipotent murine mesenchymal stem cell line C3H/10T1/2 showed featured morphology depending on the orientation state of the fibers, as well as polarized motility on aligned fibers and nonpolarized motility on random fibers. This could be a useful tool to study tumor invasion and cancer metastasis processes. These fibrous SMP substrates are T_g driven and require a relatively long time (~5 h) to complete the change in fiber orientation. T_m-driven SMPs with a similar fibrous structure have also been developed[88] and may be an effective tool for evaluating cellular responses to rapid topographic changes. Furthermore, fibrous SMP substrates, which are mold free and resemble an *in vivo* ECM structure, have attractive potential in terms of recapitulating the physiological environment.

However, in most SMP systems, temperature change is required to induce surface topography change, and the effect of temperature change on cells cannot be ignored for accurate evaluation of cell behavior and function. To address this issue, it has been necessary to identify surface topography change by photothermal conversion of highly dispersed photothermal materials in an SMP matrix. By introducing gold nanorods[89,90] or titanium nitride nanoparticles,[91] it was possible to construct a system that induces surface topography change by near-infrared light and/or sunlight, and spatially controlling the topography change and associated cell orientation change was realized by applying local light irradiation.[89] This system will be useful as a technique to further investigate the cellular response to topography changes in more detail because it does not require changes in cell culture temperature, only spatiotemporal light control. Thus the development of an SMP-based culture platform that can dynamically change its structural properties is expected to be further developed in the field of tissue engineering as a novel culture substrate for sophisticated manipulation of cells and tissues.

9.5 Conclusions and Future Perspective

In the environmental surrounding cells *in vivo*, diverse topographies exist and contribute not only to cell structural role but also to function and fate determination. Furthermore, it is becoming evident that pathological processes are accompanied by dynamic and complex topographic changes. This chapter highlighted material-based platforms that can replicate *in vivo* microenvironments. Substrates that display not only a static topography but also dynamic topographical changes on various scales can be designed by using the buckling phenomenon of PDMS for wrinkle formation, by PSR formation represented by azobenzene-containing polymers, and by shape-memory programming of SMPs. With optimization of materials and operating conditions to induce topographic changes, the topography of the substrate can be changed in a reversible or preprogrammed manner to the cells while remaining in an attached state. Cells adhered to the substrate recognize such topographic changes and dynamically change their cell morphology to adapt to the topographic changes, leading to actin reorganization and focal adhesion remodeling.

From the viewpoint of mechanobiology, it is extremely important to elucidate the sensing mechanism of topography and its changes at the molecular level. It can be inferred that the topographic change of the substrate not only plays a structural role but also applies a local force to the cell and transforms the intracellular mechanical field. However, there has been no quantitative evaluation of this, and it is thus essential to construct an evaluation and analysis system to clarify the whole picture. In addition, the materials discussed in this chapter are all very stiff and have supraphysiological mechanical properties, so there is a need to develop materials that can induce soft and dynamic topographic changes similar to those in biological tissues. Technology for materials is advancing, and it is becoming possible to realize complex and dynamic changes in structures of any scale. Although a proof-of-concept regarding the potential of materials has been demonstrated, a system that mimics the real biological processes has not yet been realized. Future research should focus on utilization of dynamic culture substrates that can change their topography to elucidate various physiological phenomena and contribute to mechanobiology research.

Acknowledgements

This work was supported by AMED-PRIME (JP19gm5810017, KU) from the Japan Agency for Medical Research and Development and Grant-in-Aid for Transformative Research Areas (A) (JP20329906, KU) from the Japan Society for Promotion of Science.

References

1. S. J. Liliensiek, P. Nealey and C. J. Murphy, *Tissue Eng., Part A*, 2009, **15**(9), 2643–2651.
2. D.-H. Kim, E. A. Lipke, P. Kim, R. Cheong, S. Thompson and M. Delannoy, *et al.*, *Proc. Natl. Acad. Sci.*, 2010, **107**(2), 565–570.
3. L.-A. Harsan, C. Dávid, M. Reisert, S. Schnell, J. Hennig and D. von Elverfeldt, *et al.*, *Proc. Natl. Acad. Sci.*, 2013, **110**(19), E1797–E1806.
4. N. Schubert, M. Axer, M. Schober, A.-M. Huynh, M. Huysegoms and N. Palomero-Gallagher, *et al.*, *Front. Neuroanat.*, 2016, **10**, 51.
5. G. Sharifi, A. M. Pajavand, S. Nateghinia, T. E. Meybodi and H. Hasooni, *Front. Hum. Neurosci.*, 2020, **13**, 472.
6. M. Modo, K. Mellodew, D. Cash, S. E. Fraser, T. J. Meade and J. Price, *et al.*, *NeuroImage*, 2004, **21**(1), 311–317.
7. H. N. Kim, A. Jiao, N. S. Hwang, M. S. Kim, D. H. Kang and D.-H. Kim, *et al.*, *Adv. Drug Delivery Rev.*, 2013, **65**(4), 536–558.

8. R. Malik, P. I. Lelkes and E. Cukierman, *Trends Biotechnol.*, 2015, **33**(4), 230–236.
9. M. W. Conklin, J. C. Eickhoff, K. M. Riching, C. A. Pehlke, K. W. Eliceiri and P. P. Provenzano, *et al.*, *Am. J. Pathol.*, 2011, **178**(3), 1221–1232.
10. P. P. Provenzano, K. W. Eliceiri, J. M. Campbell, D. R. Inman, J. G. White and P. J. Keely, *BMC Med.*, 2006, **4**(1), 38.
11. M. Klaas, T. Kangur, J. Viil, K. Mäemets-Allas, A. Minajeva and K. Vadi, *et al.*, *Sci. Rep.*, 2016, **6**(1), 27398.
12. Y. Kanzaki, Y. Yamauchi, M. Okabe, F. Terasaki and N. Ishizaka, *Circulation*, 2012, **125**(5), 738–739.
13. K. Uto, J. H. Tsui, C. A. DeForest and D.-H. Kim, *Prog. Polym. Sci.*, 2017, **65**, 53–82.
14. C. Bonnans, J. Chou and Z. Werb, *Nat. Rev. Mol. Cell Biol.*, 2014, **15**(12), 786–801.
15. R. N. Palchesko, L. Zhang, Y. Sun and A. W. Feinberg, *PLoS One*, 2012, **7**(12), e51499.
16. V. Panzetta, S. Fusco and P. A. Netti, *Proc. Natl. Acad. Sci.*, 2019, **116**(44), 22004–22013.
17. J. Fu, Y.-K. Wang, M. T. Yang, R. A. Desai, X. Yu and Z. Liu, *et al.*, *Nat Meth*, 2010, **7**(9), 733–736.
18. R. Kaunas, P. Nguyen, S. Usami and S. Chien, *Proc. Natl. Acad. Sci. U. S. A.*, 2005, **102**(44), 15895–15900.
19. L. Chen, C. Yan and Z. Zheng, *Mater. Today*, 2018, **21**(1), 38–59.
20. A. K. Patel, M. W. Tibbitt, A. D. Celiz, M. C. Davies, R. Langer and C. Denning, *et al.*, *Curr. Opin. Solid State Mater. Sci.*, 2016, **20**(4), 202–211.
21. N. Bowden, S. Brittain, A. G. Evans, J. W. Hutchinson and G. M. Whitesides, *Nature*, 1998, **393**(6681), 146–149.
22. K. Efimenko, M. Rackaitis, E. Manias, A. Vaziri, L. Mahadevan and J. Genzer, *Nat. Mater.*, 2005, **4**(4), 293–297.
23. P.-C. Lin, S. Vajpayee, A. Jagota, C.-Y. Hui and S. Yang, *Soft Matter*, 2008, **4**(9), 1830–1835.
24. W. M. Choi, J. Song, D.-Y. Khang, H. Jiang, Y. Y. Huang and J. A. Rogers, *Nano Lett.*, 2007, **7**(6), 1655–1663.
25. M.-W. Moon, S. H. Lee, J.-Y. Sun, K. H. Oh, A. Vaziri and J. W. Hutchinson, *Proc. Natl. Acad. Sci.*, 2007, **104**(4), 1130–1133.
26. N. Bowden, W. T. S. Huck, K. E. Paul and G. M. Whitesides, *Appl. Phys. Lett.*, 1999, **75**(17), 2557–2559.
27. D.-Y. Khang, H. Jiang, Y. Huang and J. A. Rogers, *Science*, 2006, **311**(5758), 208–212.
28. T. Ohzono and M. Shimomura, *Phys. Rev. B*, 2004, **69**(13), 132202.
29. F. Li, H. Hou, J. Yin and X. Jiang, *Sci. Adv.*, 2018, **4**(4), eaar5762.
30. G. M. Whitesides, E. Ostuni, S. Takayama, X. Jiang and D. E. Ingber, *Annu. Rev. Biomed. Eng.*, 2001, **3**(1), 335–373.
31. A. M. Greiner, A. Sales, H. Chen, S. A. Biela, D. Kaufmann and R. Kemkemer, *Beilstein J. Nanotechnol.*, 2016, **7**, 1620–1641.

32. M. T. Lam, W. C. Clem and S. Takayama, *Biomaterials*, 2008, **29**(11), 1705–1712.
33. M. Guvendiren and J. A. Burdick, *Adv. Healthcare Mater.*, 2013, **2**(1), 155–164.
34. P. Pholpabu, S. Kustra, H. Wu, A. Balasubramanian and C. J. Bettinger, *Biomaterials*, 2015, **39**, 164–172.
35. M. Morioka, H. Parameswaran, K. Naruse, M. Kondo, M. Sokabe and Y. Hasegawa, *et al.*, *PLoS One*, 2011, **6**(10), e26384.
36. X. Zhu, K. L. Mills, P. R. Peters, J. H. Bahng, E. H. Liu and J. Shim, *et al.*, *Nat. Mater.*, 2005, **4**(5), 403–406.
37. A. J. Ehrlicher, F. Nakamura, J. H. Hartwig, D. A. Weitz and T. P. Stossel, *Nature*, 2011, **478**(7368), 260–263.
38. P. Linke, R. Suzuki, A. Yamamoto, M. Nakahata, M. Kengaku and T. Fujiwara, *et al.*, *Langmuir*, 2019, **35**(23), 7538–7551.
39. T. Ohzono and M. Shimomura, *Langmuir*, 2005, **21**(16), 7230–7237.
40. P. Rochon, E. Batalla and A. Natansohn, *Appl. Phys. Lett.*, 1995, **66**(2), 136–138.
41. D. Y. Kim, S. K. Tripathy, L. Li and J. Kumar, *Appl. Phys. Lett.*, 1995, **66**(10), 1166–1168.
42. H. S. Kang, S. Lee and J.-K. Park, *Adv. Funct. Mater.*, 2011, **21**(23), 4412–4422.
43. H. S. Blair, H. I. Pague and J. E. Riordan, *Polymer*, 1980, **21**(10), 1195–1198.
44. S. Bian, L. Li, J. Kumar, D. Y. Kim, J. Williams and S. K. Tripathy, *Appl. Phys. Lett.*, 1998, **73**(13), 1817–1819.
45. N. C. R. Holme, L. Nikolova, S. Hvilsted, P. H. Rasmussen, R. H. Berg and P. S. Ramanujam, *Appl. Phys. Lett.*, 1999, **74**(4), 519–521.
46. T. Ikawa, T. Mitsuoka, M. Hasegawa, M. Tsuchimori, O. Watanabe and Y. Kawata, *et al.*, *J. Phys. Chem. B*, 2000, **104**(39), 9055–9058.
47. X. L. Jiang, L. Li, J. Kumar, D. Y. Kim and S. K. Tripathy, *Appl. Phys. Lett.*, 1998, **72**(20), 2502–2504.
48. A. Natansohn and P. Rochon, *Adv. Mater.*, 1999, **11**(16), 1387–1391.
49. P. S. Ramanujam, N. C. R. Holme and S. Hvilsted, *Appl. Phys. Lett.*, 1996, **68**(10), 1329–1331.
50. C. Fiorini, N. Prudhomme, G. de Veyrac, I. Maurin, P. Raimond and J. M. Nunzi, *Synth. Met.*, 2000, **115**(1), 121–125.
51. L. Frey, B. Darracq, F. Chaput, K. Lahlil, J. M. Jonathan and G. Roosen, *Opt. Commun.*, 2000, **173**(1), 11–16.
52. H. Nakano, T. Takahashi, T. Kadota and Y. Shirota, *Adv. Mater.*, 2002, **14**(16), 1157–1160.
53. H. Nakano, T. Tanino and Y. Shirota, *Appl. Phys. Lett.*, 2005, **87**(6), 061910.
54. S. Lee, H. S. Kang and J.-K. Park, *Adv. Mater.*, 2012, **24**(16), 2069–2103.
55. S. L. Oscurato, M. Salvatore, P. Maddalena and A. Ambrosio, *Nanophotonics*, 2018, **7**(8), 1387–1422.
56. S. Lee, H. S. Kang, A. Ambrosio, J.-K. Park and L. Marrucci, *ACS Appl. Mater. Interfaces*, 2015, **7**(15), 8209–8217.

57. C. M. Magin and K. S. Anseth, *Small*, 2013, **9**(4), 578–584.
58. H. Baac, J.-H. Lee, J.-M. Seo, T. H. Park, H. Chung and S.-D. Lee, *Mater. Sci. Eng., C*, 2004, **24**(1), 209–212.
59. C. F. Natale, M. Ventre and P. A. Netti, *Biomaterials*, 2014, **35**(9), 2743–2751.
60. M. Ventre, C. F. Natale, C. Rianna and P. A. Netti, *J. R. Soc., Interface*, 2014, **11**(100), 0687.
61. M. Iannone, M. Ventre, L. Formisano, L. Casalino, E. J. Patriarca and P. A. Netti, *Nano Lett.*, 2015, **15**(3), 1517–1525.
62. C. Rianna, A. Calabuig, M. Ventre, S. Cavalli, V. Pagliarulo and S. Grilli, *ACS Appl. Mater. Interfaces*, 2015, **7**(31), 16984–16991.
63. C. Rianna, L. Rossano, R. H. Kollarigowda, F. Formiggini, S. Cavalli and M. Ventre, *et al.*, *Adv. Funct. Mater.*, 2016, **26**(42), 7572–7580.
64. S. De Martino, S. Cavalli and P. A. Netti, *Adv Healthcare Mater*, 2020, **9**(13), e2000470.
65. O. Kulikovska, K. Gharagozloo-Hubmann, J. Stumpe, B. D. Huey and V. N. Bliznyuk, *Nanotechnology*, 2012, **23**(48), 485309.
66. J. S. Chua, C.-P. Chng, A. A. K. Moe, J. Y. Tann, E. L. K. Goh and K.-H. Chiam, *et al.*, *Biomaterials*, 2014, **35**(27), 7750–7761.
67. A. Puliafito, S. Ricciardi, F. Pirani, V. Čermochová, L. Boarino and N. De Leo, *et al.*, *Adv. Sci.*, 2019, **6**(14), 1801826.
68. S. De Martino, W. Zhang, L. Klausen, H.-Y. Lou, X. Li and F. S. Alfonso, *et al.*, *Nano Lett.*, 2020, **20**(1), 577–584.
69. S. Neuss, I. Blomenkamp, R. Stainforth, D. Boltersdorf, M. Jansen and N. Butz, *et al.*, *Biomaterials*, 2009, **30**(9), 1697–1705.
70. B. A. Nelson, W. P. King and K. Gall, *Appl. Phys. Lett.*, 2005, **86**(10), 103108.
71. M. Ebara, K. Uto, N. Idota, J. M. Hoffman and T. Aoyagi, *Soft Matter*, 2013, **9**(11), 3074–3080.
72. K. A. Davis, K. A. Burke, P. T. Mather and J. H. Henderson, *Biomaterials*, 2011, **32**(9), 2285–2293.
73. A. M. DiOrio, X. Luo, K. M. Lee and P. T. Mather, *Soft Matter*, 2011, **7**(1), 68–74.
74. K. Uto, K. Yamamoto, S. Hirase and T. Aoyagi, *J Controlled Release*, 2006, **110**(2), 408–413.
75. M. Ebara, K. Uto, N. Idota, J. M. Hoffman and T. Aoyagi, *Adv. Mater.*, 2012, **24**(2), 273–278.
76. D. M. Le, K. Kulangara, A. F. Adler, K. W. Leong and V. S. Ashby, *Adv. Mater.*, 2011, **23**(29), 3278–3283.
77. M. Ebara, K. Uto, N. Idota, J. M. Hoffman and T. Aoyagi, *Int. J. Nanomed.*, 2014, **9**(Suppl 1), 117–126.
78. M. Ebara, M. Akimoto, K. Uto, K. Shiba, G. Yoshikawa and T. Aoyagi, *Polymer*, 2014, **55**(23), 5961–5968.
79. D. Mosqueira, S. Pagliari, K. Uto, M. Ebara, S. Romanazzo and C. Escobedo-Lucea, *et al.*, *ACS Nano*, 2014, **8**(3), 2033–2047.
80. P. Y. Mengsteab, K. Uto, A. S. T. Smith, S. Frankel, E. Fisher and Z. Nawas, *et al.*, *Biomaterials*, 2016, **86**, 1–10.

81. T. Gong, K. Zhao, G. Yang, J. Li, H. Chen and Y. Chen, *et al.*, *Adv. Healthcare Mater.*, 2014, **3**(10), 1608–1619.
82. W. Li, T. Gong, H. Chen, L. Wang, J. Li and S. Zhou, *RSC Adv.*, 2013, **3**(25), 9865–9874.
83. D. Liu, T. Xiang, T. Gong, T. Tian, X. Liu and S. Zhou, *ACS Appl. Mater. Interfaces*, 2017, **9**(23), 19725–19735.
84. P. Yang, R. M. Baker, J. H. Henderson and P. T. Mather, *Soft Matter*, 2013, **9**(18), 4705–4714.
85. S. Sun, H. Shi, S. Moore, C. Wang, A. Ash-Shakoor and P. T. Mather, *et al.*, *ACS Appl. Mater. Interfaces*, 2020, **12**(19), 21450–21462.
86. L.-F. Tseng, P. T. Mather and J. H. Henderson, *Acta Biomater.*, 2013, **9**(11), 8790–8801.
87. J. Wang, A. Quach, M. E. Brasch, C. E. Turner and J. H. Henderson, *Biomaterials*, 2017, **140**, 150–161.
88. E. Niiyama, K. Tanabe, K. Uto, A. Kikuchi and M. Ebara, *Fibers*, 2019, **7**(3), 20.
89. Q. Shou, K. Uto, W.-C. Lin, T. Aoyagi and M. Ebara, *Macromol. Chem. Phys.*, 2014, **215**(24), 2473–2481.
90. Q. Shou, K. Uto, M. Iwanaga, M. Ebara and T. Aoyagi, *Polym. J.*, 2014, **46**(8), 492–498.
91. S. Ishii, K. Uto, E. Niiyama, M. Ebara and T. Nagao, *ACS Appl. Mater. Interfaces*, 2016, **8**(8), 5634–5640.

CHAPTER 10

Curvature Mechanobiology

RYOSUKE MATSUZAWA, MIDORI TAKEUCHI,
TAKUYA NISHINA AND TADAHIRO YAMASHITA*

Department of System Design Engineering, Faculty of Science and Technology, Keio University, 3-14-1 Hiyoshi, Kohoku, Yokohama 223-8522, Japan
*Email: yamashita@sd.keio.ac.jp

10.1 Introduction

Cells, the smallest unit of living organisms, have recently been reported to change behaviour according to the curvature of the surface to which they adhere,[1–4] as if they knew the surrounding landscape. Unlike human hikers, individual cells are currently not thought to have a visual sense sufficiently developed to measure the surrounding topography. Instead, numerous observations in mechanobiology have suggested that such cellular behaviour changes are enabled *via* intrinsic mechanical forces. This chapter introduces recent findings and perspectives on how cells sense and respond to curved surfaces at much larger than single-cell scale; these findings will very likely serve as a basis for biomaterials design in the future.

10.1.1 Motivation to Investigate Cellular Behaviours on Curved Surfaces

At a range of scales, multicellular organisms have highly organised hierarchical structures that are closely linked to the vital body functions. Supported by the proliferation of microfabrication technologies, multidisciplinary studies in the field known as mechanobiology have revealed that

Biomaterials Science Series No. 12
Material-based Mechanobiology
Edited by Jun Nakanishi and Koichiro Uto

Published by the Royal Society of Chemistry, www.rsc.org

the topographical features of a cell's surrounding environment have a significant influence on its fate.[5–7] While cellular responses to the substrate topography at the cellular and subcellular scales on planar surfaces have been thoroughly investigated, curved structures, especially those above cellular scale (*i.e.* larger than several tens of micrometres), have drawn little attention in the field for a long time. However, recent studies in the last decade have clearly demonstrated that surface curvature at such scales has a considerable influence on cellular behaviours such as migration, growth, and even differentiation (Figure 10.1A),[1–4] an influence unreported on planar surfaces. These recent studies have been motivated not only by molecular biologists but also by bioengineers aiming to guide cellular behaviours by designing structured scaffolds to produce functional tissues and organs in a self-organising way.[4,8,9]

To elucidate the general rules and mechanisms of how cells respond to such subtle curvature of the extracellular environment, several different approaches have been adopted. Some posit a biomolecular mechanism that translates the surface curvature into intracellular signalling,[2] while others account for the phenomenon based on simple physical laws.[10–13] This range of approaches to finding the basis for cellular curvature sensing arises from the fact that the curvatures of interest, ranging from micrometre to millimetre scales, are several orders of magnitude larger than individual molecules but still too small to allow us to neglect the properties and interactions of the molecular architectures within cells. Integrating knowledge from different disciplines to clarify the role of surface curvature as a biological input would contribute to an understanding of the cellular sensing mechanism in mechanobiology and also to the establishment of design strategies for bioengineering scaffolds in the field of tissue engineering. Much work still needs to be done, however.

In this chapter, we thus gather cutting-edge findings from different viewpoints on cellular responses to curved surfaces in order to give a current perspective on the cellular curvature–response mechanism. We first introduce a definition of curvature; then, following an overview of basic techniques for fabricating curved surfaces, we review several hypotheses on biomolecular mechanisms and physical frameworks accounting for the cellular curvature–response mechanism.

10.1.2 Curvature as a Fundamental Geometrical Parameter

Various terms, such as *concavity*, are used to express the curviness of a surface, yet their correspondence to mathematical formulation is sometimes ambiguous. We thus start by providing some geometrical formulations to characterise the material curvature discussed in this chapter.

When a curved line that is differentiable at any given points exists on a two-dimensional plane, curvature at an arbitrary point on the line is defined as the inverse of the radius of the osculating circle (Figure 10.1B). This definition positions curvature as a local parameter characterising how much

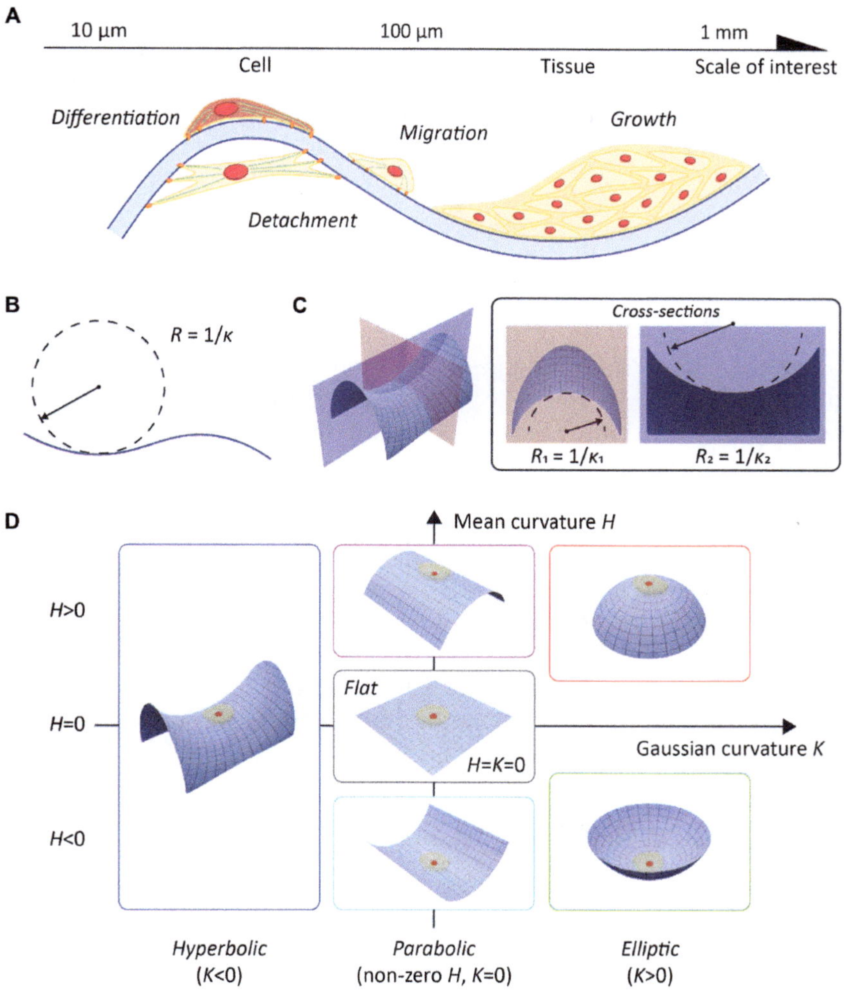

Figure 10.1 Schematics of surface curvature discussed in this chapter: (A) Overview of cellular behaviours on curved surfaces above micrometre scale. (B) Schematic representation of curvature of a line. (C) Schematic illustration of surface curvature in three-dimensional space. κ_1 and κ_2 represent the principal curvatures at the saddle point. (D) Representative local surface curvatures around the point of interest (red point) categorised according to the signs of mean curvature H and Gaussian curvature K.

the line deviates from the tangential line at a given point. This basic notion of curvature can be extended to characterise the curvature of a two-dimensional surface in a three-dimensional space (Figure. 10.1C). At a point p on the surface, the maximal and minimal curvatures $\kappa_1(p)$ and $\kappa_2(p)$ can be defined as principal curvatures if we consider all the possible contours formed by the surface and planes including p. When considering the curvature of an interface between two phases, these principal curvatures take both positive and negative values, indicating whether the contour deforms outwards or inwards at p. In this chapter, positive and negative values are used to represent the convexity and concavity of the material surface, respectively. Using these principal curvatures, the mean curvature H and the Gaussian curvature K are defined as[14]

$$H = \frac{\kappa_1 + \kappa_2}{2} \tag{10.1}$$

$$K = \kappa_1 \cdot \kappa_2 \tag{10.2}$$

H, together with K, symbolises the local landscape of the surface (Figure 10.1D). For example, the curvature at any point on a sphere is characterised by constant $H > 0$ and $K > 0$, whereas that on a spherical concave is characterised by constant $H < 0$ and $K > 0$. When a surface has at least one principal direction in which the contour tangent to the surface is straight (*e.g.* cylindrical surfaces), the local curvature is represented by nonzero H and $K = 0$. Negative K represents saddle-like geometries. At any point of a flat surface, clearly, $H = K = 0$. It should be noted that K is an identical parameter representing the intrinsic local geometry, while the sign of H depends on the side on which the surface is considered. The point p is called *elliptic* if $K(p) > 0$, *hyperbolic* if $K(p) < 0$, and *parabolic* if $K(p) > 0$ and $H(p) \neq 0$. These parameters are used to illustrate the surface curvature as necessary in the following sections.

10.2 Engineering Curved Surface

Industrial products have been adopted for cell culture on curved surfaces, by virtue of their wide availability, ease of handling, and precision in shape. Glass cylinders[15–22] and spheres[20,23] have been the most popular types. Their shapes are often transferred to other materials by replica moulding. While these commercial products serve as an affordable scaffold or mould, there are limited choices in shape, size, and composition. To widen the variety of curved surfaces to investigate cellular behaviours, various fabrication techniques have been applied (Figure 10.2, Table 10.1). This section reviews the notable techniques for engineering curved surfaces with various types of curvatures for cell culture use.

10.2.1 Basic Moulding Techniques

Moulding allows researchers to replicate structured substrates at reasonable cost. This is of practical importance for biological studies, where the use of disposable materials is desirable to avoid the risk of contamination and wear on substrate structures due to repeated use.

Replica moulding is a common technique to replicate microstructured surfaces, including curved ones,[24] thanks to its simplicity, high throughput,

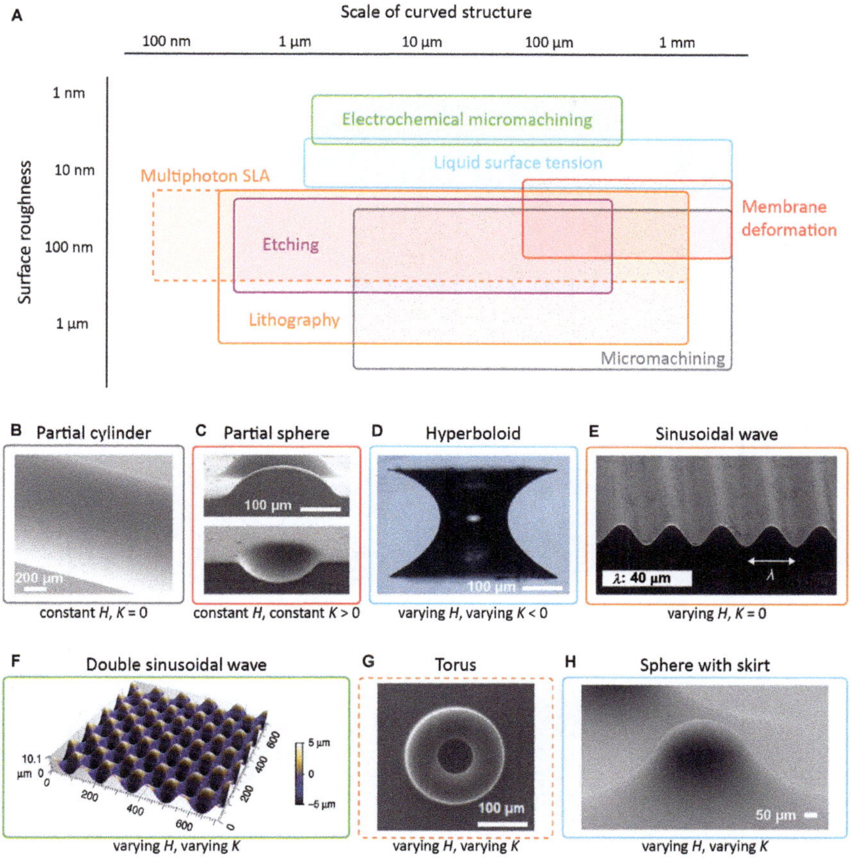

Figure 10.2 Fabrication techniques to create curved surfaces at micron scales: (A) An overview of fabrication techniques to produce curved surfaces in terms of the scale and surface roughness of the fabricable substrate. (B–H) Representative curved structures to analyse cellular behaviours: (B) Glass substrate with a semicylindrical concave (constant $H<0$, $K=0$) replicated from a metal mould fabricated by micromachining.[91] (C) Spherical concave and convex surfaces (constant H, constant $K>0$) moulded from deformed elastic membrane.[25] (D) Hyperbolic surface ($K<0$) formed by liquid surface tension.[37] (E) Sinusoidal wavy substrate (locally varying H and $K=0$) fabricated by X-ray lithography.[31] (F) Double-sinusoidal wavy substrate (locally varying H and K) fabricated by electrochemical micromachining.[44] (G) Torus structure (locally varying H and K) fabricated by multiphoton polymerisation–laser lithography.[51] (H) A sphere-with-skirt structure (SWS, locally varying H and K) fabricated *via* liquid surface tension.[35] (B) reproduced from ref. 91, with permission from Elsevier, Copyright 2016. (C) reproduced from ref. 25 with permission from the Royal Society of Chemistry. (D) reproduced from ref. 37 with permission from AAAS, Copyright 2019 The Authors. (E) reproduced from ref. 31 with permission from Elsevier, Copyright 2015. (F) reproduced from ref. 44, https://doi.org/10.1038/s41467-018-06494-6, under the terms of the CC BY 4.0 license https://creativecommons.org/licenses/by/4.0/. (G) reproduced from ref. 51 with permission from Elsevier, Copyright 2020. (H) reproduced from ref. 35 with permission from Biophysical Society, Copyright 2018.

Table 10.1 Representative fabrication methods to create curved substrates at above cellular scale for cell culture.

		Structure type [ref. number]				
Principle	Method	Spherical	Cylindrical	Sinusoid	Others	Material [ref. number]
Commercial product	Direct use for cell culture	[20,23]	[15–22]			Glass
	Replicated to another material		[28,30,36,100]	[36]		From steel to hydroxyapatite [28] From FeCrAl [30], nylon [36] and steel [100] to PDMS
Interface control	Membrane deformation	[25,33]	[32,34]			Photoresist [25] Thermoplastic polymer [32,33] Silicone membrane [34]
	Liquid surface tension	[38,39]	[26,40,54]		Sphere with skirt [35] Hyperboloid [37]	Water-glycerol [38] Photoresist [39,40,54] PDMS [26,35,37]
Machining and etching	Micromachining	[41,42]	[91]			Metal
	Electrochemical micromachining			[45]	Double sinusoid [44] flat-bottomed, round-transitioned groove [45]	Metal
	Chemical etching		[46]			Silicon
Lithography	Deep X-ray			[31]		Acrylic resin
	Stereolithography	[48]		[50]	Torus [51]	Photocurable resin

and low cost. Typically, a precursor solution is poured onto a master mould, cured, and removed from the master mould, whose surface structure it now replicates. The roughness of the replicated surface is basically determined by that of the master mould. Silicone elastomer polydimethylsiloxane (PDMS) is the most popular material for general cell culture use because it has enough transparency for microscopy, flexibility in terms of surface chemical modification, and gas permeability. Photocurable materials such as SU-8 are also used.[25,26] Elsewhere, hydroxyapatite, fabricated *via* slurry casting, is popular as an inorganic scaffold for bone tissue engineering.[27,28] While PDMS moulding is capable of transferring the master structure even at submicrometre scale, the surface roughness of slurry-cast products is usually limited by the particle size in the precursor.

While the choices in materials for simple replica moulding are limited, there are some more versatile ways to replicate a master structure. Hot embossing techniques,[29] for example, are useful in shaping a wide range of thermoplastic polymers. Though hot embossing involves technical complexity in precise tuning of the shaping conditions, it provides great flexibility in the choice of polymer compositions and allows the transfer of structures with high aspect ratios. Semicylindrical[30] and sinusoidal[31] structures ranging from ten to hundred micrometre scales have been created on polymeric materials for cell culture use.

10.2.2 Controlling Interfacial Geometries

Various types of curvature based solely on physical and geometrical laws can be seen in nature. The simplest way to create such curved surfaces is to replicate such interfaces of forces using moulding techniques.

For example, applying pressure onto a membrane with stable boundary conditions deforms it into a smooth curve. Park *et al.* combined an air pressure control and microfabricated frames to bend a thin PDMS membrane smoothly.[25] Smooth convex and concave structures for cell culture were then successfully fabricated by moulding the membrane-formed surface shape onto another material (Figure 10.2C). Pressure thermoforming is an advanced variation of this idea that has been used to shape thermoplastic polymers.[32,33] Temporal tuning of the surface curvature of the cell-adhering surface has also been demonstrated in a similar pressure deformation scheme.[34] These techniques offer the advantages of flexible tuning of curved structures with radii around several hundred micrometres.

Liquid forms a smooth meniscus under stable boundary conditions in accordance with the balance of interfacial tensions; this meniscus can be also employed as a template to fabricate curved structures. PDMS precursor has proven to be a useful material for this scheme. It is typically poured onto materials prestructured with features such as microgrooves,[26] spheres,[35] fibres,[36] and cylinder edges,[37] then thermally cured to fix the structure (Figure 10.2D, H). Though this scheme has some limitations in terms of accuracy and fabricable geometries, its simplicity provides researchers with an inexpensive way to obtain curved surfaces of very low roughness. It is thus one of the most popular techniques to fabricate curved substrates for cell culture at larger than a hundred micrometre scale. To compensate for its shortcomings in accuracy, printing techniques can be employed.[38] Thermal reflow, which applies heat in order to reshape the material surface *via* surface tension, is capable of fabricating smooth curved structures at smaller scales.[39,40]

10.2.3 Machining and Etching

Assisted by computer control, micromachining techniques enable reproducible and precise fabrication, which is thus employed to create moulds with smooth curved surfaces even at nanometre scale.[41] In addition to conventional mills, laser micromachining is also used for fabrication.[42] Metallic materials such as aluminium or steel alloy are commonly used to make master moulds for repeated use. The surface structure needs to then be transferred to other materials favourable for cell culture *via* moulding techniques (Figure 10.2B).

Electrochemical micromachining, which gradually lyses a metal immersed in electrolyte solution by applying a direct current, is an alternative technique for fabricating a mould with curved structures. The contactless principle can avoid errors arising from machining force and size effect, resulting in an extremely smooth surface of the finished product. On the other hand, it is unable to offer the level of accuracy in shaping structures at above micrometre scales that milling techniques can provide, nor can it match their throughput.[43] This technique, combined with a mass transport–limited dissolution, has been employed to fabricate a substrate double-sinusoidal wavy structures for cell culture (Figure 10.2F).[44,45]

Etching technologies are also used in conjunction with photolithography to create curved surfaces at micrometre scales. In one technique, a chemical etchant dissolves the material through mask patterns formed *via* photolithography. Because of the diffusivity of the etchant, the target surface is eroded isotropically, resulting in the rounded edges of the fabricated patterns. While the wet etching process is suitable for this type of isotropic etching, some dry etching processes, including reactive ion etching, can also be employed.[46] Silicon-based materials such as Si and SiO_2 are commonly processed using both the wet and dry etching techniques. Isotropic etching provides a way to fabricate patterns ranging from nanometre to several hundred micrometre scales, but the throughput is relatively low due to the necessity of the prior photolithography step.

10.2.4 Lithography

While photolithography is frequently employed to fabricate a supportive material for curvature fabrication,[25,26,34,39,46] it can also fabricate curved surfaces itself. Grey-tone lithography, which modulates the amount of exposure light reaching a surface using a specialised photomask patterned with intensity gradients, produces curved surfaces.[47] Deep X-ray can be also employed to fabricate a mould with curved structures (Figure 10.2E).[31] More recently, the use of a light-based 3D printing technology called stereolithography (SLA) is becoming popular in bioengineering. SLA constructs 3D objects by piling up structured layers of a photocurable resin solidified by patterned light illumination. The resolution is already enough to create smooth structures at scales around several hundred micrometres.[48] At an even smaller scale, an application of multiphoton polymerisation can achieve resolutions of several 10 nm,[49] which provides a very smooth curved surface for the characterisation of cellular behaviour (Figure 10.2G).[50,51] The throughput of SLA is lower than that of other simple fabrication methods, and the costs of the instruments and their operating environments are high. It should also be noted that unreacted resin needs to be removed at the end of SLA fabrication, and this process sometimes limits the fabricable design. However, the potential of SLA more than makes up for its limitations, and it will doubtless soon be one of the standard fabrication techniques in this field.

10.2.5 Technical Perspectives

As we have seen, fabrication of curved surfaces for cell culture has been facilitated by the application of microfabrication techniques. Simple fabrication methods exploiting liquid surface tension would appear to offer a simple and popular standard process, but such methods have intrinsic limitations in terms of fabricable scales, materials, and structures. Structural limitations are particularly apparent if we consider hierarchical structures. In this regard, however, the currently emergent additive manufacturing technologies including SLA are beginning to offer more choices. For example, metal-based additive manufacturing technologies provide versatile ways of creating porous materials with complex curvatures for practical biomedical applications.[52] It is an intrinsic problem that the more complex the material structure becomes, the more difficult it becomes to observe cells optically, due to the limited balance of resolution, brightness, and working distance of microscopy. These fabrication methods, as well as cutting-edge observation techniques optimised for observing cells positioned deep within structured materials, will be the technical basis for investigating the mechanism through which cells respond to the surface curvature, as introduced in the following sections.

10.3 Biomolecules Related to Cellular Curvature Sensing

10.3.1 What Senses Cellular-scale Curvature?

While most molecular-level insights into cells have so far been obtained from *in vitro* experiments using planar substrates, cells *in vivo* are embedded in matrices. To reveal the influence of the various hierarchical structures found *in vivo*, at scales ranging from molecular to whole-organ, numerous studies using structured surfaces have now also been conducted *in vitro*. While the influence of curvature at subcellular scales has drawn considerable attention in molecular biology, recent studies in bioengineering and mechanobiology have demonstrated that curved structures at larger-than-multicellular scales, which have been considered "flat" for individual cells, have a significant effect on cellular behaviours such as migration, proliferation, and differentiation.[1–4] Such biological effects of surface geometry cannot be neglected for practical applications in tissue engineering[4] and stem cell technology.[53] Cells seem to have a molecular mechanism that translates such subtle curvature of the extracellular environment into mechanobiological signals including YAP nuclear translocation.[54] Nonetheless, this mechanism is yet to be understood. In this section, we thus focus on the potential roles of several molecules and cellular components in surface curvature sensing (Figure 10.3A) from a biological perspective.

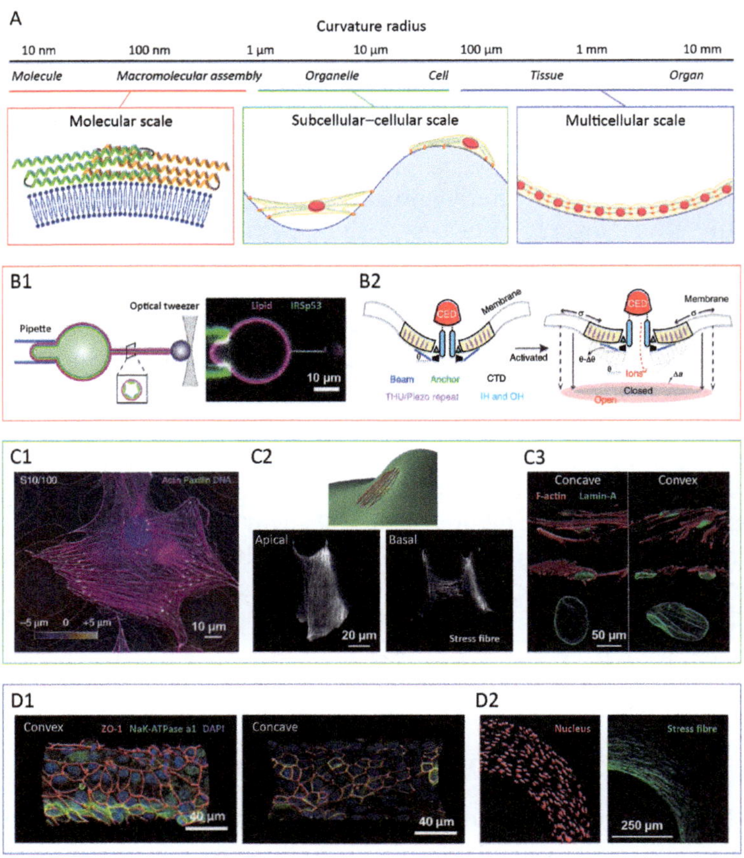

10.3.2 Curvature Recognition *via* Molecular Affinity

The geometry of intracellular compartments such as vesicles at submicrometre scale is modulated by proteins. Hence, curvature sensing *via* cellular membrane deformation is taken as one of the possible—although as yet largely unexplored—hypotheses to explain cellular curvature sensing. Recent advances in experimental techniques have enabled detailed observation of these molecules modulating the curvature of lipid bilayers.[55,56] We introduce two types of molecules that are often referred to in the context of membrane curvature – curvature-sensitive proteins and ion channels[6] – and review their roles as potential curvature mechanotransducers.

10.3.2.1 Curvature-sensitive Proteins

A cellular membrane is a physical interface distinguishing the individual cellular system from the environment, and its geometry is dynamically modulated by various curvature-responsive proteins that control the membrane deformation by forming a "coat"-like structure.[57] Dynamin[58] and clathrin[59] are typical examples of such proteins. Although their specialised response to membrane curvature plays crucial roles in vesicle formation,

Figure 10.3 Representative molecules and intracellular components speculated to be involved in cellular curvature sensing: (A) Schematic illustrations of curvature sensing mechanisms across scales. (B1) I-BAR protein IRSp53 (green) shows a preferential distribution inside the thin tube formed by pulling the lipid bilayer (purple), indicating a higher affinity for large negative curvature.[55] (B2) A potential influence of membrane curvature on ion-channel gating at nanometre scale.[68] (C1) Distribution of FAs on a double sinusoidal substrate. FAs tend to disperse densely in the convex regions to position the nucleus on the concave region.[44] (C2) Two different subpopulations of SFs in a cell adhering to a hyperbolic surface ($K<0$).[35] (C3) Observation of nucleus of mesenchymal stem cells cultured on curved surfaces. Nuclei with increased lamin-A (green) flattened by actin filaments (red), correlating to osteogenic differentiation, were observed on the convex surface.[48] (D1) Expression of tight junction protein ZO-1 (red) was enhanced on positive curvature.[26] (D2) Growing tissue confined by pores. SF networks mechanically linked by cell–cell adhesive structures maintain tissue integrity.[93] (B1) reproduced from ref. 55, https://doi.org/10.1038/ncomms9529, under the terms of the CC BY 4.0 license https://creativecommons.org/licenses/by/4.0/. (B2) reproduced from ref. 68 with permission from Elsevier, Copyright 2018. (C1) reproduced from ref. 44, https://doi.org/10.1038/s41467-018-06494-6, under the terms of the CC BY 4.0 license https://creativecommons.org/licenses/by/4.0/. (C2) reproduced from ref. 35 with permission from Elsevier, Copyright 2018. (C3) reproduced from ref. 48, https://doi.org/10.1002/advs.201600347, under the terms of the CC BY 4.0 license https://creativecommons.org/licenses/by/4.0/. (D1) reproduced from ref. 26 with permission from Elsevier, Copyright 2018. (D2) reproduced from ref. 93, https://doi.org/10.1371/journal.pone.0036336, under the terms of the CC BY 4.0 license https://creativecommons.org/licenses/by/4.0/.

these proteins are highly unlikely to be associated with cellular curvature sensing of curvatures with radii above several hundred micrometres because the scale of their polymerised structure is no larger than nanometre scale. Another class of well-known molecular curvature sensors is Bin/amphiphysin/Rvs (BAR) domain proteins.[60] Having crescent-shaped BAR domains with different properties, BAR domain proteins bind preferentially to membranes with a specific range of curvature.[61] F-BAR proteins sense positive membrane curvature at the nanometre scale and induce F-actin formation,[62] whereas I-BAR proteins sense negative membrane curvature of a few hundred nanometres (Figure 10.3B1).[55] One would naturally assume that BAR domain proteins function as curvature mechanosensors at larger scales.[2] To the best of our knowledge, however, no BAR domain proteins have yet been reported to sense curvatures with radii at or above one micrometre. These proteins remain worthy of attention as many more are yet to be discovered.[61] Septin, meanwhile, is currently known as the sole example to show a specific affinity towards micrometre-scale membrane curvature in mammalian cells.[63] Although they are not acknowledged in the context of substrate curvature recognition, septins sense curvature with their amphipathic helices[63] and show a tendency to concentrate at membranes with asymmetric curvature.[64] They could thus be proposed as one of the candidate mechanosensing molecules to detect curved geometry at above the cellular scale.

10.3.2.2 Mechanosensitive Channels

Mechanosensitive (MS) channels embedded in cellular membranes respond to numerous mechanical stimuli as molecular force sensors. Since their discovery in the 1980s,[65] their mechanosensitivity has been investigated by early structural biologists focusing on two major bacterial MS channels, MscS and MscL, and, later, on mammalian MS channels, TREK, TRAAK, and Piezo1/2.[66] Mammalian MS channels are speculated to be involved in curvature recognition[2] because they control ion flux, initiating intracellular signalling pathways. Though the opening of MS channels was originally thought to be driven by cytoskeletal tension transmitted to the membrane,[67] more recent studies suggest that it also responds to the local membrane curvature (Figure 10.3B2).[68] This insight introduced the possibility that MS channels play a role in cellular curvature sensing.

However, at the current stage, it seems unlikely that MS channels directly couple gentler membrane curvatures to ion flux. Though their sensitivity to membrane curvature at the nanometre scale has been reported,[68,69] simulation work suggests that a curvature gentler than 10 μm^{-1} (*i.e.* larger than 100 nm in radius) has little effect on the channel activity.[70] There is a slight possibility that the membrane tension modulated by physicochemical and cytomechanical factors[71] controls the gating function of MS channels *via* a curvature-induced osmotic pressure shift. However, whether directly or indirectly, the contribution of cytoskeleton and adhesive structures to membrane tension[67,72] seems to have far more influence. Given these factors, we do not currently speculate that MS channels are primarily involved in the sensing of curvatures with radii of more than a few micrometres.

10.3.3 Role of Intracellular Components

Cells sense the mechanical properties of the environment primarily through cell–matrix adhesions firmly coupled to the cytoskeleton and nucleus. Thanks to this structural integrity, cells transmit forces to intracellular components to orchestrate various cellular behaviours.[73] In order to identify the primary mechanisms translating extracellular curvature into biochemical signals, it is essential to decipher the individual roles of key intracellular components. We now explore these as candidate curvature mechanosensors.

10.3.3.1 Focal Adhesions

Focal adhesions (FAs) are essential adhesive multiprotein structures binding the cell to the extracellular matrix; they modulate cellular motion on structured surfaces.[74] Through force-dependent behavioural changes in the related molecules, FAs transduce applied forces into intracellular biochemical signalling and thus serve as the primary mechanosensors of a cell.[73] Despite their importance in cellular mechanosensing, only a few studies have discussed their contributions in the sensing of surface curvatures with radii much greater than the size of a single cell. FA size, distribution, and the applied tension are influenced differently by convex and concave curved surfaces (Figure 10.3C1).[44,48,75] Paxillin[44] and vinculin,[48] the intracellular components of FAs required for their maturation, tend to be observed in research to date, but their contribution to cellular curvature sensing remains unexplored. On the other hand, very few studies have focused on the differential roles of integrin heterodimers[76] in curvature sensing. Though cell migration speed is obviously influenced by substrate curvature,[25,31,44,77,78] integrin traffic and the roles of integrin subtypes have not been investigated. It is interesting that some cells form a bridge or spider-like shape with a limited adhesive area on concaves;[48,78,79] this evokes the idea that a highly tensile force is applied to FAs in such areas. The contribution of surface curvature to activation of integrin β_1,[80] and the elevated mRNA expression of integrin α_2 and α_5[79] has been reported for substrates with heterogeneous curvature. These observations, as well as the accumulated knowledge in the field, suggest the necessity of FAs in cellular-scale curvature sensing. However, further quantitative and systematic characterisation of their morphology and turnover is required to assess their contribution in detail.

10.3.3.2 Stress Fibres

Stress fibres (SFs) are bundles of contractile actin filaments firmly associated with cellular mechanosensitivity.[81] Fibrous elements including SFs were shown to align longitudinally along the axis of cylindrical curvature in early studies, which led to a theory by Dunn and Heath,[15] explaining that cells elongate in a direction minimising the cytoskeletal strain imposed by curvature. This theory was supported by later observations of fibroblasts cultured on cylindrical surfaces.[16,17] It was, however, proven insufficient by the presence of short SFs aligning in the circumferential direction.[18,19] The solution to this disagreement was only found with the development of other numerical theories, as introduced in the following section.

From the biological side, further insights into cellular orientation on curved surfaces was provided by the identification of two different SF subpopulations: long "apical" SFs locating above the nucleus, which align along the cylinder axis, and short "basal" SFs sitting below the nucleus, which align in the circumferential direction (Figure 10.3C2).[22,35] Variations in the balance of these SFs resulted in a shift in cellular alignment,[22] a phenomenon consistent with an improved theory by Biton and Safran considering both the strain and contractility of SFs.[10] So far, the molecular mechanism controlling the complex SF arrangement to minimise the bending penalty or contractility remains unclear.[45] Nevertheless, the mechanism here could perhaps be mediated by potential mechanosensitive components constituting SFs such as actin filament[82] and myosin II.[75,83] These studies show the significant role of actin cytoskeletons as a major component supporting cell bodies and highlight the necessity to understand the biomolecular basis of cytoskeletal organisation to elucidate the cellular curvature-sensing mechanism.

10.3.3.3 Nuclei

The cell nucleus, the largest and stiffest organelle, has attracted attention for its potential role as mechanosensor.[84] At the middle of the cell body, the nucleus senses the extracellular mechanical environment through forces transmitted *via* actin cytoskeletons and the linker of the nucleocytoskeleton complex.[85] The nucleus changes its structural and chemical organisation in response to loading force[86] and dynamically alters cellular behaviour by regulating gene expression.

Observations of lamin-A, a nucleoskeletal protein giving mechanical strength to nuclei, have suggested the involvement of the nucleus in curvature-dependent cellular behaviours. Numerous studies have reported curvature-induced cellular differentiation.[23,48,79,87,88] One study has reported that convex surfaces can cause significant deformation of the nucleus and enhanced lamin-A expression on convex surfaces (Figure 10.3C3) and that these phenomena correlate to the osteogenic differentiation of mesenchymal stem cells.[48] Another study demonstrated a unique cellular migration mode, for which it coined the term "curvotaxis", on a wavy curved surface;[44] curvotaxis was suppressed under lamin-A knockdown or actomyosin inhibition. These studies highlight the importance of loading force upon the nucleus as an initial trigger of curvature-dependent cellular behaviours. The downstream mechanism of nuclear deformation and its connection to nuclear mechanotransduction pathways[89] in terms of curvature sensing remain unclear and need to be quantitatively investigated.

10.3.4 Role of Intercellular Connections

Recent studies have suggested the importance of cell–cell adhesion in controlling collective cellular behaviour on curved surfaces. For example, cellular alignment[22] and the arrangement of cortical actin[90] on curved surfaces have been sensitised by matured cell–cell adhesion. Collective cellular tension can cause catastrophic detachment of the tissue from curved surfaces.[54,91,92] Also, a curvature-dependent expression change in cadherin,[90] which forms adherens junctions and in ZO-1 (Figure 10.3D1)[26] and occludin,[39] which form tight junctions, implies the involvement of intercellular connection to control the cellular behaviours on curved surfaces. On the other hand, some studies report a trivial change in cadherin expression on convex surfaces[39] and a negligible contribution of cadherin to curvature sensing at curvature radii above 40 μm.[30] These fragmentary pieces of information might indicate that intercellular junctions amplify their response to the surface curvature, potentially *via* orchestrated cytoskeletal forces across the linked cells (Figure 10.3D2),[9,54,93] but they do not necessarily indicate that the junctions are involved in the intrinsic curvature-sensing mechanism.

10.3.5 Current Perspectives

Recent studies have shown that mechanically coupled intracellular elements, including FAs, actin cytoskeletons, and nuclei, are involved in determining the cellular response to surface curvature. However, the molecular mechanism underlying it remains unclear, and the involvement of membrane-curvature-sensitive proteins has not yet been confirmed. While intercellular connections are likely to sensitise collective behaviour to the surface curvature, their role in fundamental curvature sensing needs further investigation. Most of the curvature-transducing candidates discussed in this section seem to be influenced by extracellular curvature *via* tension, in either direct or indirect ways, due to their tight functional interdependence. Hence, further technical improvements in spatiotemporal observation of intracellular force/strain in living cells would help us to grasp their individual contribution. Despite the difficulties, revealing the molecular mechanism of cellular curvature sensing would be a key step in understanding cellular behaviours on structured materials.

10.4 Modelling Cellular Behaviours on Curved Surfaces

Aiming to better understand the mechanobiological basis of cellular behaviours, numerous *in silico* models have been proposed. Such studies aim to provide a physical basis for our understanding of the mechanism underlying various vital phenomena by treating cells, organelles, and molecular complexes as physical components.[94] This methodology has also been used to elucidate the basis of cellular behaviours on curved surfaces. These trials are also expected to contribute to the design of artificial scaffolds for tissue engineering[95] by providing insights and methodologies to quantitatively predict how cells behave on steric scaffolds and how they form tissues in a self-organising way.[96] As a fundamental geometrical parameter, curvature has drawn much attention in cellular models. This section introduces several different approaches to modelling the behaviour of cells and tissues on curved surfaces (Figure 10.4). For general *in silico* modelling approaches, extensive reviews are mentioned.[97,98]

10.4.1 Continuum Modelling Approaches

Continuum mechanics, which basically handles the target as a uniform body and solves the problem in equilibrium, has been applied with the aim of understanding cellular behaviours on curved surfaces. This discipline also allows coupled analysis with a range of other factors, including interfacial tension and chemical potential, to capture behaviour either at single-cell or at tissue scale. This section reviews the recent trials to draw physical

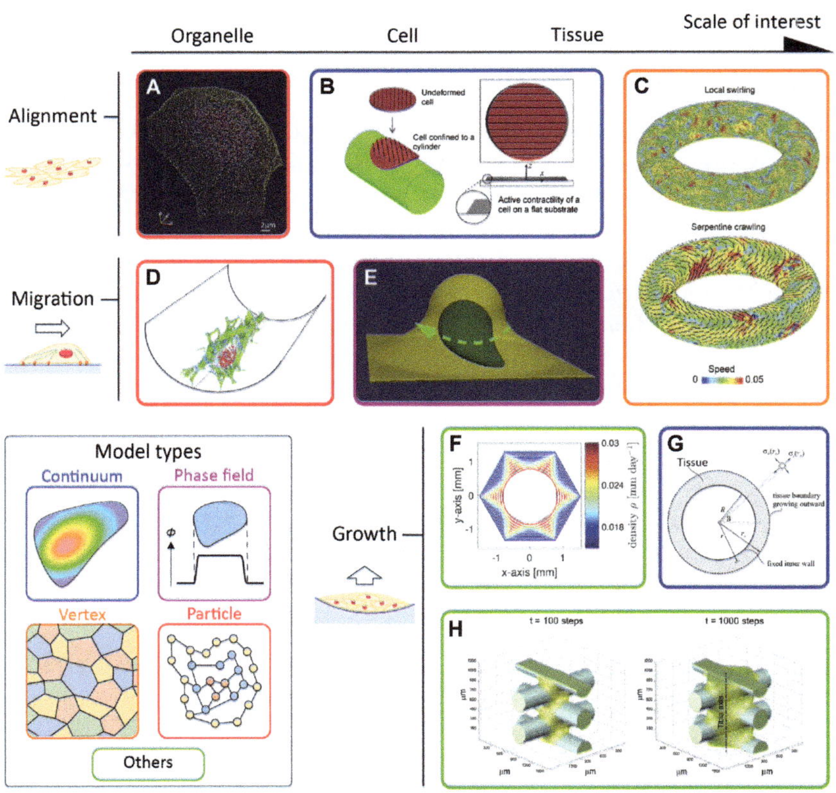

principles from the complex cellular and tissue behaviours seen with the continuum approach, especially focusing on cellular alignment, locomotion, and growth. The contributions of related disciplines, such as phase-field modelling, are also introduced where relevant.

10.4.1.1 Cellular Alignment Analysed Based on Internal Stress

Cellular alignment is an essential factor that determines the functionality of tissues and organs. While cellular alignment on curved surfaces has been under investigation for several decades, the theoretical basis was only established relatively recently.

To understand the physical aspects of the optimal orientation angle of cells on cylindrical surfaces, Biton and Safran focused on the balance between the strain energy of a bent SF and the actomyosin-generated shear stress and on how it determines the optimal angle of the fibre alignment to minimise the total energy (Figure 10.4B).[10] Their description accords well with the conventional observation that different types of fibroblast and

Figure 10.4 (A) A particle-based model calculating physical interactions between rigid intracellular components accounts for the contribution of each component to the cellular alignment on curved surfaces.[116] (B) A continuum model explaining cellular orientation angle by competition of the shear stress and bending stiffness of the SFs.[10] (C) A vertex model predicts the cellular alignment and the transition of several different migration modes on curved surfaces.[13] (D) A discrete 3D mechanical model reproduces the cellular shape and migration on convex and concave surfaces by considering the total energy of deformable cell membrane, solid spherical nucleus, and string-like cytoskeleton.[117] (E) A phase field model implementing a lamellipodium-based cellular migration captures the characteristic migration modes on curved surfaces with various Gaussian curvatures.[101] (F) An interface-based model predicts tissue growth kinetics in a confined space by considering the change in cell density due to the contraction or expansion of the local surface area.[12] (G) A tissue growth model combining continuum mechanics, thermodynamics, and interfacial tension accounts for the different tissue growth kinetics on concave and convex surfaces.[107] (H) The prediction from a curvature-driven 3D tissue growth model accords well with an *in vivo* endogenous soft tissue formation within the scaffolds.[102] (A) reproduced from ref. 116, https://doi.org/10.1007/s10237-017-0888-4, under the terms of the CC BY 4.0 license https://creativecommons.org/licenses/by/4.0/. (B) reproduced from ref. 10 with permission of IOP Publishing, Copyright 2009. (C) reproduced from ref. 13 with permission from the Royal Society of Chemistry. (D) reproduced from ref. 117 with permission of IOP Publishing, Copyright 2017. (E) reproduced from ref. 101, https://doi.org/10.1038/s42005-019-0185-x, under the terms of the CC BY 4.0 license https://creativecommons.org/licenses/by/4.0/. (F) reproduced from ref. 12 with permission from Elsevier, Copyright 2017. (G) reproduced from ref. 107 with permission from Elsevier, Copyright 2013. (H) reproduced from ref. 102 with permission from Elsevier, Copyright 2017.

epithelial cells show different optimal angles on a cylindrical convex curve.[17] This basic scheme has been further extended to account for the elongation and alignment of epithelial cell sheets on cylindrical concaves by considering the influence of cell–cell adhesion.[30] The stress field induced on curved surfaces with a complex geometry can be solved by employing numerical analysis techniques such as the finite element method (FEM). Sanz-Herrera *et al.* investigated how the cell-contractility-derived stress field of cytoskeletons determines the cellular orientation on various types of curved surfaces,[99] which accorded well with the prediction by Dunn and Heath.[15] The maximum principal stress analysed using FEM explained the alignment of confluent cell sheets on cylindrical surfaces,[100] underlining the importance of the stress field in cellular alignment.

Continuum models have been able to explain the basic tendencies of cellular alignment on curved surfaces at different scales because they are unencumbered by the details of the stress-generating unit. However, they have less applicability to describing nonstationary deformations or those involving intracellular events. As a complementary approach, a phase field model simulating a lamellipodium-based cellular migration on curved surfaces was developed to consider the degree of local parallel ordering of actin filaments (Figure 10.4E).[101] This framework captured unique cellular migration modes on cylindrical surfaces (constant H, $K=0$),[21] as well as on a sphere-with-skirt structure (varying H and K, Figure 10.2H).[35] These complementary approaches, including other discrete models, could be useful in describing cellular locomotion involving a complex rearrangement of intracellular components.[22]

10.4.1.2 Multidisciplinary Tissue Growth Models

Modelling of tissue growth kinetics under geometrical constraints has been one of the core interests in the field. Such modelling is particularly relevant to certain tissue formation processes both *in vitro* and *in vivo*, such as bone tissue engineering[8,102] and wound healing.[103,104] The simulation of tissue formation processes is complex in that it must take into account the mechanical equilibrium at a stationary state together with a time-dependent volume increase. Both continuum approaches and other phenomenological and interface-based modelling approaches have been employed to elucidate the physical rules of tissue growth kinetics with a viscoelastic nature. We now review notable models to clarify the extent to which geometry-guided tissue formation process can currently be predicted.

The growth kinetics of osteoblasts filling a pore was first characterised in a simple phenomenological model in two dimensions.[27] Adopting a general rule of physical systems containing interfaces, the growth speed of tissues inside pores with different geometries was shown to be proportional to the local curvature, highlighting the importance of scaffold geometry as a principal determinant of the tissue formation process. This simple framework was then extended to another chord-based tissue growth model explicitly depicting how the local curvature and cellular contractile elements cooperatively induce local contraction.[93] The chord-based model accorded well with the observation of growing tissues within pores with various geometries.[105] Although some features, such as a slowdown in the growth speed at the later stage remained unexplained, these models successfully reproduced the initial stage of the geometry-dependent tissue formation process.

A phenomenological approach was also taken to a relevant mechanical theorem. Nelson *et al.* reported that cellular growth patterns arise from the stress distribution induced by the geometrical constraints on planar surfaces.[106] Extending Nelson's notion to three dimensions, Dunlop *et al.* proposed a theoretical model that explains tissue growth in terms of "growth stress" caused by the chemical potential increasing the tissue volume and the elastic resistance of the tissue body to deformation.[11] This combined theory of continuum mechanics and thermodynamics demonstrated the linear relationship between the local curvature and the growth stress at the growing front, supporting the assumption of the phenomenological models.[27,93] Furthermore, the different tissue growth kinetics on concave and convex surfaces were explained by considering the tissue surface tension (Figure 10.4G).[107] As an alternative approach, Alias and Buenzli proposed an interface-based model considering the curvature-dependent change in the spatial distribution of cellular density (Figure 10.4F).[12] Decrease and diffusion of cells at the growth front, which continuum mechanics models have fundamental difficulty implementing, explained the slowdown in the tissue growth at the late stage. A reaction-diffusion model also derived the curvature-dependent tissue growth, accounting for the change in cellular density based on cellular migration and proliferation.[108] It is remarkable that these continuum-based and interface-based models are able to explain tissue growth dynamics simply as consequences of geometrical conditions without considering the cellular biological properties such as phenotype, despite the obvious heterogeneity of the biological environment within the growing tissue.[104,109]

These models thus have considerable potential to predict the tissue formation process on artificial scaffolds with complex geometries. As an example, the structure of trabecular bone is characterised by negative Gaussian curvature, and the average mean curvature is close to zero.[110] This is known to be one of the minimal surfaces that combines mechanical strength with smooth mass transportation, and it has thus been of interest in the design of optimal scaffolds for bone tissue engineering.[111,112] To predict a realistic tissue formation on such complex structures, several technical improvements have been made. A scanning mask scheme[102,113] and a versatile level-set method[114] were adopted in the tissue growth models, which reproduced the features of the initial tissue formation process on microstructured surfaces both *in vitro*[113,114] and *in vivo* (Figure 10.4H).[102] Though practical tuning is still required, these *in silico* models are expected to be a tool for the practical design of tissue engineering scaffolds.

10.4.2 Discrete Modelling Approaches

Discrete models express the target as an assembly of discrete elements and predict the target motion as a consequence of the momentum of all the components such as particles or vertices forming polygonal units. These approaches are suitable for reconstructing anisotropic events such as protrusion formation at the cellular and tissue scales because they express structural components explicitly and are flexible in that they can adopt various types of interaction between the components. On the other hand, discrete models demand great care in defining a new interaction and setting parameters because this scheme needs to translate all the biological events into simplified physical interactions between the elements, and there are often difficulties in quantitative comparison between simulations and experiments.[98,115]

Particle-based cellular models, in which various physical interactions of intracellular components treated as rigid elements are considered, have been proposed as a means of quantitatively analysing the mechanical impact of surface curvature on intracellular components and the mechanism by which curvature induces anisotropic alignment of single cells (Figure 10.4A and D).[116,117] Implementing a feature of lamellipodium-based cellular migration, the model clarified that cellular migration follows a preceding nuclear displacement, whereas nuclear stiffness contributes to the directionality,[118] accounting for the basic tendency of curvotaxis.[44] Another model, considering cell–substrate adhesion with the effect of cytoskeletal polymerisation, contraction, and protrusion, explained the tendency for cylindrical concaves to direct persistent cellular migration in the axial direction[31] and also to induce more anisotropic spread than convex surfaces.

At multicellular scale, a vertex model, expressing cellular clusters as connected polygons, was recently employed to predict collective cellular behaviour on curved surfaces.[13] Surface curvature, together with local cellular alignment and contact inhibition of locomotion, was predicted to control various global migration modes such as local swirling and serpentine crawling (Figure 10.4C). Some features, including the migratory direction and the cellular area independent of curvature, partially resembled a recent observation of confluent epithelial cells cultured on a torus substrate (Figure 10.2G).[51] However, current experimental platforms are still limited in their ability to verify such predictions; further technical advances would be desirable in order to characterise such complex collective cellular behaviours.

10.4.3 Future Subjects

Various curvature-induced cellular behaviours have been reproduced by assuming relatively simple laws *in silico*. It is notable that several models have managed to identify the dominant parameters to capture the basic features of alignment[10] and migration,[101,118] though technical difficulties still remain in measuring some of the proposed parameters experimentally. There are increasing numbers of models demonstrating tissue growth under geometrical constraints,[12,93] yet there is still room for fine-tuning in some aspects, such as the late stage of tissue formation and correspondence to the real event in terms of timescale. Some major challenges remain in predicting the total tissue formation process *in silico* for practical scaffold design. First, nonstationary deformations of tissue, such as detachment and spheroid formation,[54,91,119] which have been neglected by current models, need to be considered. Formulation of these motions, perhaps in a discrete modelling approach, and their integration into growth models would widen the applicability of the models. Second, formulation of physicochemical interactions between the tissue and the environment, taking into account the porosity and viscoelasticity of the scaffold,[120] might be a key to predicting the total tissue growth because such environmental conditions slowly but steadily change as tissue fills spaces. While micro-/nanobiofabrication technologies are rapidly progressing, there is still only a limited theoretical framework supporting design strategies for bioconstructs. If they are able to overcome the challenges, these modelling approaches will be of increasing importance for tissue engineering in the near future.

10.5 Conclusions

In this chapter, we have introduced various techniques and investigations pursuing the mechanisms that drive the unique cellular behaviours found on curved surfaces at cellular and tissue scales. Technical advances in microfabrication have enabled precise fabrication of smooth curved surfaces for cell culture. Biological observations have confirmed that curved surfaces at above cellular scale have a significant influence on cellular properties, including phenotype. Meanwhile, modelling approaches have accounted for the basic tendencies in curvature-dependent behaviours such as migration and tissue growth from simple physical assumptions, without considering the biological properties of cells. In order to obtain a complete picture of the mechanism through which surface curvature controls cellular behaviours, phenomenological observations need to be incorporated into our understanding of the interaction between macroscopic curvature and the biomolecular signalling pathways that determine the macroscopic behaviour of cells following physical laws. Future challenges may rely on the development of quantitative measurement techniques to observe the interface where biomolecules, force, and geometrical constraints meet. Clarifying the mechanisms underlying curvature mechanobiology is an important challenge that, if mastered, will allow us to predict cellular behaviours on structured surfaces and thus contribute to the future design strategies of biomaterials.

References

1. R. K. Assoian, N. D. Bade, C. V. Cameron and K. J. Stebe, *Open Biol.*, 2019, **9**, 190155.

2. D. Baptista, L. Teixeira, C. van Blitterswijk, S. Giselbrecht and R. Truckenmüller, *Trends Biotechnol.*, 2019, **37**, 838–854.
3. S. J. P. Callens, R. J. C. Uyttendaele, L. E. Fratila-Apachitei and A. A. Zadpoor, *Biomaterials*, 2020, **232**, 119739.
4. M. Werner, N. A. Kurniawan and C. V. C. Bouten, *Materials*, 2020, **13**, 963.
5. M. Nikkhah, F. Edalat, S. Manoucheri and A. Khademhosseini, *Biomaterials*, 2012, **33**, 5230–5246.
6. V. Vogel and M. Sheetz, *Nat. Rev. Mol. Cell Biol.*, 2006, **7**, 265–275.
7. K. Uto, E. Fisher, H.-N. Kim, C. H. Seo and D.-H. Kim, *Integrative Mechanobiology: Micro-and Nano-Techniques in Cell Mechanobiology*, 2015, p. 69.
8. A. A. Zadpoor, *Biomater. Sci.*, 2015, **3**, 231–245.
9. J. Foolen, T. Yamashita and P. Kollmannsberger, *J. Phys. D: Appl. Phys.*, 2015, **49**, 053001.
10. Y. Y. Biton and S. A. Safran, *Phys. Biol.*, 2009, **6**, 046010.
11. J. W. C. Dunlop, F. D. Fischer, E. Gamsjäger and P. Fratzl, *J. Mech. Phys. Solids*, 2010, **58**, 1073–1087.
12. M. A. Alias and P. R. Buenzli, *Biophys. J.*, 2017, **112**, 193–204.
13. S.-Z. Lin, Y. Li, J. Ji, B. Li and X.-Q. Feng, *Soft Matter*, 2020, **16**, 2941–2952.
14. W. Kühnel, *Differential Geometry: Curves—Surfaces—Manifolds*, American Mathematical Soc., 3rd edn, 2015.
15. G. A. Dunn and J. P. Heath, *Exp. Cell Res.*, 1976, **101**, 1–14.
16. P. E. Fisher and C. Tickle, *Exp. Cell Res.*, 1981, **131**, 407–410.
17. Y. A. Rovensky and V. I. Samoilov, *J. Cell Sci.*, 1994, **107**, 1255–1263.
18. T. M. Svitkina, Y. A. Rovensky, A. D. Bershadsky and J. M. Vasiliev, *J. Cell Sci.*, 1995, **108**, 735–745.
19. E. M. Levina, L. V. Domnina, Y. A. Rovensky and J. M. Vasiliev, *Exp. Cell Res.*, 1996, **229**, 159–165.
20. S. J. Lee and S. Yang, *Rev. Sci. Instrum.*, 2012, **83**, 094302.
21. H. G. Yevick, G. Duclos, I. Bonnet and P. Silberzan, *Proc. Natl. Acad. Sci. U. S. A.*, 2015, **112**, 5944–5949.
22. N. D. Bade, R. D. Kamien, R. K. Assoian and K. J. Stebe, *Sci. Adv.*, 2017, **3**, e1700150.
23. S. J. Lee and S. Yang, *Biotechnol. J.*, 2017, **12**, 1700360.
24. M. A. Rose, J. J. Bowen and S. A. Morin, *Chem. Phys. Chem.*, 2019, **20**, 909–925.
25. J. Y. Park, D. H. Lee, E. J. Lee and S.-H. Lee, *Lab Chip*, 2009, **9**, 2043–2049.
26. S.-M. Yu, J. M. Oh, J. Lee, W. Lee-Kwon, W. Jung, F. Amblard, S. Granick and Y.-K. Cho, *Acta Biomater.*, 2018, **77**, 311–321.
27. M. Rumpler, A. Woesz, J. W. C. Dunlop, J. T. van Dongen and P. Fratzl, *J. R. Soc., Interface*, 2008, **5**, 1173–1180.
28. M. Pilia, T. Guda, S. M. Shiels and M. R. Appleford, *J. Biol. Eng.*, 2013, **7**, 23.

29. H. N. Hansen, R. J. Hocken and G. Tosello, *CIRP Ann.*, 2011, **60**, 695–714.
30. A. K. Yip, P. Huang and K.-H. Chiam, *Biophys. J.*, 2018, **114**, 1707–1717.
31. K. H. Song, S. J. Park, D. S. Kim and J. Doh, *Biomaterials*, 2015, **51**, 151–160.
32. I. Hebeiss, R. Truckenmüller, S. Giselbrecht and U. Schepers, *Lab Chip*, 2012, **12**, 829.
33. D. Baptista, L. M. Teixeira, Z. T. Birgani, S. van Riet, T. Pasman, A. Poot, D. Stamatialis, R. J. Rottier, P. S. Hiemstra, P. Habibović, C. van Blitterswijk, S. Giselbrecht and R. Truckenmüller, *Biomaterials*, 2021, **266**, 120436.
34. T. Yamashita, T. Nishina, I. Matsushita and R. Sudo, *Anal. Sci.*, 2020, **36**, 1015–1019.
35. N. D. Bade, T. Xu, R. D. Kamien, R. K. Assoian and K. J. Stebe, *Biophys. J.*, 2018, **114**, 1467–1476.
36. V. Hosseini, P. Kollmannsberger, S. Ahadian, S. Ostrovidov, H. Kaji, V. Vogel and A. Khademhosseini, *Small*, 2014, **10**, 4851–4857.
37. S. Ehrig, B. Schamberger, C. M. Bidan, A. West, C. Jacobi, K. Lam, P. Kollmannsberger, A. Petersen, P. Tomancak, K. Kommareddy, F. D. Fischer, P. Fratzl and J. W. C. Dunlop, *Sci. Adv.*, 2019, **5**, eaav9394.
38. S. Surdo, A. Diaspro and M. Duocastella, *Appl. Surf. Sci.*, 2017, **418**, 554–558.
39. D. A. Soscia, S. J. Sequeira, R. A. Schramm, K. Jayarathanam, S. I. Cantara, M. Larsen and J. Castracane, *Biomaterials*, 2013, **34**, 6773–6784.
40. B. Sun, K. Xie, T.-H. Chen and R. H. W. Lam, *RSC Adv.*, 2017, **7**, 6788–6794.
41. T. Liu, C.-C. Chien, L. Parkinson and B. Thierry, *ACS Appl. Mater. Interfaces*, 2014, **6**, 8090–8097.
42. L. Sun, C. B. Danoux, Q. Wang, D. Pereira, D. Barata, J. Zhang, V. LaPointe, R. Truckenmüller, C. Bao, X. Xu and P. Habibovic, *Acta Biomater.*, 2016, **42**, 364–377.
43. K. K. Saxena, J. Qian and D. Reynaerts, *Int. J. Mach. Tools Manuf.*, 2018, **127**, 28–56.
44. L. Pieuchot, J. Marteau, A. Guignandon, T. Dos Santos, I. Brigaud, P.-F. Chauvy, T. Cloatre, A. Ponche, T. Petithory, P. Rougerie, M. Vassaux, J.-L. Milan, N. Tusamda Wakhloo, A. Spangenberg, M. Bigerelle and K. Anselme, *Nat. Commun.*, 2018, **9**, 3995.
45. P. Rougerie, L. Pieuchot, R. S. Dos Santos, J. Marteau, M. Bigerelle, P.-F. Chauvy, M. Farina and K. Anselme, *Sci. Rep.*, 2020, **10**, 14784.
46. M. Nikkhah, J. S. Strobl and M. Agah, *Biomed. Microdevices*, 2009, **11**, 429–441.
47. B. Wagner, H. J. Quenzer, W. Henke, W. Hoppe and W. Pilz, *Sens. Actuators, A*, 1995, **46**, 89–94.
48. M. Werner, S. B. G. Blanquer, S. P. Haimi, G. Korus, J. W. C. Dunlop, G. N. Duda, D. W. Grijpma and A. Petersen, *Adv. Sci.*, 2017, **4**, 1600347.
49. S. Waheed, J. M. Cabot, N. P. Macdonald, T. Lewis, R. M. Guijt, B. Paull and M. C. Breadmore, *Lab Chip*, 2016, **16**, 1993–2013.

50. D. Cheng, R. K. Jayne, A. Tamborini, J. Eyckmans, A. E. White and C. S. Chen, *Biofabrication*, 2019, **11**, 021001.
51. S.-M. Yu, B. Li, F. Amblard, S. Granick and Y.-K. Cho, *Biomaterials*, 2020, **265**, 120420.
52. X. Wang, S. Xu, S. Zhou, W. Xu, M. Leary, P. Choong, M. Qian, M. Brandt and Y. M. Xie, *Biomaterials*, 2016, **83**, 127–141.
53. M. Bao, J. Xie and W. T. S. Huck, *Adv. Sci.*, 2018, **5**, 1800448.
54. K. E. Broaders, A. E. Cerchiari and Z. J. Gartner, *Integr. Biol.*, 2015, **7**, 1611–1621.
55. C. Prévost, H. Zhao, J. Manzi, E. Lemichez, P. Lappalainen, A. Callan-Jones and P. Bassereau, *Nat. Commun.*, 2015, **6**, 8529.
56. Y. Chen, J. Wang, X. Li, N. Hu, N. H. Voelcker, X. Xie and R. Elnathan, *Adv. Mater.*, 2020, **32**, e2001668.
57. H. T. McMahon and J. L. Gallop, *Nature*, 2005, **438**, 590–596.
58. K. Takei, P. S. McPherson, S. L. Schmid and P. D. Camilli, *Nature*, 1995, **374**, 186–190.
59. B. M. Pearse, *Proc. Natl. Acad. Sci. U. S. A.*, 1976, **73**, 1255–1259.
60. K. S. Cannon, B. L. Woods and A. S. Gladfelter, *Trends Biochem. Sci.*, 2017, **42**, 961–976.
61. M. Simunovic, E. Evergren, A. Callan-Jones and P. Bassereau, *Annu. Rev. Cell Dev. Biol.*, 2019, **35**, 111–129.
62. H.-Y. Lou, W. Zhao, X. Li, L. Duan, A. Powers, M. Akamatsu, F. Santoro, A. F. McGuire, Y. Cui, D. G. Drubin and B. Cui, *Proc. Natl. Acad. Sci. U. S. A.*, 2019, **116**, 23143–23151.
63. K. S. Cannon, B. L. Woods, J. M. Crutchley and A. S. Gladfelter, *J. Cell Biol.*, 2019, **218**, 1128–1137.
64. H. Kang and D. J. Lew, *Curr. Genet.*, 2017, **63**, 75–77.
65. F. Guharay and F. Sachs, *J. Physiol.*, 1984, **352**, 685–701.
66. S. S. Ranade, R. Syeda and A. Patapoutian, *Neuron*, 2015, **87**, 1162–1179.
67. B. D. Matthews, C. K. Thodeti, J. D. Tytell, A. Mammoto, D. R. Overby and D. E. Ingber, *Integr. Biol.*, 2010, **2**, 435–442.
68. X. Liang and J. Howard, *Curr. Biol.*, 2018, **28**, R357–R359.
69. A. J. Patel, M. Lazdunski and E. Honoré, *Curr. Opin. Cell Biol.*, 2001, **13**, 422–428.
70. O. Bavi, C. D. Cox, M. Vossoughi, R. Naghdabadi, Y. Jamali and B. Martinac, *Membranes*, 2016, **6**, 14.
71. A.-L. Le Roux, X. Quiroga, N. Walani, M. Arroyo and P. Roca-Cusachs, *Philos. Trans. R. Soc., B*, 2019, **374**, 20180221.
72. S. G. Brohawn, Z. Su and R. MacKinnon, *Proc. Natl. Acad. Sci. U. S. A.*, 2014, **111**, 3614–3619.
73. F. Martino, A. R. Perestrelo, V. Vinarský, S. Pagliari and G. Forte, *Front. Physiol.*, 2018, **9**, 824.
74. C. Oakley and D. M. Brunette, *J. Cell Sci.*, 1993, **106**, 343–354.
75. T. Ozdemir, L.-C. Xu, C. Siedlecki and J. L. Brown, *Integr. Biol.*, 2013, **5**, 1407–1416.

76. M. Michael and M. Parsons, *Curr. Opin. Cell Biol.*, 2020, **63**, 31–37.
77. M. Werner, N. A. Kurniawan, G. Korus, C. V. C. Bouten and A. Petersen, *J. R. Soc., Interface*, 2018, **15**, 20180162.
78. M. Werner, A. Petersen, N. A. Kurniawan and C. V. C. Bouten, *Adv. Biosyst.*, 2019, **3**, e1900080.
79. Y.-P. Lo, Y.-S. Liu, M. G. Rimando, J. H.-C. Ho, K.-H. Lin and O. K. Lee, *Sci. Rep.*, 2016, **6**, 21253.
80. Z. Li, W. Wang, X. Xu, K. Kratz, J. Zou, L. Lysyakova, M. Heuchel, A. Kurtz, M. Gossen, N. Ma and A. Lendlein, *J. Mater. Chem. B*, 2017, **5**, 7415–7425.
81. A. Livne and B. Geiger, *J. Cell Sci.*, 2016, **129**, 1293–1304.
82. V. I. Risca, E. B. Wang, O. Chaudhuri, J. J. Chia, P. L. Geissler and D. A. Fletcher, *Proc. Natl. Acad. Sci. U. S. A.*, 2012, **109**, 2913–2918.
83. H. Elliott, R. S. Fischer, K. A. Myers, R. A. Desai, L. Gao, C. S. Chen, R. S. Adelstein, C. M. Waterman and G. Danuser, *Nat. Cell Biol.*, 2015, **17**, 137–147.
84. G. R. Fedorchak, A. Kaminski and J. Lammerding, *Prog. Biophys. Mol. Biol.*, 2014, **115**, 76–92.
85. N. Wang, J. D. Tytell and D. E. Ingber, *Nat. Rev. Mol. Cell Biol.*, 2009, **10**, 75–82.
86. K. Anselme, N. T. Wakhloo, P. Rougerie and L. Pieuchot, *Adv. Healthcare Mater.*, 2018, **7**, e1701154.
87. A. Marino, C. Filippeschi, G. G. Genchi, V. Mattoli, B. Mazzolai and G. Ciofani, *Acta Biomater.*, 2014, **10**, 4304–4313.
88. X. Lin, S. Romanazzo, K. Lin, C. Kelly, J. J. Gooding and I. Roohani, *Materialia*, 2020, **14**, 100870.
89. M. Maurer and J. Lammerding, *Annu. Rev. Biomed. Eng.*, 2019, **21**, 443–468.
90. W.-H. Jung, K. Elawad, S. H. Kang and Y. Chen, *Cells*, 2019, **8**, 813.
91. T. Yamashita, P. Kollmannsberger, K. Mawatari, T. Kitamori and V. Vogel, *Acta Biomater.*, 2016, **45**, 85–97.
92. F. A. Maechler, C. Allier, A. Roux and C. Tomba, *J. Cell Sci.*, 2019, **132**, jcs222372.
93. C. M. Bidan, K. P. Kommareddy, M. Rumpler, P. Kollmannsberger, Y. J. M. Bréchet, P. Fratzl and J. W. C. Dunlop, *PLoS One*, 2012, **7**, e36336.
94. J. Sharpe, *Development*, 2017, **144**, 4214–4225.
95. L. Geris, T. Lambrechts, A. Carlier and I. Papantoniou, *Curr. Opin. Biomed. Eng.*, 2018, **6**, 92–98.
96. P. F. Egan, S. J. Ferguson and K. Shea, *J. Mech. Des.*, 2017, **139**, 061401.
97. P. Kollmannsberger, C. M. Bidan, J. W. C. Dunlop and P. Fratzl, *Soft Matter*, 2011, **7**, 9549–9560.
98. B. A. Camley and W.-J. Rappel, *J. Phys. D: Appl. Phys.*, 2017, **50**, 113002.
99. J. A. Sanz-Herrera, P. Moreo, J. M. García-Aznar and M. Doblaré, *Biomaterials*, 2009, **30**, 6674–6686.
100. C. Liu, J. Xu, S. He, W. Zhang, H. Li, B. Huo and B. Ji, *J. Mech. Behav. Biomed. Mater.*, 2018, **88**, 330–339.

101. B. Winkler, I. S. Aranson and F. Ziebert, *Commun. Phys.*, 2019, **2**, 1–11.
102. M. Paris, A. Götz, I. Hettrich, C. M. Bidan, J. W. C. Dunlop, H. Razi, I. Zizak, D. W. Hutmacher, P. Fratzl, G. N. Duda, W. Wagermaier and A. Cipitria, *Acta Biomater.*, 2017, **60**, 64–80.
103. P. Joly, G. N. Duda, M. Schöne, P. B. Welzel, U. Freudenberg, C. Werner and A. Petersen, *PLoS One*, 2013, **8**, e73545.
104. P. Kollmannsberger, C. M. Bidan, J. W. C. Dunlop, P. Fratzl and V. Vogel, *Sci. Adv.*, 2018, **4**, eaao4881.
105. C. M. Bidan, K. P. Kommareddy, M. Rumpler, P. Kollmannsberger, P. Fratzl and J. W. C. Dunlop, *Adv. Healthare. Mater.*, 2013, **2**, 186–194.
106. C. M. Nelson, R. P. Jean, J. L. Tan, W. F. Liu, N. J. Sniadecki, A. A. Spector and C. S. Chen, *Proc. Natl. Acad. Sci. U. S. A.*,, 2005, **102**, 11594–11599.
107. E. Gamsjäger, C. M. Bidan, F. D. Fischer, P. Fratzl and J. W. C. Dunlop, *Acta Biomater.*, 2013, **9**, 5531–5543.
108. P. R. Buenzli, M. Lanaro, C. S. Wong, M. P. McLaughlin, M. C. Allenby, M. A. Woodruff and M. J. Simpson, *Acta Biomater.*, 2020, **114**, 285–295.
109. C. M. Bidan, P. Kollmannsberger, V. Gering, S. Ehrig, P. Joly, A. Petersen, V. Vogel, P. Fratzl and J. W. C. Dunlop, *J. R. Soc., Interface*, 2016, **13**, 20160136.
110. H. Jinnai, H. Watashiba, T. Kajihara, Y. Nishikawa, M. Takahashi and M. Ito, *Bone*, 2002, **30**, 191–194.
111. S. Torquato, S. Hyun and A. Donev, *Phys. Rev. Lett.*, 2002, **89**, 266601.
112. S. C. Kapfer, S. T. Hyde, K. Mecke, C. H. Arns and G. E. Schröder-Turk, *Biomaterials*, 2011, **32**, 6875–6882.
113. C. M. Bidan, F. M. Wang and J. W. C. Dunlop, *Comput. Methods Biomech. Biomed. Eng.*, 2013, **16**, 1056–1070.
114. Y. Guyot, I. Papantoniou, Y. C. Chai, S. Van Bael, J. Schrooten and L. Geris, *Biomech. Model. Mechanobiol.*, 2014, **13**, 1361–1371.
115. S. Alt, P. Ganguly and G. Salbreux, *Philos. Trans. R. Soc., B*, 2017, **372**, 20150520.
116. M. Vassaux and J. L. Milan, *Biomech. Model. Mechanobiol.*, 2017, **16**, 1295–1308.
117. X. He and Y. Jiang, *Phys. Biol.*, 2017, **14**, 035006.
118. M. Vassaux, L. Pieuchot, K. Anselme, M. Bigerelle and J.-L. Milan, *Biophys. J.*, 2019, **117**, 1136–1144.
119. D. Gonzalez-Rodriguez, K. Guevorkian, S. Douezan and F. Brochard-Wyart, *Science*, 2012, **338**, 910–917.
120. G. Lutzweiler, A. Ndreu Halili and N. Engin Vrana, *Pharmaceutics*, 2020, **12**, 602.

CHAPTER 11

Dynamic Hydrogel

TAKESHI UEKI*[a,b] AND RYOTA TAMATE[c]

[a] Research Center for Functional Materials, National Institute for Materials Science (NIMS), 1-1 Namiki, Tsukuba, Ibaraki 305-0044, Japan; [b] Graduate School of Life Science, Hokkaido University, Kita 10, Nishi 8, Kita-ku Sapporo 060-0810, Japan; [c] Center for Green Research on Energy and Environmental Materials, National Institute for Materials Science (NIMS), 1-1 Namiki, Tsukuba, Ibaraki 305-0044, Japan
*Email: ueki.takeshi@nims.go.jp

11.1 Hydrogel Classification and Switchable Hydrogel Design Principles

This section summarises the recent mechanobiological research on hydrogel-containing cell scaffolds, focusing on the molecular design and action mechanism of hydrogel substrates that provide mechanical stimulation to cells. In living organisms, cells adhere to a viscoelastic substrate called the extracellular matrix (ECM). The contractile force generated by the cells themselves plays a fundamental role in the assembly of the cytoskeleton, which significantly affects the activation of biochemical signalling pathways, gene transcription, migration, proliferation, cell adhesion, apoptosis, and differentiation. To date, much effort has been directed at the imitation of the ECM in living organisms, and numerous *in vitro* cell scaffold materials have been proposed. Soft biomaterials such as hydrogels, which offer the benefits of hydrophilicity and the capacity to mimic the aspects of different soft tissues have emerged as a prominent platform for the *in vitro* studies of diseases, allowing for the multidimensional culturing and regeneration of human cells. Synthetic hydrogels intended for use as cell

Biomaterials Science Series No. 12
Material-based Mechanobiology
Edited by Jun Nakanishi and Koichiro Uto

Published by the Royal Society of Chemistry, www.rsc.org

scaffolds should feature hydrophilic and water-swellable polymer networks. Based on their three major components (polymer network, crosslinking point, and solvent), polymer gels are generally classified according to (1) the network-forming polymer, (2) crosslinking form, (3) gel dimensions, and (4) solvent.

Classification schemes (1) and (2) are particularly important for the design of dynamic polymer hydrogels as prospective cell scaffolds. The network-forming polymers can be classified into synthetic (*e.g.* polyethylene glycol [PEG] and polyacrylamide) and naturally derived ones, with the latter further subdivided into proteins (*e.g.* collagen, gelatin, Matrigel) and polysaccharides (*e.g.* alginic acid, hyaluronic acid, chitosan, dextran). Synthetic polymers offer the advantages of design freedom, the possibility of reproducible gel scaffold preparation without interbatch variation, facile mechanical property tuning, ease of processing, and low immunological/contamination risks, but they do not have matrix metalloproteinase- (MMP-) cleavable sites that decompose in response to cell secretions or cell adhesion sites such as RGD. In many cases, naturally derived proteins have adhesion ligands and can be used as scaffolds for adhesive cells without any chemical treatment. As polysaccharides bear functional groups (*e.g.* –COOH), they can be easily modified with RGD motifs and biodegradable structures such as MMP-cleavable sites *via* well established methods such as Michael addition, Schiff base formation, and Huisgen cyclisation. However, biological proteins possess high immune risks and suffer from large batch-to-batch differences and potential contamination.

Crosslinking points can be classified into those relying on permanent chemical bonds (chemical crosslinking) and those relying on relatively weak exchangeable intermolecular interactions such as hydrogen bonds, ionic bonds, polymer chain entanglement, and association of polymer (physical crosslinking). Hydrogels with chemical crosslinking feature a completely frozen microscopic structure and are therefore macroscopically elastic; *i.e.* they are quite different from viscoelastic hydrogels. Viscoelastic hydrogels such as the ECM may contain exchangeable crosslinking. To produce an adaptive *in vitro* cell scaffold, one should design a mechanism for the break-up of chemical crosslinking in response to external stimuli, cell secretion, or the force exerted by the cells themselves or introduce breakable bonds into the matrix polymer. Further details pertaining to the recent trends and design strategies for cell culture–oriented dynamic hydrogels or adaptive hydrogels are found elsewhere.[1–5]

11.2 Hydrogels Exhibiting One-way Stiffness Changes

11.2.1 Design of Switchable Hydrogels with Decreasable Elastic Modulus

According to the classical rubber elasticity theory, the storage elastic modulus G' of a polymer gel depends on the number of crosslinkages per unit volume (n) (crosslinking density) at a given timescale, the Boltzmann constant (k), and absolute temperature (T):

$$G' = nkT. \tag{11.1}$$

Therefore, at a constant temperature, the polymer gel can be softened *via* the reduction of crosslinking density within the polymer network using a nanoscopically preinstalled molecular mechanism. From this perspective, it is preferable to choose permanently crosslinked gels rather than physically crosslinked gels, as the latter are capable of interchangeable crosslinking within a given time scale. A better strategy of preparing gels exhibiting a one-way stiffness decrease is the introduction of cleavable bonds within a stable covalent network (Table 11.1). Anseth *et al.*[6] incorporated a photocleavable nitrobenzyl group at a bioinert PEG terminal to afford a photocleavable bi-functional oligomer (Figure 11.1A). The resulting gel softened upon irradiation at 365–405 nm because of the irreversible cleavage of the nitrobenzyl group, and the extent of softening was found to depend on light wavelength and intensity. These authors encapsulated human mesenchymal stem cells (hMSCs) into a photodegradable three-dimensional (3D) cell scaffold and observed cell elongation in response to network photodegradation upon irradiation. Moreover, the RGD motif was incorporated into the polymer network *via* a nitrobenzyl group, and the chondrogenic behaviour of fibrosarcoma cells upon the photoinduced cleavage of the RGD motif from the gel network was investigated in detail. The photodegradable gels were claimed to allow the manipulation of material properties or chemistry in real time, providing dynamic environments with the scope to answer the fundamental questions pertaining to the material regulation of live cell functions. Subsequently, the authors probed the effects of hMSC differentiation on the decrease in gel elastic modulus using the same nitrobenzene-modified photocleavable PEG hydrogel (Figure 11.1B, C).[7] In this case, hMSCs were cultured on the surface of stiff gels (Young's modulus (E) $\approx$ 10 kPa, $E = 2G(1+\nu)$, where ν is Poisson's ratio) for 10 days and then cultured on soft gels ($E \approx 2$ kPa, prepared by irradiation of stiff gels) for 1, 5, or 10 days. Interestingly, the nuclear localisation of RUNX2 and YAP (which are essential transcriptional [co]factors during bone differentiation) was observed for all hMSCs cultured on the soft gel surface. As all hMSCs were initially cultured on a stiff interface, this result indicates that the cells "remembered" past mechanical stimuli that affected their status. Previous experimental designs required the cells to be treated with trypsin and transferred from a stiff activating substrate (*e.g.* a polystyrene dish) to a compliant deactivating substrate (*e.g.* a soft hydrogel), which introduced several inherent complications possibly detrimental to result interpretation. This photodegradable gel system provides a unique experimental platform for the further study of mechanical memory and dosing *via* the investigation of the dynamic response of stem cells when the modulus of the underlying substrate is switched from activating ($E \approx 10$ kPa) to deactivating ($E \approx 2$ kPa) at selected time points. In addition, the authors employed the same

Table 11.1 Summary of dynamic hydrogels changing their elastic moduli in response to external stimuli.

Reference	Trigger	Elastic modulus	Comment	Cell line
6	Light	Decrease	Photocleavable nitrobenzene-bearing crosslinker	hMSCs, fibrosarcoma cells
7	Light	Decrease	Photocleavable nitrobenzene-bearing crosslinker	hMSCs
8	Light	Decrease	Photomediated addition-fragmentation chain transfer process	hMSCs
9	Light	Increase	Photoinitiated polymerization of crosslinkable hyaluronic acid	hMSCs
10	Light	Increase	Photoinitiated polymerization of crosslinkable hyaluronic acid	hMSCs
11	Light	Increase	Photodimerization of 8-arm PEG-conjugated anthracene	Cardiac fibroblasts
12	Light (VIS)	Increase	Photodimerization of 4-arm PEG-conjugated electron-rich anthracene	hMSCs (for viability check)
13	Light (VIS)	Increase	Photoisomerisation of azobenzene-bearing crosslinker	hMSCs
14	pH	Reversible	pH-responsive gel made by ABA triblock copolymer self-assembly	C2C12
15	pH	Reversible	pH-responsive gel made by ABA triblock copolymer self-assembly	hMSCs
16	Ions (Ca^{2+}, Na^{+})	Reversible	Collagen-alginate interpenetrate network (IPN) gel partially crosslinked with ionic bond	C3T/10T/1/2
17	Redox reaction	Reversible	Protein folding-unfolding triggered by redox-responsive reversible disulfide cleavage reaction	N/A
18	Ions (Ca^{2+})	Reversible	Protein folding-unfolding coupled with reversible chelation of Ca^{2+}	NIH3T3

Table 11.1 (*Continued*)

Reference	Trigger	Elastic modulus	Comment	Cell line
19	Enzymatic reaction	Reversible	SrtA enzyme–reactive peptide sequence conjugate with PEG network	hMSCs, COLO-357
20	Chemicals	Reversible	Inclusion complex formation between β-CD and adamantane in different polymer network	MIN6 β-cell
21	Chemicals	Reversible	Inclusion complex formation between β-CD and adamantane in same polymer network	C2C12
18	Light	Reversible	Protein folding-unfolding triggered by photoresponsive protein (LOV-2)	NIH3T3
22	Light	Reversible	Photoisomerization of azobenzene incorporated into polymer network	NIH3T3 (for viability check)
23	Light	Reversible	Polymer network of hydrophilicity changes depending on photoisomerization state of azobenzene	VICs
24	Light	Reversible	Photocleavable nitrobenzene and photocrosslinkable methacrylated hyaluronic acid	MCF-7
26	Light	Reversible	Photoresponsive reversible conformation changes of engineered cyanobacterial Chytochrome1	hMSCs
27	Light (VIS)	Reversible	Photocleavable nitrobenzene and photocrosslinkable methacrylated hyaluronic acid	hMSCs
28	Light (VIS)	Reversible	Photodimerization reaction of 4-arm PEG-conjugated styrylpyrene	hMSCs (for viability check)
29	Magnetic field	Reversible	Alignment of iron particle covalently conjugated polymer network	hMSCs

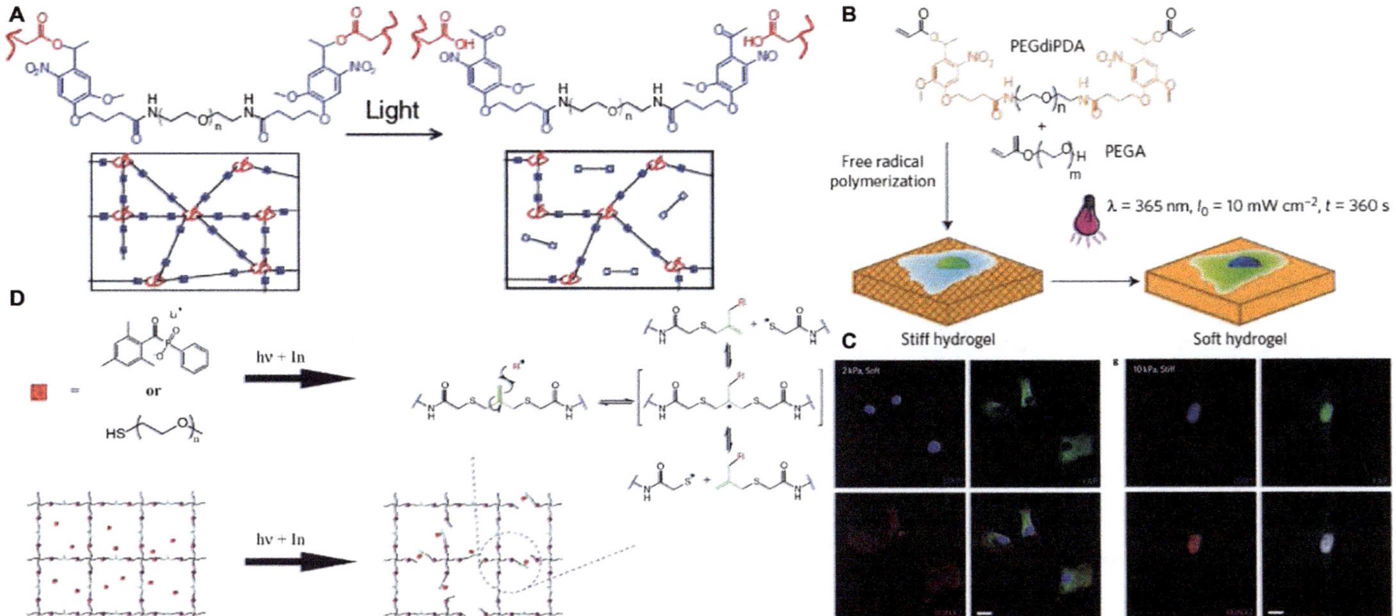

Figure 11.1 (A) Structure of the photocleavable nitrobenzene-containing crosslinker and a schematic illustration of the polymer network responsible for the decrease in gel elastic modulus.[6] (B) Cell culturing on the surface of the photodegradable hydrogel. (C) Nuclear colocalisation of YAP and RUNX2 in a 10 kPa hydrogel.[7] (D) Light-triggered degradation mechanism based on addition-fragmentation chain transfer.[8] Reproduced from ref. 6 with permission from National Academy of Sciences. Reproduced from ref. 7 with permission from Springer Nature, Copyright 2014. Reproduced from ref. 8 with permission from John Wiley and Sons, Copyright © 2017 WILEY-VCH Verlag GmbH & Co. KGaA, Weinheim.

mechanism of lowering crosslinking density to realise different molecular design concepts. For example, they used the cleavage of an allyl sulphide group *via* an addition-fragmentation chain transfer process in the presence of phenyl-2,4,6-tri-methylbenzoylphosphinate or thiol-decorated methoxy-PEG, which generates free radicals in response to light (Figure 11.1D).[8]

11.2.2 Design of Switchable Hydrogels with Increasable Elastic Modulus

A completely opposite strategy aims to realise hydrogels that stiffen in response to external stimuli due to the increase in the number of cross-linkages within the hydrogel polymer network. Matrix stiffening may be more relevant for many biological events such as tissue development, wound healing, and the progression of diseases (*e.g.* fibrosis and tumour formation). To imitate such stiffening, Burdick *et al.*[9] used hydrogels that stiffened in the presence of cells because of the *in situ* polymerisation–induced formation of stable chemical crosslinking (Figure 11.2A–C). Hyaluronic acid, a polysaccharide with multiple hydroxyl groups, was modified with methacrylic anhydride (MA) and RGD, and crosslinking *via* the Michael addition of dithiothreitol afforded a single-network gel. Subsequent irradiation generated reactive radicals from the photoinitiator (Irgacure 2959) incorporated into the polymer network and thus induced the polymerisation of vinylidene groups. This treatment resulted in a sudden increase in crosslinking density and afforded stiffened gels within several minutes. Then, hMSCs were cultured on the surface of these hydrogels, and the cell projected area was correlated with the increase in substrate elastic modulus. An elastic modulus increase of 3–30 kPa resulted in an increase in cell area and traction force on the cell adhesion surface within several hours. The elastic modulus of the hydrogel scaffold was increased at selected time points during the 14 day differentiation induction period. hMSCs that had experienced elastic modulus increases at an early stage showed bone differentiation, whereas those that had experienced elastic modulus increases at a late stage showed adipogenic differentiation. Thus stem cell differentiation was strongly influenced by the timing of the elastic modulus increase, and irreversible bone differentiation was promoted by longer periods of culturing on a stiff substrate. Subsequently, the same strategy was applied to a 3D cell scaffold hydrogel with an elastic modulus that could be increased at an arbitrary time (Figure 11.2D, E).[10] In this case, hMSCs were encapsulated inside a hyaluronic acid–based hydrogel using two-step chemical crosslinking. When cells were cultured inside MMP-degradable hydrogel, a switch from bone differentiation to adipogenic differentiation occurred if polymer network degradation in the vicinity of the cell was suppressed. A detailed discussion of this work from a biological perspective is given in Chapter 13, which focuses on 3D scaffold cell culturing systems.

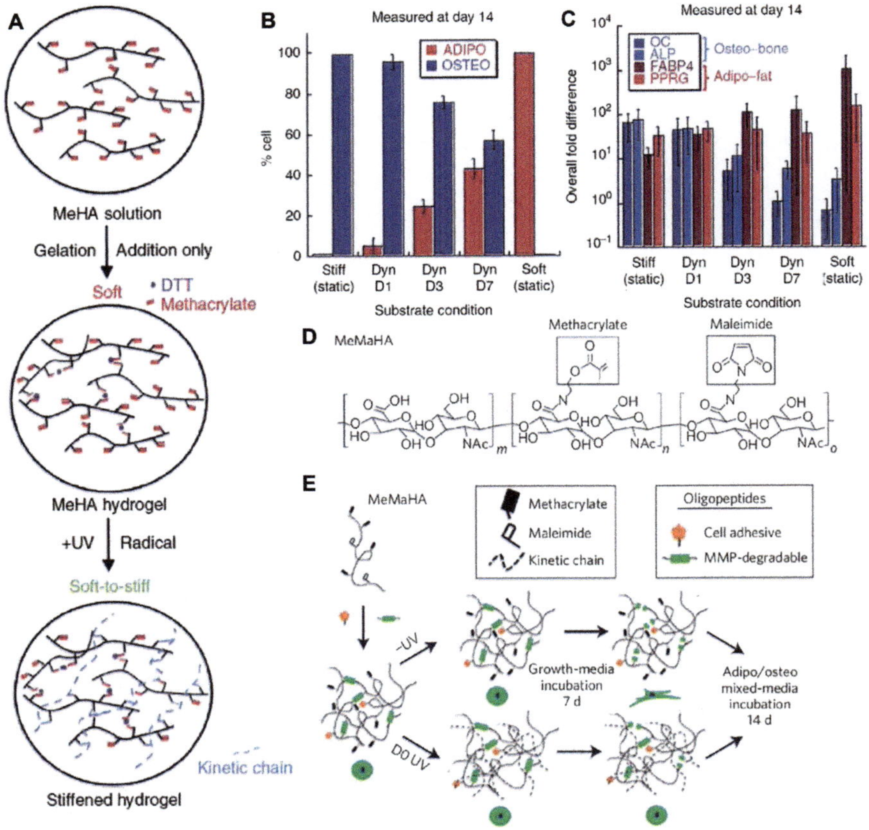

Figure 11.2 (A) Schematics of crosslinking using methacrylated hyaluronic acid. (B) Mean percentages of hMSCs stained positive for ALP (osteo) and Oil Red (adipo) as functions of dynamic stiffening time. The total duration of cell culturing equalled 14 days. Error bars indicate the standard deviation (s.d.; $n = 3$). (C) Overall fold difference of osteogenic (CC, ALP) and adipogenic (FABP4, peroxisome proliferator-activated receptor gamma [PPRG]) gene expression obtained from RT-PCR experiments as a function of stiffening time. Error bars indicate the s.d. ($n = 3$).[9] (D) Structure of hyaluronic acid possessing methacrylate and maleimide groups (MeMaHA, $m = 0.755$, $n = 0.14$, $o = 0.105$). (E) Schematics of the sequential crosslinking of MeMaHA *via* primary addition and secondary radical polymerisation to create -UV and D0 UV hydrogels, respectively.[10] Reproduced from ref. 9 and 10 with permission from Springer Nature, Copyright 2012, 2013.

Specific chemical structures such as anthracene, coumarin (CA), thymine, and cynnamoylones can undergo dimerization upon irradiation and have therefore been employed to directly increase the number of crosslinkages in polymer gels.[11] Hydrogels prepared from eight-arm PEG terminally modified with anthracene exhibited a stepwise increase in stiffness upon multiple instances of irradiation because of the direct increase in crosslinking density due to the [4 + 4] cycloaddition between anthracene units. Using this cell scaffold hydrogel, Anseth *et al.*[11] investigated the role of dynamic stiffness changes in intracellular calcium signal transduction. Forsyth *et al.*[12] attached an electron-donating group to the anthracene backbone to realise a polymer gel capable of changing its elastic modulus in response to illumination with low-phototoxicity visible light (Figure 11.3A–D). Richardson

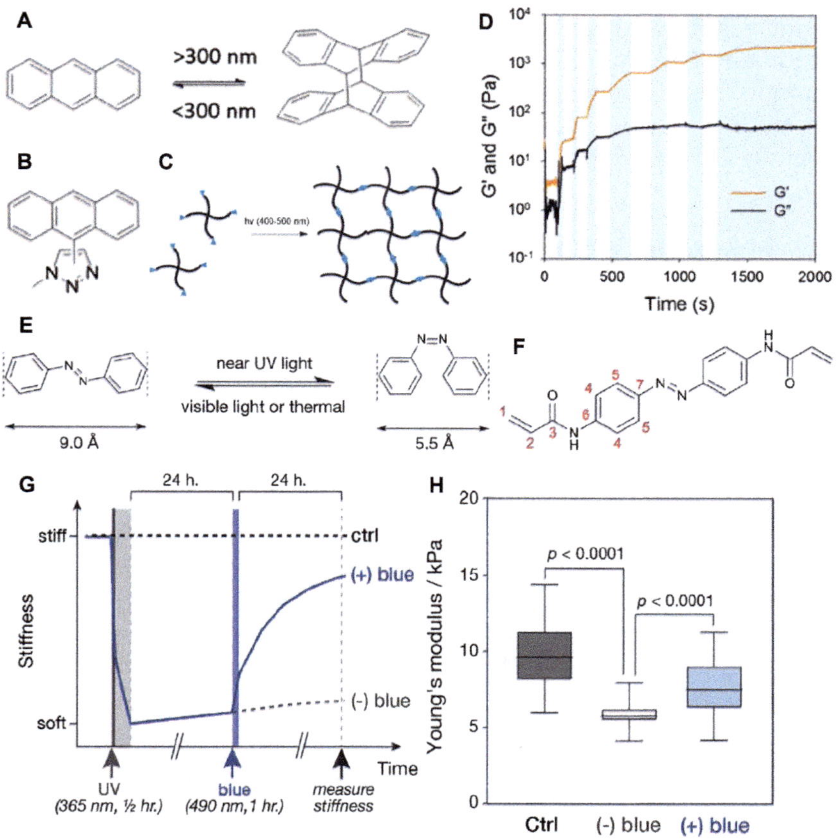

Figure 11.3 (A) Photochemical dimerisation of anthracene. (B) Structure of a representative electron-rich anthracene derivative. (C) Schematics of polymer crosslinking under visible light. (D) Effect of irradiation time on the rheological behaviour of a four-arm PEG attached to the electron-rich anthracene derivative shown in (B) (3 mM). Polymerisation was stopped when the light was turned off.[12] (E) Structure and photoisomerisation of azobenzene. (F) Structure of an azobenzene derivative containing a bifunctional crosslinker (4,4′-di(acrylamide)azobenzene). (G) Schematic overview of experimental design showing the treatments of photoresponsive hydrogels. The hydrogels were softened by exposure to UV (365 nm) irradiation for 30 min; 24 h later, (+) blue gels were exposed to blue (490 nm) light for 1 h, whereas (−) blue gels were not irradiated (*i.e.* left in the dark). (H) Plot of hydrogel stiffness obtained by AFM. (−) Blue gels were significantly softer than controls (5.9 *versus* 9.7 kPa, respectively), while (+) blue gels were much stiffer (7.6 kPa) than their (−) counterparts.[13] Reproduced from ref. 12 and 13 with permission from American Chemical Society, Copyright 2017 and 2018.

et al.[13] prepared a photoresponsive stiffening hydrogel using the well known photochromic reaction of azobenzene (Figure 11.3E–H). A special crosslinker agent bearing two vinyl groups at both ends of the azobenzene molecule was synthesised and used to prepare an acrylamide-based hydrogel. When the gel was irradiated with UV light before cell culturing, the azobenzene moieties were isomerised into their *cis*-state to afford a soft gel. Irradiation of the *cis*-gel with visible (blue) light induced a transition to the *trans*-state and hence increased the elastic modulus. This strategy avoided the exposure of cells to potentially cytotoxic UV light. Although the photoisomerisation of azobenzene itself is known to be reversible, the authors evaluated cellular behaviour only for a one-way *cis*-to-*trans* change induced by irradiation with cell-friendly visible light in view of the potential cytotoxicity of UV light.

11.3 Hydrogels Capable of Reversible Stiffness Changes

Many living tissues are exposed to periodic stress changes, *e.g.* cardiomyocytes are constantly exposed to heartbeat, and blood vessels are subjected to periodic stretching and compression synchronised with blood flow pulsation. Consequently, much effort has been directed at the fabrication of materials more closely mimicking the dynamic properties of a living body *via* the realisation of cell scaffold hydrogels with mechanical properties reversibly changing in response to external stimuli. Compared to that used in the case of unidirectional mechanical property changes, the latter material design is more complicated. Herein, we introduce studies on the realisation of reversible stiffness changes classified according to the external stimuli, namely chemical stimuli (*e.g.* chemical substances, pH, ions) and physical stimuli (*e.g.* temperature, light, magnetic field, electric field).

11.3.1 Reversible Stiffness Changes in Response to Chemical Stimuli

Yoshikawa *et al.*[14] used the self-assembly of an ABA-type triblock copolymer to realise hydrogels capable of reversible elastic modulus changes in response to pH as a typical chemical stimulus (Figure 11.4A, B). The well defined ABA-type triblock copolymer consisting of poly(2-(methyacryloyloxy)ethyl phosphorylcholine) (PMPC) as the middle block and poly(2-(diisopropylamino)ethyl methacrylate) (PDPA) as end blocks was prepared by atom-transfer radical polymerisation. PMPC is known to be biocompatible, while PDPA is a pH-responsive polymer with a pK_a of 6.2. The protonation state along the PDPA main chain drastically changes in response to extremely subtle changes around $pH \approx 7$, which is a condition suitable for cell culturing. At higher pH ($pH \approx 8$), the PDPA chains become more deprotonated and hence more hydrophobic, which leads to stronger interchain interactions and affords a highly elastic, physically crosslinked hydrogel substrate. On the other hand, the hydrophobic interaction between the partially charged PDPA blocks (and thus the physical crosslinking) becomes much weaker at lower pH (pH$\approx$7), which results in a softer gel substrate. These authors investigated the cellular response of mouse myoblasts (C2C12) to mechanical environment changes using the pH-responsive hydrogel and observed changes in the cell projected area corresponding to reversible changes in elastic modulus. A hydrogel capable of reversible stiffness change in response to subtle pH variation was further developed for the study of hMSC expansion (Figure 11.4C–G).[15] hMSCs cultured on the surface of the triblock copolymer gel maintained the expression of STRO-1 even when culturing was performed for 20 days, which suggested multiple lineage potentials. Furthermore, a change of substrate elasticity every two days suppressed the proliferation of hMSCs by up to 90% without any loss of multiple lineage potential. This behaviour implies that rhythmic changes of the elastic modulus result in an oscillation of the hMSC differentiation signal and the external mechanical environment, suggesting that the multipotency of stem cells is maintained upon the application of a periodic mechanical stimulus.

Sia *et al.*[16] used a hydrogel comprising collagen as a well-known natural ECM and alginate as a polysaccharide to prepare a synthetic scaffold material capable of reversible elastic modulus changes in response to ion species (Figure 11.5A). As both collagen and alginate bear –COOH groups, they can form chelating structures in the presence of polyvalent cations such as Ca^{2+}. The formation of these structures increases the elastic modulus of the hydrogel by increasing crosslinking density. In contrast, when monovalent cations such as Na^{+} are introduced, the physical crosslinking of the chelate structure break up, and the elastic modulus decreases. As a result, the authors demonstrated reversible elastic modulus changes in response to ion species coexisting with the hybrid gel. Although this pioneering work realised an *in vitro* cell scaffold material capable of reversible elastic modulus changes, it might be possible to accidentally modulate the intracellular signal transduction with Ca^{2+} cations added from the external environment, which act as a trigger to form physical crosslinking.

Stimuli-responsive hydrogels utilising biological phenomena such as protein folding/unfolding and enzymatic reactions have also been actively studied (Figure 11.5B),[17] *e.g.* Li *et al.* proposed incorporating protein-folding switches into tandem modular proteins to construct protein hydrogels and thus achieve dynamic mechanical property regulation. Under the concept of protein engineering, these authors prepared mutually exclusive proteins that changed their conformations in response to redox reactions. The newly synthesised protein became unfolded when its disulphide bond was cleaved under reducing conditions, whereas the folded state was recovered upon the formation of a cystine structure under oxidising conditions. As expected, the hydrogel crosslinked with the engineered protein swelled and became softer under reducing conditions and shrank and became stiffer under oxidising conditions, exhibiting an up to fivefold increase in E (Young's modulus). These mechanical property changes were fully reversible. DeForest *et al.*[18]

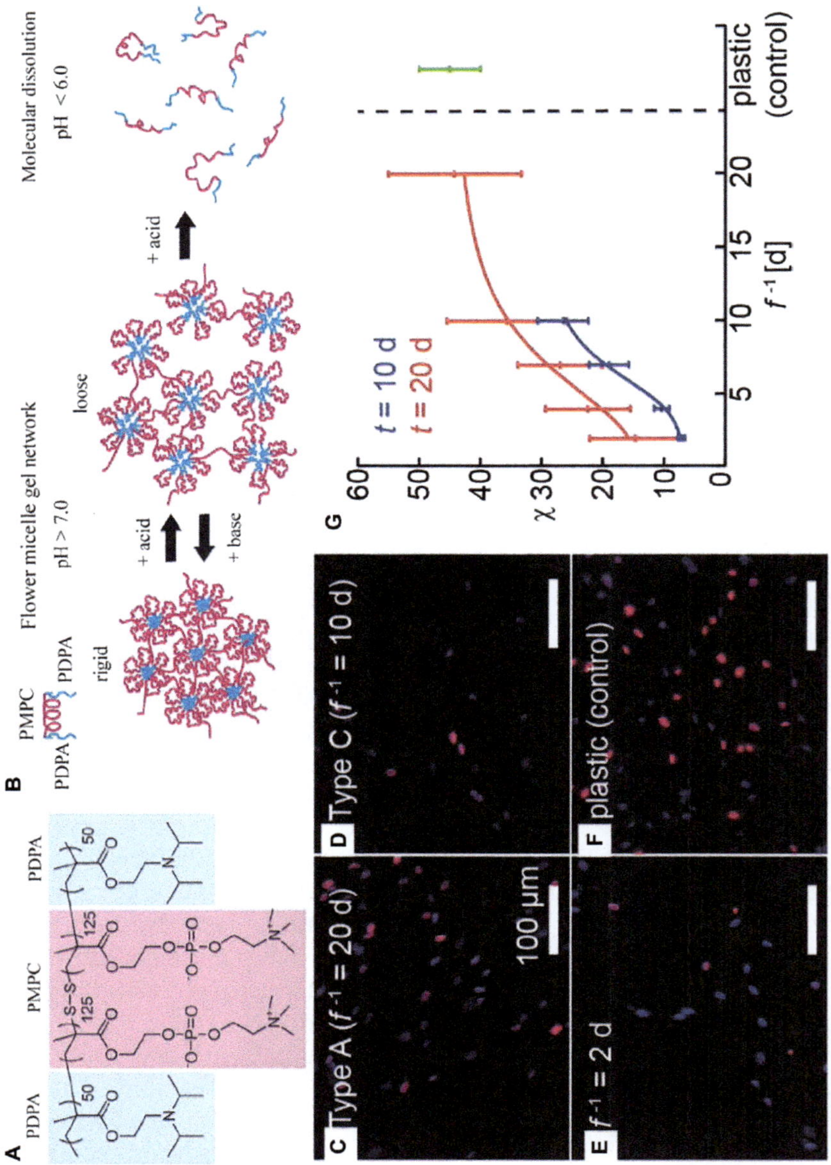
A
PDPA
PMPC
PDPA
B
PMPC
PDPA
PDPA
Flower micelle gel network
pH > 7.0
rigid
loose
+ acid
+ base
+ acid
Molecular dissolution
pH < 6.0
C Type A (f⁻¹ = 20 d)
100 µm
D Type C (f⁻¹ = 10 d)
E f⁻¹ = 2 d
F plastic (control)
G
t = 10 d
t = 20 d
χ
f⁻¹ [d]
plastic (control)

prepared a chemical stimulus–responsive hydrogel capable of reversible stiffness changes using protein conformation changes. Specifically, the authors synthesised a hydrogel featuring a tetra-arm PEG combined with calmodulin (CaM), a Ca^{2+}-responsive protein that adopts a compact conformation in the presence of Ca^{2+} because of the intramolecular binding of these ions by the M13 peptide *via* chelation. On the other hand, Ca^{2+} ions are chelated away in the absence of Ca^{2+}, and the CaM structure and the intramolecular association relax. CaM was attached to the chemically crosslinked branched-arm PEG, and the reversible conformational changes induced by the addition of Ca^{2+} were found to result in reversible stiffening. In a similar attempt, Lin *et al.*[19] used a sortase A (SrtA)-catalysed transpeptidation reaction to fabricate a chemically responsive hybrid protein-PEG hydrogel capable of reversible stiffening (Figure 11.5C). Specifically, a linear peptide crosslinker flanked by SrtA sequences was introduced into the hydrogel, and the pendant SrtA substrates could be covalently ligated in the presence of SrtA to form secondary crosslinking and thus increase the crosslinking density. The stiffened gel could be softened by the cleavage of the secondary crosslinked structure upon incubation with additional SrtA and the soluble glycine amide. In this case, the stiffening and softening processes were fully reversible.

In 2014, Lin *et al.*[20] reported that the chemical stimulus–triggered formation of a supramolecular β-cyclodextrin (β-CD) inclusion complex induced a reversible change in elastic modulus (Figure 11.6A). The authors incorporated β-CD into a permanently crosslinked PEG hydrogel using click chemistry. When adamantane (Ad)-modified four-arm PEG was added to the matrix of the β-CD–modified PEG hydrogel, the elastic modulus increased because of the increase in crosslinking density inside the polymer network. In view of the high equilibrium constant of Ad–β-CD complex formation, the addition of the branched Ad-PEG induced the formation of a polymer complex inside the β-CD–containing matrix and thus increased crosslinking density. The elastic modulus of the hydrogel returned to the original stiffness range when the formation of the β-CD–Ad inclusion complex was interrupted by the addition of free β-CD or phosphate-buffered saline (PBS).

Figure 11.4 (A) Structure of pH-sensitive PDPA50-PMPC250-PDPA50 ABA-type triblock copolymer. (B) Schematic representation of reversible structural changes occurring in the polymer network upon pH adjustment.[14] Dual staining (DAPI and anti-BrdU) of hMSCs (culturing time $t = 20$ days) experiencing (C) no change, (D) $f^{-1} = 10$ days, (E) $f^{-1} = 2$ days, and (F) the results obtained for hMSCs cultured on plastic dishes (control). (G) Plots of the fraction of anti-BrdU positive (proliferating) cells χ *versus* the frequency of mechanical step f^{-1} at 10 days (lower line) and 20 days (upper line). Each data point represents mean values ± s.d. for $n > 30$ cells.[15] Reproduced from ref. 14 with permission from American Chemical Society, Copyright 2011. Reproduced from ref. 15, https://doi.org/10.1038/srep24264, under the terms of the CC BY 4.0 license https://creativecommons.org/licenses/by/4.0/.

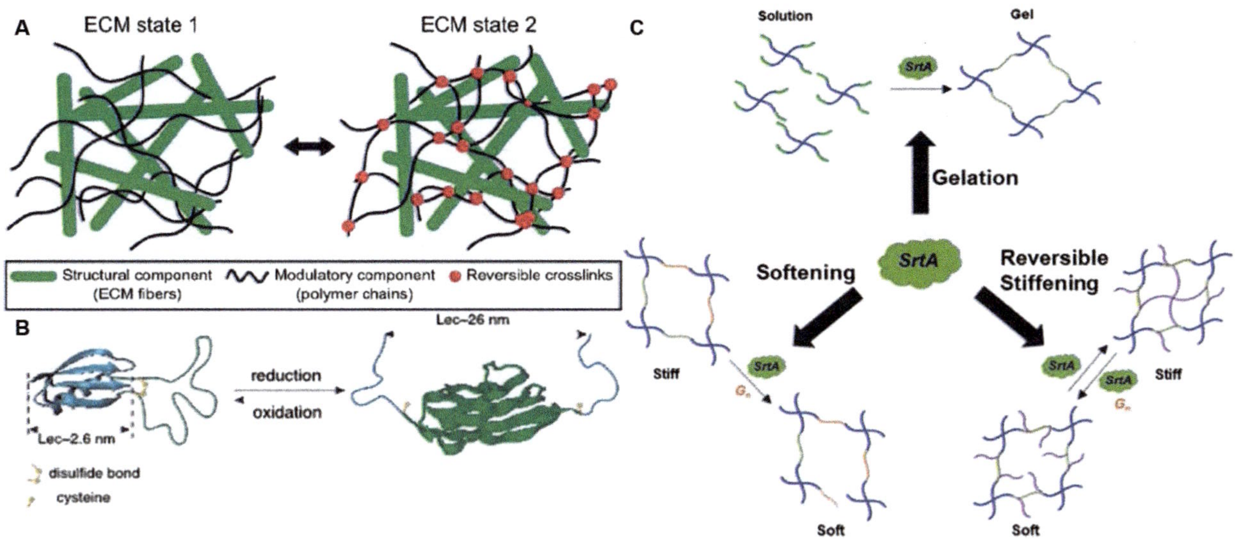

Figure 11.5 (A) Schematics of changes in the 3D microenvironment due to Ca^{2+} concentration switching. In the two-component ECM, one material serves as a structural scaffold (ECM fibres), while the other acts as a modulatory component (polymer chains) that can be reversibly crosslinked *via* chelation of multivalent ions.[16] (B) Schematics of the mutually exclusive protein-based redox-responsive folding switch.[17] (C) SrtA-mediated gelation of PEG peptide hydrogels.[19] Reproduced from ref. 16 and 17 with permission from John Wiley & Sons, Copyright 2010, 2014 Wiley-VCH Verlag GmbH & Co. KGaA, Weinheim. Reproduced from ref. 19 with permission from Elsevier, Copyright 2019.

On the other hand, Harada *et al.*[21] prepared an acrylamide-based copolymer comprising β-CD and Ad groups in identical polymer chains (Figure 11.6B, C). As β-CD and Ad groups along the polymer side chain formed an inclusion complex to afford physical crosslinkages, the polymer network created a host–guest gel at an appropriate polymer concentration and a suitable β-CD : Ad group ratio. Upon the addition of free β-CD, the elastic modulus of the host–guest hydrogel significantly decreased because of the drop in crosslinking density (as in the work of Lin *et al.*) but immediately recovered when β-CD was removed because of the re-formation of the pseudo-crosslinking from β-CD and Ad groups. Owing to the reversible nature of host–guest interactions, the magnitude of hydrogel softening and stiffening could be modulated by varying the concentrations of free host molecules (β-CD) in solution. The Young's modulus of the hydrogel ranged from 4 kPa to 11 kPa, and the projected areas of C2C12 on the hydrogels exhibited a clear correlation with substrate stiffness. The fitting of the experimental results with the empirical Hill equation indicated that hydrogel elasticity covered the optimal range for myoblast regulation. Furthermore, the dynamic mechanoresponse of C2C12 was fully reversible upon the softening and stiffening of the host–guest hydrogel substrate.

11.3.2 Reversible Stiffness Changes in Response to Physical Stimuli

Hydrogels utilising the cyclodextrin host–guest chemistry and undergoing reversible elastic modulus changes in response to physical stimuli are also available. Rosales *et al.*[22] used azobenzene and β-CD to fabricate a supramolecular hydrogel in which the number of inclusion complexes changes in response to light stimuli (Figure 11.6D). This host–guest supramolecular hydrogel was obtained by mixing a solution of azobenzene-modified hyaluronic acid (AzoHA) and β-CD–modified hyaluronic acid (CDHA). In this case, azobenzene was in a thermally stable *trans*-state; however, a thermally metastable *cis*-state was formed upon irradiation of the mixed solution with UV light. The equilibrium constant for the complexation of *cis*-azobenzene by β-CD (277 M^{-1}) was approximately 2.5 times lower than that for the complexation of *trans*-azobenzene (760 M^{-1}). Therefore, reversible elastic modulus changes due to the changes in the number of supramolecular crosslinking were realised by irradiation of a gel comprising AzoHA and CDHA. Note that the physical gel with dynamic crosslinking points constructed utilising host–guest chemistry featured a strongly frequency-dependent elastic modulus, especially when both azobenzene and CD were loaded. The photoisomerisation of azobenzene was also used to prepare a CD-free gel capable of reversible stiffness changes.[23] In this case, an azobenzene-containing bifunctional peptide was attached to the terminal of a multiarm PEG *via* Michael addition. Reversible changes in the hydrogel elastic modulus were due to the conformational changes in junction points

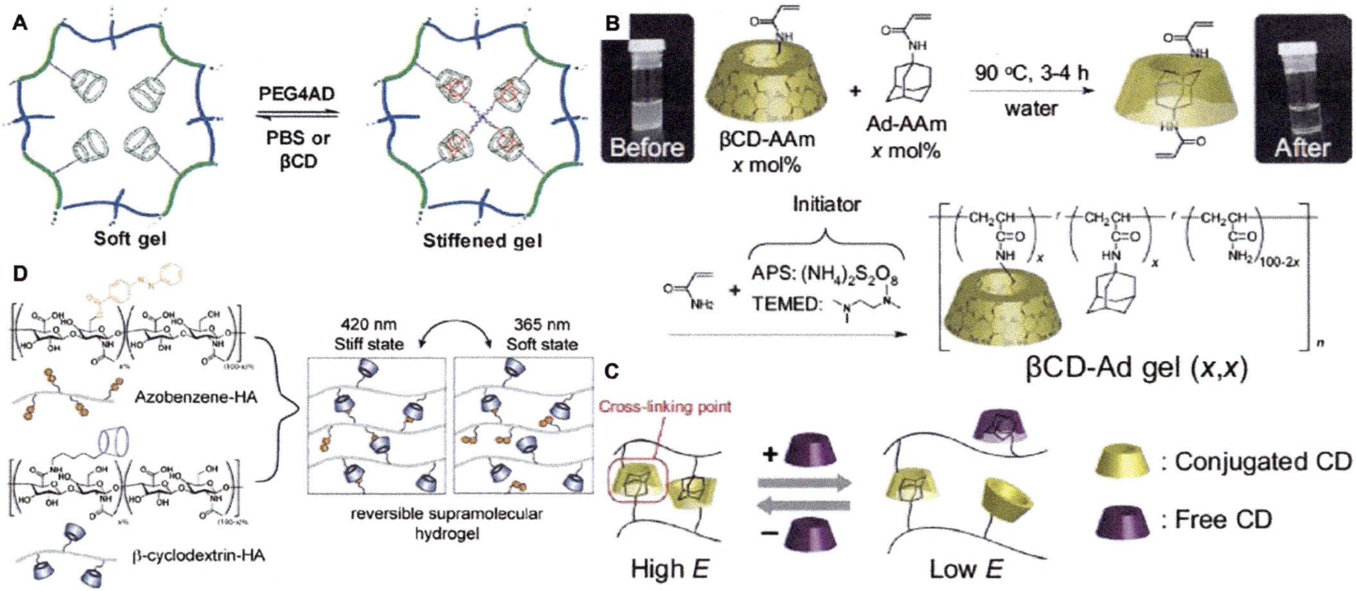

Figure 11.6 (A) Reversible elastic modulus changes of a β-CD-containing polymer network hybridised with Ad-modified tetra-branched PEG. Gel softening was achieved by incubating the stiffened gel in PBS or a solution of freely diffusing β-CD.[20] (B) Preparation and structure of β-CD–Ad gels (x, x), where x represents the contents of β-CD and Ad units (mol%). (C) Concept of the reversible switching of β-CD–Ad bonds connected to the polymer chains depending on β-CD concentration.[21] (D) Schematic illustration of azobenzene isomerisation within networks upon irradiation at 365 nm and the resulting decrease in the β-CD–azobenzene binding constant and crosslinking density. The isomerisation could be reversed by irradiation with visible light (420 nm).[22] Adapted from ref. 20 with permission from the Royal Society of Chemistry. Reproduced from ref. 21, https://doi.org/10.1038/s41598-017-07934-x, under the terms of the CC BY 4.0 license https://creativecommons.org/licenses/by/4.0/. Reproduced from ref. 22 with permission from American Chemical Society, Copyright 2018.

induced by the photoisomerisation of azobenzene. This photoisomerisation was not directly connected to the decrease in the number of crosslinkages, and the expansion and contraction degrees of the polymer chain connected to the photoisomerisable moiety were small, as the change in the distance due to the preceding isomerisation in the fluctuating gel network was also small. Thus the change in elastic modulus was as small as 10%. Ueki and Nakanishi *et al.*[24] used the photoisomerisation of azobenzene to change the hydrophilicity of the entire azobenzene-bearing polymer network and thus to realise another design of hydrogel scaffolds capable of reversible stiffness changes (Figure 11.7A, B). Copolymers comprising hydrophobic acrylate moieties with azobenzene side chains and hydrophilic dimethylacrylamide moieties exhibit lower critical solution temperature (LCST) type of phase separation temperatures dependent on the photoisomerisation state of azobenzene.[25] For this reason, in the temperature region between the LCST of the *trans*-form and that of the *cis*-form, the hydrophilicity of the entire azobenzene-containing polymer changes even at a constant temperature, depending on the photoisomerisation state. The authors synthesised a copolymer hydrogel with an optimal content of azobenzene and dimethylacrylamide monomers and found that the hydrogel swelling ratio, which is inversely proportional to the elastic modulus, reversibly changed upon irradiation with UV or visible light at a cell culture temperature of 37 °C. The hydrogel does not change the number of the crosslinking point itself inside the polymer network but does change the crosslinking density of the network by changes in its volume. The elastic modulus decreased by 38% upon the UV (365 nm, 58.5 J cm^{-2}) light-induced *trans*- to *cis*-isomerisation. As the initial elastic modulus and the extent of its change during photoisomerisation can be precisely tuned by optimising the hydrogel preparation conditions, the obtained hydrogel, which can stiffen or soften in response to the same stimulus (*i.e.* light), is expected to be useful for more patient-specific personal disease modelling.

Two different strategies of modulating crosslinking density were employed to achieve a hydrogel cell scaffold capable of reversible elastic modulus changes at a user-defined time point (Figure 11.7C).[26] A hyaluronic acid-based scaffold material was modified with an adhesion ligand (RGD), photocleavable nitrobenzyl groups, and vinylidene groups. The acrylate units were subsequently crosslinked with dithiothreitol. Under the condition of base-catalysed Michael addition, the acrylate units were nucleophilically attached in preference to methacrylate units because of the more electrophilic β-carbon of the former. Thus a hydrogel scaffold was formed by hyaluronic acid coupling. Irradiation of the resultant gels at 365 nm induced nitrobenzyl group cleavage and hence decreased the elastic modulus. Finally, a photoinitiator was introduced, and the hydrogel was irradiated at 400–500 nm to crosslink the methacrylate moieties along with the polymer backbone and thus increase the elastic modulus. As the methacrylate moieties were already present in the polymer network, there was no need for an additional crosslinking agent, which allowed the allocation of the

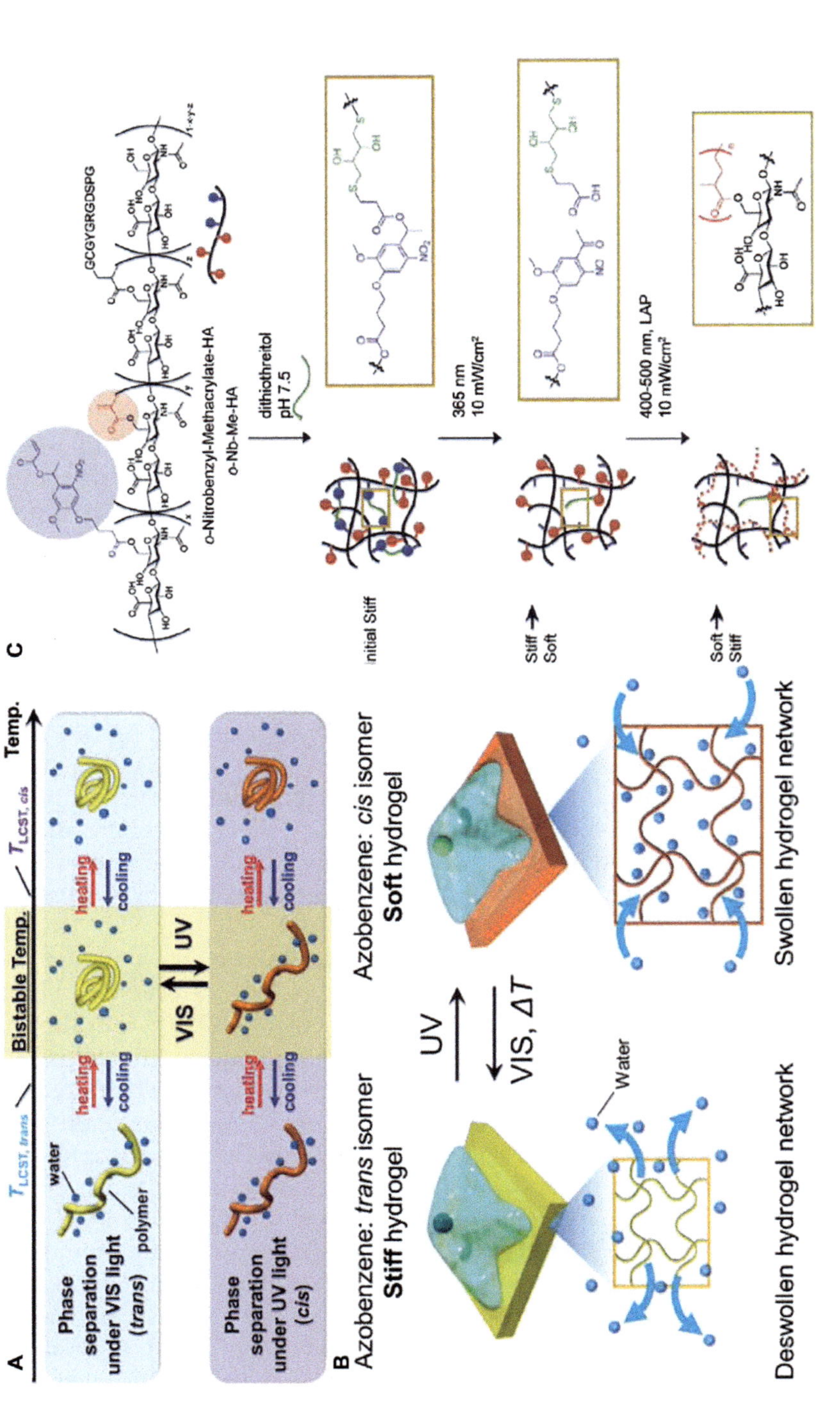
A
Phase separation under VIS light (trans)
Phase separation under UV light (cis)
water
polymer
heating
cooling
Bistable Temp.
Temp.
VIS
UV
B
Azobenzene: trans isomer
Stiff hydrogel
Azobenzene: cis isomer
Soft hydrogel
UV
VIS, ΔT
Water
Deswollen hydrogel network
Swollen hydrogel network
C
GCGYGRGDSPG
o-Nitrobenzyl-Methacrylate-HA
o-Nb-Me-HA
dithiothreitol
pH 7.5
Initial Stiff
365 nm
10 mW/cm²
Stiff → Soft
400-500 nm, LAP
10 mW/cm²
Soft → Stiff

macroscopic inhomogeneity of the elastic modulus caused by the later addition of crosslinking agents to the polymer network. Although only one softening–stiffening cycle was possible, a uniform reversible change in the elastic modulus was achieved in the entire region of the gel. This gel was subsequently used as a scaffold material for hMSCs, and changes in projected area, circularity, and nuclear localisation of YAP/TAZ following the change in hydrogel stiffness were investigated.

Numerous material designs have been developed to realise reversible elastic modulus changes in response to physical stimuli, *e.g.* protein conformation changes,[27] reversible photodimerisation reactions,[28] and magnetic fields.[29] Despite the difference in the input of physical and chemical stimuli, the molecular-level mechanism behind reversible stiffness changes is the same as that described in the previous section. For further details, the readers are referred to the original paper.

11.4 Cell Scaffolds Capable of Stress Relaxation

The elastic modulus of synthetic ECM scaffolds can significantly affect cell function and fate. In previous sections, we introduce the design principles and examples of hydrogel scaffolds that can dynamically switch the static component of mechanical properties and elastic modulus. On the other hand, time-dependent mechanical parameters such as viscoelasticity can also significantly impact cell response. Cooper-White *et al.*[30] were the first to systematically investigate the effect of substrate viscoelasticity on cell behaviour and phenotype (hMSCs). Through careful variation of the initial combination of an acrylamide monomer and a methylene bis(acrylamide) crosslinker, the authors produced a range of gels with a nearly constant elastic modulus, while the loss modulus (related to the viscous component of the substrate) varied over two orders of magnitude. An increase in the loss modulus of polyacrylamide hydrogels increased the hMSC spreading area and proliferation but decreased the size and maturity of focal adhesions. The ability of hMSCs to differentiate towards both osteogenic and adipogenic lineages was also enhanced on substrates with a high loss modulus. Mooney

Figure 11.7 Conceptual illustration of (A) the photoinduced phase transition of the azobenzene-containing polymer and (B) the corresponding crosslinked hydrogel exhibiting reversible mechanical property changes upon irradiation with UV or visible light.[24] (C) Structure and conceptual illustration of a hydrogel undergoing photochemical reactions. The initially stiff hydrogel was obtained using a Michael-type addition reaction with dithiothreitol and was softened *via* the cleavage of the *o*-nitrobenzyl group upon irradiation at 365 nm. Restiffening was induced by the radical polymerisation of methacrylate units upon irradiation at 400–500 nm in the presence of a photoinitiator.[26] Reproduced from ref. 24 with permission from Elsevier, Copyright 2021. Reproduced from ref. 26 with permission from John Wiley & Sons, Copyright 2017 Wiley-VCH Verlag GmbH & Co. KGaA, Weinheim.

et al.[31] systematically studied the mechanism of cellular responses to scaffold stress relaxation using an adaptive hydrogel that undergoes stress relaxation *via* ion crosslinking based on the alginate modification (Table 11.2). Computational simulations and experimental investigations suggested that at low initial elastic moduli and high ligand densities, more pronounced cell spreading should be observed for cells adhering to viscoelastic substrates that exhibit stress relaxation than for elastic substrates. Subsequently, the same researchers fabricated a stress relaxation hydrogel exhibiting independent elastic modulus changes from 9 kPa to 17 kPa and intrinsic stress relaxation times of 40–3300 s by changing the molecular weight of the crosslinked alginate.[32] When the stress relaxation hydrogel-based scaffold material was used for the 3D culturing of hMSCs, osteogenic differentiation was promoted inside the scaffold material with a shorter intrinsic stress

Table 11.2 Summary of adaptive hydrogels undergoing stress relaxation.

Reference	Origin of dissipation	Comment	Cell line
30	Viscous nature of gel	Same elastic gel possessing different loss moduli	hMSCs
31	Ionic bond	Both ionically and covalently crosslinked hydrogels	3T3, U2OS
32	Ionic bond	Stress relaxation control with ionically and covalently crosslinked hydrogels	3T3, hMSCs
33	Ionic bond	Stress relaxation control with ionically and covalently crosslinked hydrogels	Bovine chondrocyte cell
34	Dynamic covalent bond	Control of stress relaxation based on aldehyde–hydrazine dynamic covalent bond	hMSCs
35	Ionic bond	Independent control of stress relaxation and initial stiffness of alginate–PEG hydrogels	3T3, D1 mouse MSCs
36	Dynamic covalent bond	Control of stress relaxation based on aldehyde–hydrazine dynamic covalent bond	C2C12
37	Dynamic covalent bond	Fast stress relaxation (within ~10 s) based on boronate dynamic covalent bond	hMSCs
38	Dynamic covalent bond	Fast stress relaxation (within ~10 s) based on boronate dynamic covalent bond	MC3T3-E1
39	Polymer physical association	Stress relaxation switching from viscoelastic gel to elastic gel	HeLa

relaxation time (60 s) even at a similar initial modulus (17 kPa). As matrix remodelling occurs more readily in materials with shorter stress relaxation times, the clustering of ligands such as RGD is expected to occur on a shorter timescale, creating a matrix more susceptible to osteogenesis. This explanation is consistent with the fact that the bone differentiation of hMSCs prevents culturing inside nondegradable crosslinked elastic hydrogels.[10] Chaudhuri *et al.* comprehensively investigated the effects of stress relaxation on cell behaviour (including chondrocyte differentiation) using the precise control of stress relaxation time based on changes in alginate molecular weight,[33] control of the rate constant of the dynamic hydrazine–aldehyde covalent bond formation (Figure 11.8A, B),[34] and detailed studies on the independent control of the initial modulus and stress relaxation time by changing the number and molecular weight of PEG grafting to alginate (Figure 11.8C).[35] Anseth *et al.* produced a hydrogel scaffold material capable of stress relaxation using tetra- and octabranched PEG end-modified with functional groups capable of dynamic covalent bonding instead of using biopolymer-derived materials such as polysaccharides. Reversible hydrazine–aldehyde covalent bonding,[36] as well as the reversible formation of boronates from boronic acids and diols (*e.g.* catechol)[37] (Figure 11.8D, E), were used to achieve rapid stress relaxation. Sun *et al.*[38] synthesised a hydrogel with a stress relaxation time of <10 s using type I collagen (which encapsulates the adhesion factors for cells and fibrillates) as the main matrix and composites of dialdehyde-functionalised PEG (which form dynamic crosslinking by reacting with the collagen amino groups). Stress relaxation gels were used to culture mouse pre-osteoblast MC3T3-E1 cells, and fast stress relaxation effectively promoted osteodifferentiation *via* the activation of TRPV4 ion channels.

Tamate and Ueki *et al.*[39] precisely synthesised an ABA-type triblock copolymer capable of photoswitchable stress relaxation using reversible addition-fragmentation chain transfer (RAFT) polymerisation (Figure 11.9). The ABA-type triblock copolymer (P(NIPAAm-*r*-CA)-*b*-PEG-*b*-P(NIPAAm-*r*-CA)) could be switched between the states of a viscoelastic liquid (viscoelastic hydrogel) capable of stress relaxation and a viscoelastic solid (elastic hydrogel) not capable of stress relaxation. Each end of the block copolymer comprised a statistical polymer with temperature-responsive poly(*N*-isopropylacrylamide) (PNIPAAm) moieties and an acrylate with a coumarine side chain (CA) that underwent photodimerisation, while the centre comprised a bioinert PEG. As the polymer segments at both ends exhibited temperature responsiveness derived from NIPAAm, they formed a physically crosslinked structure because of the hydrophobic interactions between the polymers above the phase transition temperature, which resulted in gelation. The formation of physical crosslinking by the association of polymer aggregates allowed the viscoelastic hydrogel to exhibit stress relaxation. The stress relaxation time corresponded to the time when the polymer chains were pulled out from the polymer physical association (physically crosslinked structure). Irradiation at 365 nm induced a photodimerisation

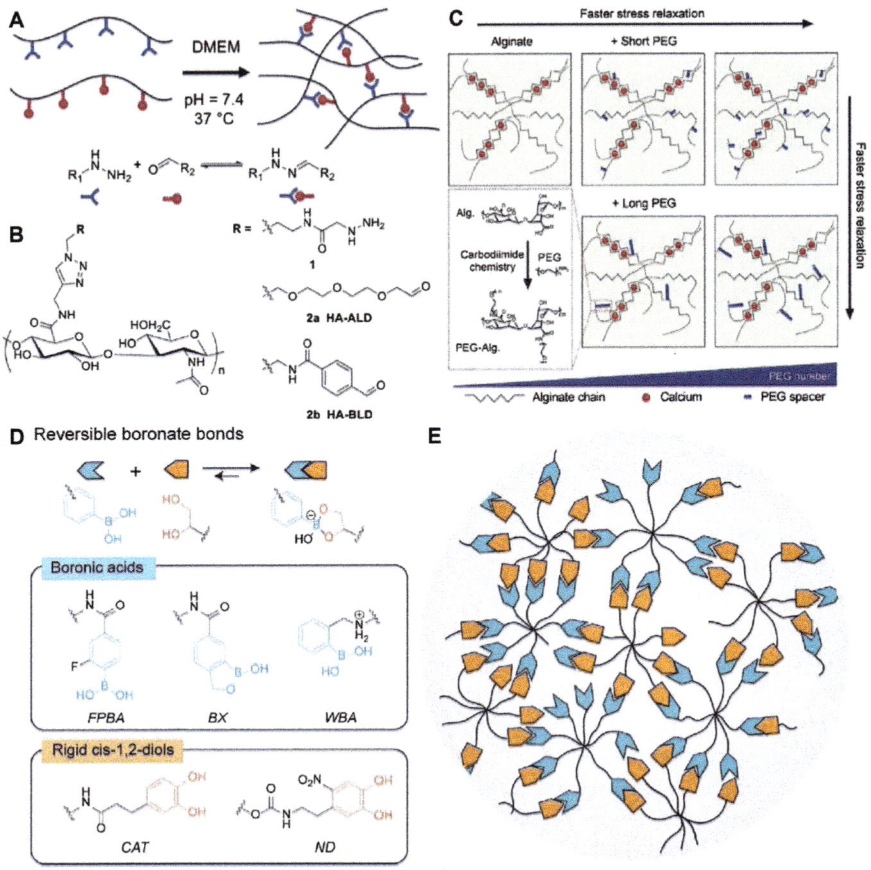

Figure 11.8 (A) Formation of a viscoelastic hyaluronic acid–based single-network hydrogel *via* dynamic covalent hydrazone crosslinking. (B) Chemical structures of hyaluronic acid modified with an aliphatic aldehyde (2a, HA-ALD) and benzaldehyde (2b, HA-BLD).[34](C) Schematic illustration depicting the molecular structure of the alginate-PEG hydrogel. Precise control over stress relaxation time was achieved by grafting alginate polymers with PEG chains of various lengths and numbers through carbodiimide chemistry.[35] (D) Reversible formation of boronates from boronic acids and *cis*-1,2-diols. Chemical structures of several boronic acids (2-fluorophenylboronic acid [FPBA], boroxole [BX], Wulff-type-like boronic acid [WBA]) and *cis*-1,2-diols (catechol [CAT] and nitrodopamine [ND]). (E) Schematic illustration of a dynamic network based on boronates derived from boronic acids and *cis*-1,2-diols functionalised with eight-arm PEGs.[37] Reproduced from ref. 34 and 35 with permission from Elsevier, Copyright 2018, 2019. Reproduced from ref. 37, https://doi.org/10.1002/advs.201800638, under the terms of the CC BY 4.0 license https://creativecommons.org/licenses/by/4.0/.

reaction between CA moieties inside the physical crosslinking, thus introducing permanent covalent bonds inside these linkages and affording a chemically crosslinked gel incapable of stress relaxation. When HeLa cells

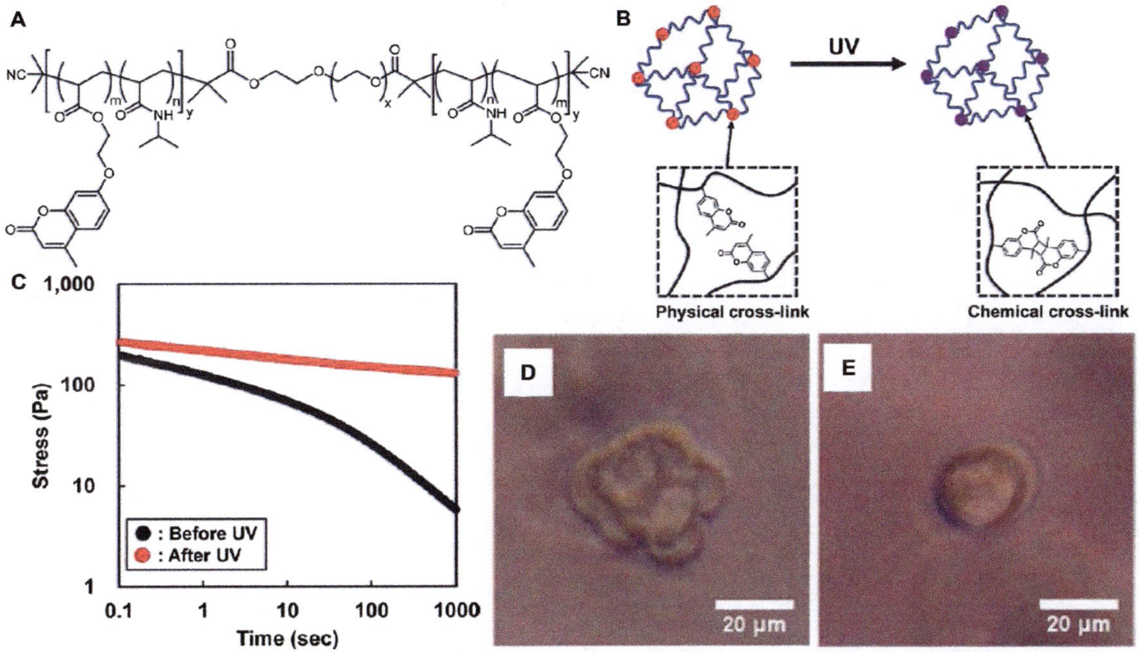

Figure 11.9 (A) Structure of an ABA-type triblock copolymer, P(NIPAAm-*r*-CA)-*b*-PEG-*b*-P(NIPAAm-*r*-CA). (B) Coumarine (CA) photodimerisation induces a change from a stress relaxation hydrogel (viscoelastic liquid) to an elastic hydrogel that does not undergo stress relaxation (viscoelastic solid). (C) Results of stress relaxation tests performed for the hydrogel before and after UV irradiation at 37 °C. Representative phase contrast images acquired after 2 days of incubation for HeLa cells cultured inside hydrogels (D) capable and (E) incapable of stress relaxation. Reproduced from ref. 39 with permission from American Chemical Society, Copyright 2016.

were cultured in this hydrogel, a clear difference in cell shape was observed before and after the switching of stress relaxation. Prior to irradiation (viscoelastic liquid capable of stress relaxation), the cells showed a spreading morphology, whereas after irradiation (viscoelastic solid showing no stress relaxation), the cells retained their rounded shape. Despite several examples of changes in static mechanical properties (such as elastic modulus) in response to stimuli, this material platform is unique, as it allows stress relaxation to be controlled freely and at any time point in response to light stimuli using precisely designed synthetic polymers.

11.5 Conclusions and Future Perspectives

In this chapter, the recent showcasing research in the fabrication of dynamic hydrogels for materials-based mechanobiology were summarized. In the first part, we introduced switchable hydrogels that could change their stiffness in response to external trigger from the outside the cell scaffold, such as a cell secretion or artificial trigger from the external environment. In this context, photoillumination triggers are useful and frequently used in the design of switchable hydrogels (Table 11.1). Contactless and spatiotemporally controllable photostimuli are highly desirable for the design of stimuli-responsive mechanobiological materials. On the other hand, short-wavelength light is known to show cell cytotoxicity. Therefore, the construction of a more sophisticated light-responsive mechanobiological cell scaffold requires the establishment of a fundamental technology for photochromic reactions that can be driven to completion using long-wavelength low-intensity light. In recent years, the magnitude of the influence of the dissipative component, such as stress relaxation, of the polymer gel on biological phenomena has been clarified. We introduced adaptive hydrogel that could express stress relaxation in the final part of this chapter. A recent theoretical research impressively confirmed that the viscous feature of viscoelastic substrate significantly affects cell behaviour: cell spreading speed depends on the stress relaxation of cell scaffold only in the case when the elastic modulus of gel is low. In the case of such a soft substrate, cell spreading maximizes when the relaxation times of the cells and of the substrate material match.[40] From this perspective, to gain deeper insight into cell behaviour against the viscoelastic substrate, a materials design strategy of dynamic gels is expected that can be tuned to the elastic modulus and viscous modulus of gel independently. To elucidate the effect of the dissipative nature of substrate on cellular behaviour, research on the behaviour of cells at the hydrophobic liquid interface has also been actively conducted.[41,42] In principle, however, the elastic modulus of gels can be controlled only by the effective polymer chain participating in the crosslinking, however, dissipative features of gels are dominated not only by the elastic polymer matrix but also by the viscous components of other building blocks of gel, such as noncrosslinked polymer residue (dangling chain, looping back chain, *etc.*), and the viscosity of liquid entrapped within a polymer network. In view of this knowledge, the construction of dynamic gels independently of a controllable static mechanical feature (*e.g.* Young's modulus, *etc.*) and a dynamic feature (*e.g.* stress relaxation, dissipative nature of gel *etc.*), focusing on the role of solvents, might become an interesting research field in the future.

References

1. J. A. Burdick and W. L. Murphy, *Nat. Commun.*, 2012, **3**, 1269.
2. A. M. Rosales and K. S. Anseth, *Nat. Rev. Mater.*, 2016, **1**, 15012.
3. K. Uto, J. H. Tsui, C. A. DeForest and D. H. Kim, *Prog. Polym. Sci.*, 2017, **65**, 53–82.
4. P. J. LeValley and A. M. Kloxin, *ACS Macro Lett.*, 2018, **8**, 7–16.
5. O. Chaudhuri, J. Cooper-White, P. A. Janmey, D. J. Mooney and V. B. Shenoy, *Nature*, 2020, **584**, 535–546.
6. A. M. Kloxin, A. M. Kasko, C. N. Salinas and K. S. Anseth, *Science*, 2009, **324**, 59–63.
7. C. Yang, M. W. Tibbitt, L. Basta and K. S. Anseth, *Nat. Mater.*, 2014, **13**, 645–652.
8. T. E. Brown, I. A. Marozas and K. S. Anseth, *Adv. Mater.*, 2017, **29**, 1605001.
9. M. Guvendiren and J. A. Burdick, *Nat. Commun.*, 2012, **3**, 792.
10. S. Khetan, M. Guvendiren, W. R. Legant, D. M. Cohen, C. S. Chen and J. A. Burdick, *Nat. Mater.*, 2013, **12**, 458–465.
11. K. A. Gunay, T. L. Ceccato, J. S. Silver, K. L. Bannister, O. J. Bednarski, L. A. Leinwand and K. S. Anseth, *Angew. Chem., Int. Ed.*, 2019, **58**, 9912–9916.
12. V. X. Truong, F. Li and J. S. Forsythe, *ACS Macro Lett.*, 2017, **6**, 657–662.
13. I. N. Lee, O. Dobre, D. Richards, C. Ballestrem, J. M. Curran, J. A. Hunt, S. M. Richardson, J. Swift and L. S. Wong, *ACS Appl. Mater. Interfaces*, 2018, **10**, 7765–7776.
14. H. Y. Yoshikawa, F. F. Rossetti, S. Kaufmann, T. Kaindl, J. Madsen, U. Engel, A. L. Lewis, S. P. Armes and M. Tanaka, *J. Am. Chem. Soc.*, 2011, **133**, 1367–1374.
15. V. Frank, S. Kaufmann, R. Wright, P. Horn, H. Y. Yoshikawa, P. Wuchter, J. Madsen, A. L. Lewis, S. P. Armes, A. D. Ho and M. Tanaka, *Sci. Rep.*, 2016, **6**, 24264.
16. B. M. Gillette, J. A. Jensen, M. Wang, J. Tchao and S. K. Sia, *Adv. Mater.*, 2010, **22**, 686–691.
17. N. Kong, Q. Peng and H. Li, *Adv. Funct. Mater.*, 2014, **24**, 7310–7317.
18. L. Liu, J. A. Shadish, C. K. Arakawa, K. Shi, J. Davis and C. A. DeForest, *Adv. Biosyst.*, 2018, **2**, 1800240.
19. M. R. Arkenberg, D. M. Moore and C. C. Lin, *Acta Biomater.*, 2019, **83**, 83–95.
20. H. Shih and C. C. Lin, *J. Mater. Chem. B*, 2016, **4**, 4969–4974.
21. M. Horning, M. Nakahata, P. Linke, A. Yamamoto, M. Veschgini, S. Kaufmann, Y. Takashima, A. Harada and M. Tanaka, *Sci. Rep.*, 2017, **7**, 7660.
22. A. M. Rosales, C. B. Rodell, M. H. Chen, M. G. Morrow, K. S. Anseth and J. A. Burdick, *Bioconjugate Chem*, 2018, **29**, 905–913.

23. A. M. Rosales, K. M. Mabry, E. M. Nehls and K. S. Anseth, *Biomacromolecules*, 2015, **16**, 798–806.
24. K. Homma, A. C. Chang, S. Yamamoto, R. Tamate, T. Ueki and J. Nakanishi, *Acta Biomater.*, 2021, **132**, 103–113.
25. R. Kröger, H. Menzel and M. L. Hallensleben, *Macromol. Chem. Phys.*, 1994, **195**, 2291–2298.
26. A. M. Rosales, S. L. Vega, F. W. DelRio, J. A. Burdick and K. S. Anseth, *Angew. Chem., Int. Ed.*, 2017, **56**, 12132–12136.
27. M. Horner, K. Raute, B. Hummel, J. Madl, G. Creusen, O. S. Thomas, E. H. Christen, N. Hotz, R. J. Gubeli, R. Engesser, B. Rebmann, J. Lauer, B. Rolauffs, J. Timmer, W. W. A. Schamel, J. Pruszak, W. Romer, M. D. Zurbriggen, C. Friedrich, A. Walther, S. Minguet, R. Sawarkar and W. Weber, *Adv. Mater.*, 2019, **31**, e1806727.
28. V. X. Truong, F. Li, F. Ercole and J. S. Forsythe, *ACS Macro Lett.*, 2018, **7**, 464–469.
29. A. A. Abdeen, J. Lee, N. A. Bharadwaj, R. H. Ewoldt and K. A. Kilian, *Adv. Healthcare Mater.*, 2016, **5**, 2536–2544.
30. A. R. Cameron, J. E. Frith and J. J. Cooper-White, *Biomaterials*, 2011, **32**, 5979–5993.
31. O. Chaudhuri, L. Gu, M. Darnell, D. Klumpers, S. A. Bencherif, J. C. Weaver, N. Huebsch and D. J. Mooney, *Nat. Commun.*, 2015, **6**, 6364.
32. O. Chaudhuri, L. Gu, D. Klumpers, M. Darnell, S. A. Bencherif, J. C. Weaver, N. Huebsch, H. P. Lee, E. Lippens, G. N. Duda and D. J. Mooney, *Nat. Mater.*, 2016, **15**, 326–334.
33. H. P. Lee, L. Gu, D. J. Mooney, M. E. Levenston and O. Chaudhuri, *Nat. Mater.*, 2017, **16**, 1243–1251.
34. J. Lou, R. Stowers, S. Nam, Y. Xia and O. Chaudhuri, *Biomaterials*, 2018, **154**, 213–222.
35. S. Nam, R. Stowers, J. Lou, Y. Xia and O. Chaudhuri, *Biomaterials*, 2019, **200**, 15–24.
36. D. D. McKinnon, D. W. Domaille, J. N. Cha and K. S. Anseth, *Adv. Mater.*, 2014, **26**, 865–872.
37. S. Tang, H. Ma, H.-C. Tu, H.-R. Wang, P.-C. Lin and K. S. Anseth, *Adv. Sci.*, 2018, **5**, 1800638.
38. A. Liu, K. Wu, S. Chen, C. Wu, D. Gao, L. Chen, D. Wei, H. Luo, J. Sun and H. Fan, *Biomacromolecules*, 2020, **21**, 3745–3755.
39. R. Tamate, T. Ueki, Y. Kitazawa, M. Kuzunuki, M. Watanabe, A. M. Akimoto and R. Yoshida, *Chem. Mater.*, 2016, **28**, 6401–6408.
40. Z. Gong, S. E. Szczesny, S. R. Caliarie, E. E. Charrier, O. Chaudhuri, X. Cao, Y. Ling, R. L. Mauck, P. A. Janmey, J. A. Burdick and V. B. Shenoy, *Proc. Natl. Acad. Sci. U. S. A.*, 2018, **115**, E2686–E2695.
41. D. Kong, L. Peng, S. Di Cio, P. Novak and J. E. Gautrot, *ACS Nano*, 2018, **12**, 9206–9213.
42. X. Jia, K. Minami, K. Uto, A. C. Chang, J. P. Hill, J. Nakanishi and K. Ariga, *Adv. Mater.*, 2020, **32**, e1905942.

CHAPTER 12

Manipulation of Durotaxis on a Matrix with Cell-scale Stiffness Heterogeneity

SATORU KIDOAKI

Lab of Biophysical and Biomedical Chemistry, IMCE, Kyushu University, CE41-204, 744 Motooka, Nishi-ku, Fukuoka 819-0395, Japan
Email: kidoaki@ms.ifoc.kyushu-u.ac.jp

12.1 Introduction

Durotaxis, directional cell movement toward the stiffer region of the extracellular environment, was first reported in 2000.[1] In early studies on durotaxis, the term mechanotaxis was also used.[2,3] The prefix to the root "-taxis" indicates the driving force of taxis, such as in chemotaxis, phototaxis, geotaxis, topotaxis, *etc.* Thus mechanical stimuli-induced taxis should generally be referred to as mechanotaxis. On the other hand, "duro-" refers to stiffness. Nowadays, the term "durotaxis" is especially used for substrate stiffness gradient–induced taxis, while "mechanotaxis" is applied mainly to shear-stress-induced taxis.[4–6] Under this definition, durotaxis is a type of substrate adhesion-based tactic behavior, along with thigmotaxis,[7] haptotaxis,[8] topotaxis,[9] plithotaxis,[10] ratchetaxis,[11] and curvotaxis.[12] To investigate durotaxis, it is critical to rigorously separate these various tactic behaviors from stiffness gradient–induced taxis. This review focuses on only pure durotaxis, while carefully excluding the effects of other factors related to thigmotactic cues of surface chemistry and topography, in order to precisely discuss the effect of stiffness of the extracellular environment on taxis from the view of materials-based mechanobiology.

Biomaterials Science Series No. 12
Material-based Mechanobiology
Edited by Jun Nakanishi and Koichiro Uto

Published by the Royal Society of Chemistry, www.rsc.org

Durotaxis has attracted considerable interest from mechanobiologists since this phenomenon is deeply related to the mechanosensory activity of cells. The recognition of mechanosensory cell migration opened the way for research on matrix stiffness-dependent modulation of cellular functions such as proliferation[13–16] and differentiation.[17–20] Efforts to identify the mechanism for such cellular responses have enhanced our understanding of the mechanosensitive growth of focal adhesions and cytoskeletons and of the dynamics of intracellular traction stress.[21,22] In this sense, it would not be an exaggeration to state that studies on durotaxis have been critical for the development of mechanobiology. Despite this importance, general views on durotaxis have not yet been established due to the difficulty of systematically comparing the characteristics of durotaxis reported by different studies, each of which used a different experimental system, *i.e.* differences in the range of the stiffness gradient, the material properties of the matrix, cell type, *etc.* This review presents general conditions to induce durotaxis and proposes methods for the manipulation and application of durotaxis.

12.2 Threshold Stiffness Gradient Required to Induce Durotaxis

12.2.1 Fabrication of a Stiffness Gradient Substrate

Adherent cells sense a stiffness gradient in the culture environment and process the mechanical signals from the extracellular milieu to exhibit directed cell movement toward the stiffer region. In the first report on this motility, Lo *et al.* prepared a hydrogel substrate with a sharp elasticity boundary between a stiff region (Young's modulus: 30 kPa) and a soft region (14 kPa) by placing pieces of stiff and soft polyacrylamide (PAAm) gels, with chemically immobilized collagen on the top surface, adjacent to each other, and observing the movement of fibroblast cells between the gels.[1] They found that the cells approached the boundary and moved from the soft region to the stiff region but did not return from the stiff region to the soft region. They called this behavior durotaxis. In their experiments, the stiffness gradient was limited to a *ca.* 15 kPa jump in the elastic modulus at the sharp boundary due to the fabrication method (adjoining two pieces of gel with different stiffness values). Although this method can provide a sharp stiffness boundary and a steep jump in the elastic modulus, it is difficult to ensure topographic continuity between the two pieces of gels. Hydrogels with different stiffness values generally exhibit different degrees of swelling, which should, in principle, produce a discrepancy in the thickness of the gels. A topographic bump should affect natural cell movement around the boundary and interfere with the induction of durotaxis, so the pure effect driven by the stiffness distribution in the induction of durotaxis must be checked after the topographic disturbance is excluded. In early studies, the difficulty of systematically preparing various stiffness gradient conditions on a hydrogel surface with smooth topographic continuity hindered the precise determination of the threshold stiffness gradient required to induce durotaxis. Therefore, in subsequent research on durotaxis, various methods have been used to fabricate materials with a stiffness gradient, such as a gradient maker,[23,24] molding/layering patterning,[25] photolithography,[3,26–33] *etc.*

Gradient makers or generators that use a microfluidic device to prepare a concentration gradient of gelators solved the issue of the systematic preparation of a stiffness gradient on hydrogel with a smooth topography.[23,24] The device is equipped with several inlet channels to introduce a monomer solution with different concentrations into the chamber. The chamber is composed of multibranched paths of convoluted microfluidic channels connected to one another. The monomer solutions are mixed at different ratios while flowing inside the channels and are then finally ejected from different adjoining outlets. Monomer solutions with different concentrations generate laminar flow, which can be photocrosslinked, and this generates a stiffness gradient on the gel surface. On the other hand, this method provides only a diffuse stiffness gradient field over a millimeter scale. The typical gradient strength tends to be weak, such as *ca.* 15 Pa μm^{-1}. In that system, the weak tendency of cell repositioning over a long time-scale was observed, and the induction of durotaxis was very weakly limited.[23,24] It was difficult to perform a systematic cell trajectory analysis for a number of cells.

Molding/layering patterning is another good option for preparing a sharp stiffness boundary with a flat surface.[25] In this method, a stiffness boundary is designed on a micropatterned mold of a glass substrate covered first with a thick layer and then a second thin layer of polydimethylsiloxane (PDMS). The thick PDMS layer is formed on a glass trench and produces the soft region, while the thin PDMS layer on the support substrate makes the stiff region. The surface is almost completely flat where the stiffness boundary becomes sharp. Though this method offers a good balance between a sharp stiffness jump and smooth topographic continuity, systematic setting of the local thickness of each layer, *i.e.* the local depth of the trench support, is required to achieve a desired distribution of the elastic modulus and strength of the stiffness gradient. Fabrication of the support substrate is needed to precisely tune the local stiffness, which is difficult for hydrogels with values under several tens of kPa.

To achieve the systematic tunability of stiffness in the physiological range, the designability of a sharp stiffness jump, and topographic continuity, photographic microstiffness patterning is advantageous for fabrication of the culture substrate to investigate durotaxis. Several options are available for making such a platform. Materials that are frequently used for studies on durotaxis include PDMS and PAAm, chemically modified with cell-adhesive extracellular factors such as fibronectin and a collagen layer on the top surface. Typical tunable ranges of the elastic modulus are MPa for PDMS and 0.1–100 kPa for PAAm; thus PAAm is more suitable for investigating durotaxis under physiological conditions. On the other hand, a collagen layer immobilized on the surface of PAAm gels may have a certain instability in its modified state. For example, Yip *et al.* pointed out that the lateral strain of the collagen layer affected the cell adhesion state and that the methods used for modification, such as the type of immobilization reagent, should affect the lateral mobility of the collagen molecules.[34] Such lateral deformability of a thin polymer layer on the top surface of the substrate has been revealed to critically affect cell spreading, thus interfering with the lineage specification of cultured stem cells.[35,36] In this sense, a noncell-adhesive substrate modified with collagen is not necessarily an ideal system for investigating the mechanosensory activity of cells, though this system has been extensively used in this research field.

To solve these problems, one of the useful potential candidates is gelatin, a completely denatured form of collagen. Cell adhesion to gelatin is rather simple and well defined, and it is based solely on the interaction between integrin and adhesive ligands adsorbed from the culture medium, such as fibronectin and vitronectin. In the case of collagen-immobilized gels, the surface binding-induced partial denaturation of the triple helix in collagen should be an issue. The collagen triple helix includes the GFOGER sequence, which is a binding site to integrin.[37] Since integrin is a biomechanical sensor that serves as a hub in transmitting signals between cells and the culture substrate, uncertainty regarding the conformation and content of integrin binding sites complicates the attempt to characterize cell behaviors. Gelatin can be easily chemically modified because of the abundant chemical groups suitable for bioconjugation, such as amino- and carboxy-groups, which can be modified with a photoreactive moiety to prepare photocurable gelatins for photolithographic position-dependent crosslinking reactions. The resulting photocurable gelatins can form a single bulky layer of hydrogels, so additional unstable effects of the lateral deformability of a different polymer layer such as collagen should be excluded. Therefore, we have established a surface elasticity micropatterning for photocurable styrenated gelatin gels, a cell adhesion hydrogel with a variable surface modulus, and have investigated the conditions of the gradient intensity at the elastic boundary required to induce and control durotaxis.

12.2.2 Determination of the Threshold Stiffness Gradient Required to Induce Durotaxis

The first systematic characterization of the threshold stiffness gradient required to induce durotaxis was performed on microelastically patterned photocrosslinkable gelatins,[3,27] which showed that the induction of durotaxis by fibroblasts requires a certain threshold gradient of elastic modulus within a single-cell adhesion area and depended on both the elastic modulus in the soft region and the matrix density (gelatin concentration in the gels). For example, on 40 wt% gelatin gels, durotaxis occurred with a stiffness gradient of 40 kPa per 50 μm on a soft base of a few kPa, and with a stiffness gradient of 100–200 kPa per 50 μm on a soft base of 20 kPa.[3] On the other hand, on 30% gelatin gels, a sharp stiffness gradient of 300–400 kPa per 50 μm on a soft base of 10 kPa is required.[27] Below the threshold stiffness gradient, durotaxis did not occur. As exemplified in these early studies, threshold stiffness gradients have been reported with large variations. We had considered that such variations were due to differences in the experimental systems, such as the matrix conditions (kind of matrix; collagen/PAAm, gelatin, PDMS, density of matrix) and the kind of cells tested. A systematic summary of the threshold stiffness conditions can be found in a comprehensive review.[38] Although these factors significantly affect the threshold gradient required to induce durotaxis, we found that the threshold value itself may be different in some cases even if the kind of materials, density condition, and cell type are the same. That was the effect from absolute elasticity in the soft base region as discussed next.[33]

To clarify the principles for the manipulation of durotaxis, we focused on the absolute stiffness of the soft region and evaluated its effect on the threshold gradient (TG) required to induce durotaxis. Microelastically patterned gels that differed with respect to both the absolute stiffness of the soft region and the strength of the stiffness gradient were photolithographically prepared using photocrosslinkable gelatins, and the TG for mesenchymal stem cells (MSCs) was examined systematically for each condition. As a result, the TG values for soft regions with elastic moduli of 2.5, 5.0, and 10 kPa were 0.14, 1.0, and 1.4 kPa μm^{-1}, respectively; *i.e.* TG markedly increased with an increase in the absolute stiffness of the soft region. An analysis of the area and the long-axis length for focal adhesions revealed that the adhesivity of MSCs was more stable on a stiffer soft region. These results suggested that the initial location of cells starting durotaxis plays an essential role in determining the TG values, and furthermore the relationship between the position-dependent TG and intrinsic stiffness gradient (IG) of the culture substrate should be carefully considered for inducing durotaxis; the IG must be higher than TG ($IG \geq TG$) (Figure 12.1). This principle provides a fundamental guide for designing biomaterials to manipulate cellular durotaxis. Therefore, when researchers consider durotaxis in their system, the most important point is that the TG sensitively varies depending on the stiffness at the initial location and cannot be determined without defining the initial position of the cell.

12.3 Manipulation of Durotaxis

12.3.1 Rectified Durotaxis

As clarified previously, the strength and efficiency of durotaxis depend on the absolute stiffness of the softer region where the cells start to move and on the position-dependent threshold stiffness gradient. Durotaxis can be induced only on a limited surface that satisfies the condition threshold gradient > intrinsic gradient. This means that durotaxis is just a transient

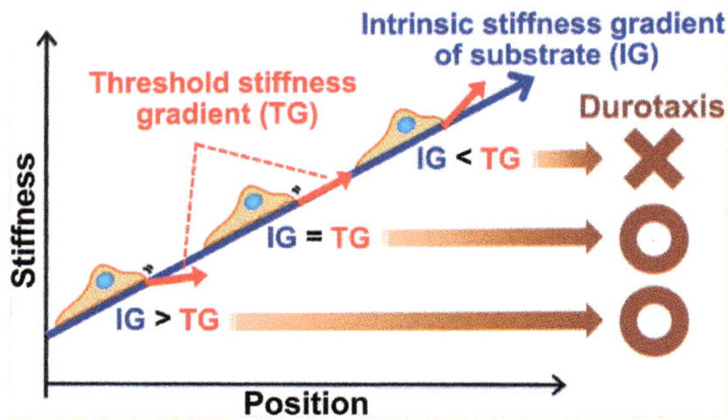

Figure 12.1 General conditions for a stiffness gradient required to induce durotaxis. Since the threshold stiffness gradient (TG) sensitively depends on the stiffness of the soft region or initial position of the moving cells, durotaxis occurs in the region where TG is greater than an intrinsic stiffness gradient (IG). In the region of TG < IG, cells exhibit random crawling rather than durotaxis. Reproduced from ref. 33 with permission from American Chemical Society, Copyright 2019.

local behavior that should finally disappear in stiffer regions. On a large substrate with a broad milliscale stiffness gradient, most of the matrix should not be involved in the induction of durotaxis. Thus for high-throughput statistical analysis for a sufficient number of durotactic cells, the culture substrate should have multiple units of local stiffness gradients with certain dense locations, *i.e.* microstiffness patterning. Typical stiffness patterns reported included square domains[2,27] and stripes,[39,40] which made it possible to obtain cell trajectories for enough numbers of cells to determine the induction efficiency of durotaxis.

On the other hand, even if the micropatterning-based spatial design of stiffness gradients is adopted, the transient nature of durotaxis itself is not altered, and all the cells on the substrate finally concentrate in the microscopic stiff region, leading the cells to a stable equilibrium state, since durotaxis is just a transient phenomenon. The final state is the same as that when cells are cultured on the usual stiff plain substrate, which does not produce any further application. To apply durotaxis to general cell manipulation technology, a culture platform is needed so that durotaxis can be induced repeatedly without equilibrated halting.

For this issue, based on the exact application of the threshold stiffness gradient, it is possible to design a platform that allows for forced repeating of durotaxis over a wide range of culture substrates. To attain this behavior, it should be noted that durotaxis occurs over a certain threshold gradient and disappears below the threshold and that the cells show random crawling. This characteristic of durotaxis provides an idea for the use of asymmetric patterning of a stiffness gradient. For example, saw-like asymmetric stiffness gradients, in which a couple of slopes have different gradient strengths but the same absolute elastic modulus peak, can be used to rectify durotaxis. Specifically, by introducing a saw-like gradient field as shown in Figure 12.2A, cells are expected to receive constant selective motile pressure

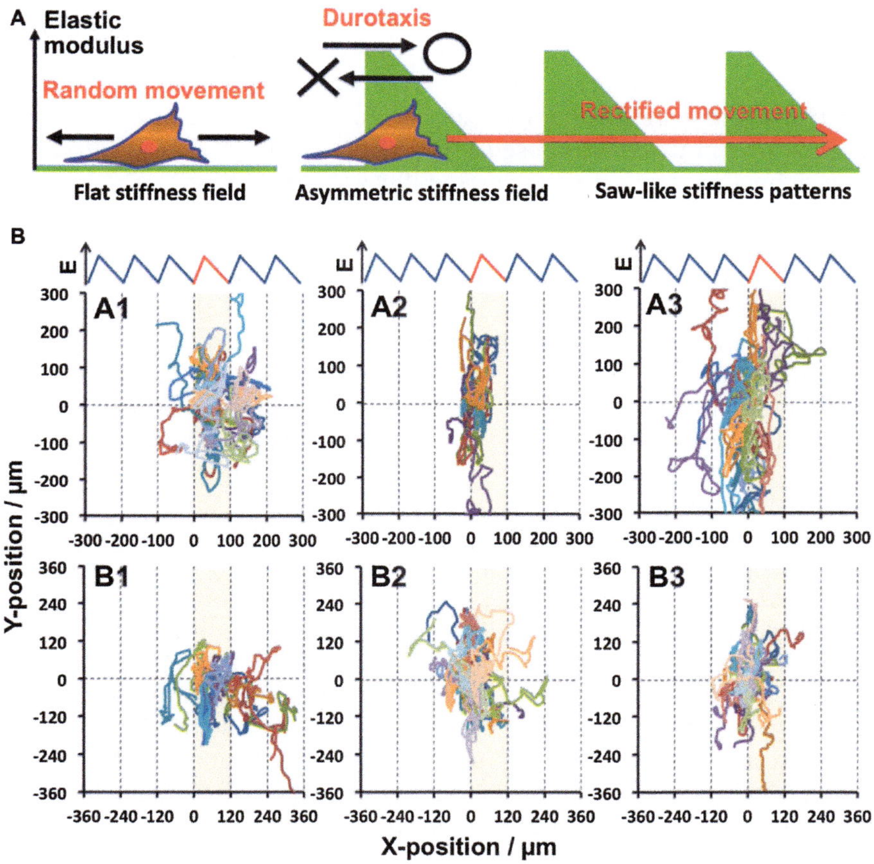

Figure 12.2 (A) Scheme of the durotactic rectification of cell crawling on saw-like patterns with asymmetric stiffness-gradient teeth. Durotaxis occurs on the side of the sharp boundary with a stiffness gradient above the threshold stiffness gradient but not in the region with a weaker stiffness gradient. (B) Migration trajectories of cells cultured on saw-like asymmetric stiffness patterns. The widths of an asymmetric tooth unit in series A and B are 100 µm and 120 µm, respectively. Series 1, 2, and 3 have a peak elastic modulus of *ca.* 100 kPa, 300 kPa, and 500 kPa, respectively. The ratio of ascending:descending stiffness gradients in each unit is 1 : 2, *i.e.* the X-position of the peak elastic modulus is located at *ca.* 30 µm and 40 µm in series A and B, respectively. Only the conditions in B1 can induce the durotactic rectification of crawling cells. Reproduced from ref. 28, https://doi.org/10.1371/journal.pone.0078067, under the terms of the CC BY 4.0 license https://creativecommons.org/licenses/by/4.0/.

toward the stiff region at each elastic boundary, which would induce unidirectional rectified long-range movement.

Rectified movement was actually observed under a somewhat narrow window of conditions. Different types of saw-like asymmetric stiffness gradient patterns were prepared on photocurable gelatin with unit lengths of

100 and 120 μm on a soft region of 100 kPa and with stiffness peaks in the stiff region of 100–500 kPa.[28] Here, the asymmetricity was designed so that the stiffness peak was located at a 1 : 2 position within the unit pattern. When fibroblasts were seeded on these asymmetric gradient gels at 1.25×10^3 cells per cm^2 and time-lapse observations were performed every 15 min for 30 h, a rightward rectification of cell motility was observed with asymmetric gradient gels with a unit length of 120 μm and a hard region peak of 100 kPa (Figure 12.2B, B1). In contrast, the cells on the saw-like gels with stiff regions of 300 kPa and 500 kPa showed strong biased movement in the *Y*-axis direction (Figure 12.2B, A2, A3, B2, B3). It was suggested that the optimal condition of an asymmetric gradient gel to rectify cell motions was a stiffness peak of 100 kPa. On asymmetric gradient gels with stiffness peaks of 300 and 500 kPa, the gradient strength was sufficiently high on both sides of the steep and slow elastic gradients to induce durotaxis, suggesting that the cells were easily trapped in the stiff region, which was unfavorable for the rectification of durotaxis.

To further improve the rectification efficiency and drive cell motility over longer distances, the stiffness peak should be as high as possible, the soft base should be as soft as possible, and the unit length of the asymmetric tooth should be short enough to allow cells to reach the stiffness boundary in a shortened time. On the other hand, trapping on the *Y*-axis, as previously described, is a critical inhibitor. To eliminate trapping, the introduction of soft lanes vertical to the asymmetric saw-like pattern should be effective. The best condition to rectify durotaxis (unit tooth length: 90 μm; stiffness peak: 500 kPa; width of vertical soft lane: 20 μm) markedly induced long-distance directional cell movement up to 500 μm for 30 h.[28] Attempts to rectify cell movement using a saw-like asymmetric field have drawn attention as ratchetaxis,[11] and the activation of duro-ratchetaxis would be valuable as a platform for research on the mechanics of cell movement. In addition, the application of this platform to cell chromatography on various types of cells with different phenotypes of motility would also be possible.

12.3.2 Reverse Durotaxis

While the stiffness gradient conditions required to induce durotaxis were becoming increasingly clear, this knowledge was limited to situations with a linear stiffness boundary perpendicular to the direction of the stiffness gradient, *i.e.* a boundary field of a parallel uniaxial stiffness gradient (Figure 12.3A, B). However, this situation is so rare in actual living organisms that the obtained gradient conditions may not necessarily support actual durotactic behavior *in vivo*. Living tissues and organs are generally microscopically heterogeneous in stiffness and the spatial distribution of the stiffness of the extracellular environment should be nonlinear. In particular, typical biological organs are comprised of curved tissues with various curvatures. Based on such realistic stiffness distributions and nonlinear curved structural properties in living organisms, we thought to evaluate the

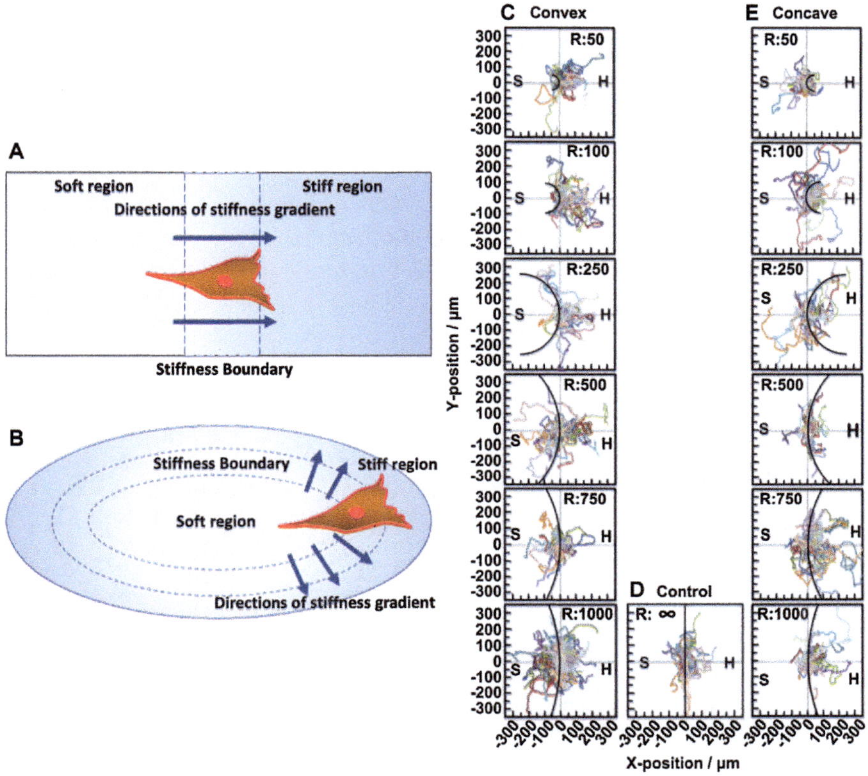

Figure 12.3 Scheme of the spatial distribution of the stiffness gradient: (A) Linear stiffness boundary that has been conventionally used. The directions of the stiffness gradient are parallel. (B) Model of a nonlinear curved stiffness boundary that should be microscopically generated in living tissues. The directions of the stiffness gradient are nonparallel within a single cell's adhered area. (C–E) Migration trajectories of cells cultured on curved stiffness boundaries with a different radius of curvature (R unit = μm). (C) Convex boundaries toward the stiff region; (D) control linear boundary; and (E) concave boundaries toward the stiff region. Reverse durotaxis can be observed for R50 in the concave boundary. Reproduced from ref. 41 with permission from Elsevier, Copyright 2015.

influence of the curvature of stiffness boundaries on the efficiency of induction of durotaxis.[41] Through this characterization, we were the first to experimentally find reverse durotaxis, *i.e.* movement from a stiff region to a soft region.

Various shapes of curved stiffness boundaries were prepared on circularly stiffness-patterned gels with curvatures with radii of 100, 250, 500, and 750 μm, with both convex and concave boundaries toward the stiff region (stiff region: 300 kPa; soft region: 10 kPa). Fibroblasts were seeded on these gels at a density of 1.5×10^3 cells per cm^2 and incubated in an incubator for 8 h. Next, time-lapse observations were performed at 15 min intervals for 24 h. The cell trajectory was analyzed for 12 h for cells that interacted with the elastic boundary (Figure 12.3C–E). At convex boundaries toward the stiff region, we found that large radii of curvature (500 and 750 μm) did not induce durotaxis, even though the elastic gradient was sufficient to induce durotaxis in fibroblast cells with the usual linear elasticity boundary. Durotaxis was strongly induced at a radius of curvature below 250 μm. Interestingly, the induction efficiency of durotaxis was increased in the order 1000 μm radius convex, linear, and 1000 μm radius concave, which clearly indicated that the durotaxis response should depend on the shape of the elastic boundary. A finding that highlighted

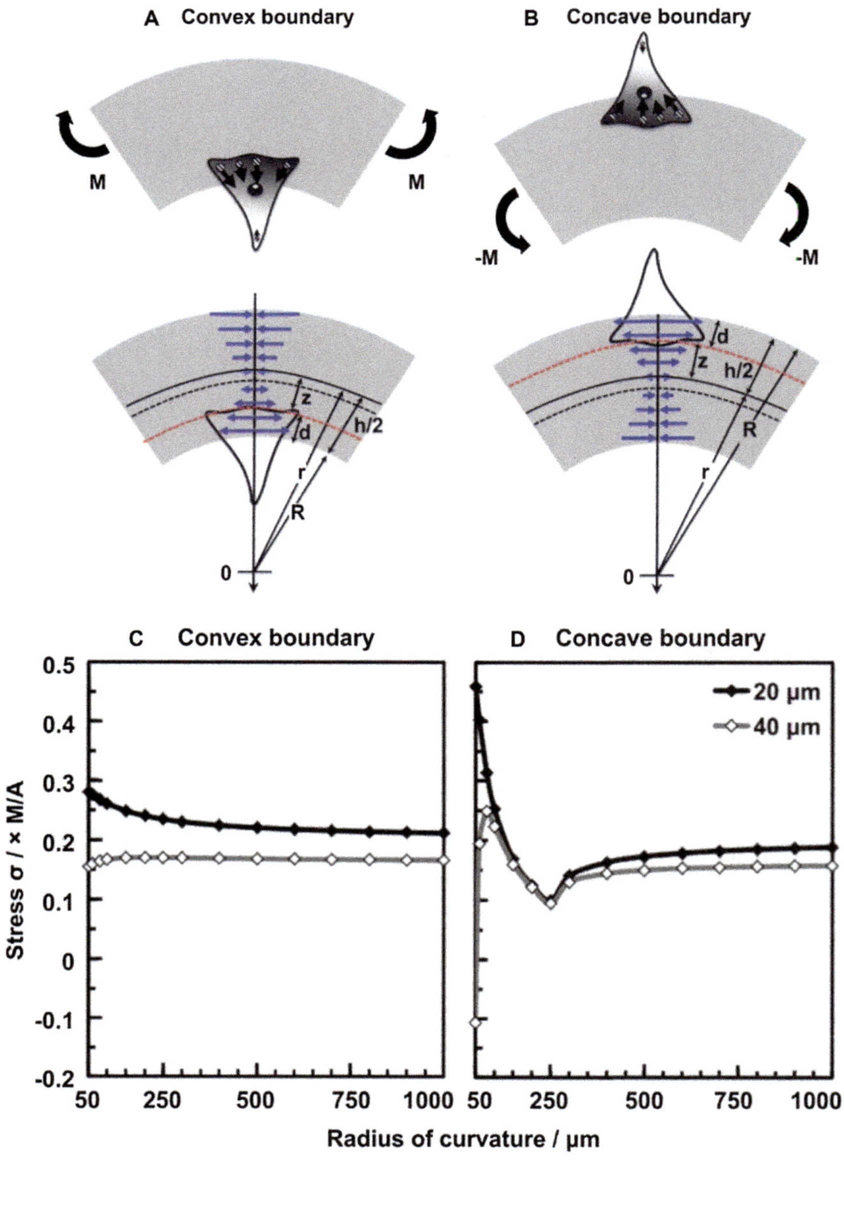

the effect that a nonlinear stiffness gradient was obtained for a concave boundary with a radius curvature smaller than 50 μm; *i.e.* movement from the stiff region to the soft region emerged. This was the first experimental demonstration of reverse durotaxis.[41]

We have considered the mechanism of such a curvature-dependent modulation in durotaxis from the view of stress distribution around the elasticity boundary as the cells move across it. As general features for curved materials, the direction of the inner bending stress changes from extension to compression across the neutral axis when the bending moment acts to decrease or increase the curvature of the curved materials,[42] A field with extensive stress is considered to be more advantageous for the maturation of focal adhesions, while a field with compressive stress is unfavorable for cell spreading. To estimate the stress distribution around curved boundaries, we applied mechanics theory on a curved beam to curved gel bands with different R values and calculated the bending stresses around the boundary caused by the traction forces generated by adhered cells (Figure 12.4A, B). For the induction of reverse durotaxis, a characteristic distribution of bending stresses around concave boundaries with very small R values was found; *i.e.* the bending stress at the stiff region of 20 μm inside the small concave boundaries ($R<100$ μm) is highly extensive, while the bending stresses at 40 μm inside the boundaries are predicted to be highly compressive and local stress steeply declines (Figure 12.4C, D). This analysis suggests that the complete invasion and spreading of cells becomes unfavorable at stiff regions of a concave boundary located further from the boundary, and thus cells may even be excluded from stiff regions to soft regions in acutely curved boundaries. This analysis for the stress distribution on a curved beam is quite simplified as it assumes that the load from

Figure 12.4 (A, B) Schematic representations of the mechanical model used to calculate the distribution of bending stress loaded by cells migrating across curved boundaries. The gray-shaded area indicates the stiff region. Positive and negative signs of the bending moment M are defined according to the direction of decreasing or increasing curvature of the boundary when a cell load's traction forces across the convex are (A) concave or (B) boundary, respectively. The solid and dashed curves shown in the gray-shaded region represent the centroid and neutral axis, respectively. Extensive and compressive bending stresses perpendicular to the moving radius are shown as lateral arrows. Stress levels were calculated at a distance d of 20 or 40 μm ahead of the boundary. (h, width of the gray-shaded stiff region; z, distance from the centroid axis to an arbitrary position; r, distance between the centroid axis and the center of curvature; and R, radius of curvature.) (C, D) Calculated distributions of bending stresses (s) at 20 μm or 40 μm ahead of the convex (c) and concave (d) boundaries with different Rs. In the concave boundary with R 50 μm, a steep increase in σ (extensive direction) appears at 20 μm ahead of the boundary, but a marked decrease in σ (compressive direction) emerges 40 μm ahead of the boundary, making focal adhesion unstable, which can induce reverse durotaxis. Reproduced from ref. 41 with permission from Elsevier, Copyright 2015.

cell traction on the gels is the bending moment M of convoluted force contribution in a coarse-grained manner. To verify these interpretations more precisely and comprehensively, traction force microscopy for a curved stiffness boundary with a small radius of curvature was also performed. Traction force distributions well corresponded with the predicted distribution of bending stresses around the curved elasticity boundary.[43] A nonlinear distribution of local bending stresses was found to play a critical role to induce reverse durotaxis.

12.4 Combined Durotaxis-induced Nomadic Cell Migration

Though conventional durotaxis is only a transitional phenomenon, based on the previously mentioned methodologies of microstiffness patterning of a cell culture matrix, it is possible to induce continuous durotaxis by using the rectification and reversal of durotaxis through the design of the stress distribution of the extracellular mechanical field. An essential factor for the manipulation of durotaxis is the cell-scale stiffness heterogeneity of the extracellular matrix (ECM); *i.e.* durotaxis in principle occurs because stiffness gradients within the adhered area of a single cell of around tens of microns can induce significant anterior–posterior polarity of the cell.[44] In this sense, only cell-sized stiffness patterning can effectively manipulate continuous durotaxis, rather than subcellular nano-/micron-sized patterning. For example, on a matrix with subcellular-scale stiffness heterogeneity, micron-scale stiff spots that are favorable for the formation of stable focal adhesion should be preferentially occupied with focal adhesions (FAs), while soft spots are associated with the poor development of FAs. These behaviors of FAs are not significantly different from the situation with a plain stiff matrix. On the other hand, on a matrix with cell-scale stiffness heterogeneity, FAs with different degrees of maturation can distribute within the adhered interface of a single cell where different regions of stiffness exist within a range of several tens of microns, which can play a role in inducing the polarized distribution of FAs. Therefore, the most basic approach that gives rise to variations in manipulating durotaxis is to carve cell-scale stiffness heterogeneity onto the matrix.

Since cell-scale stiffness heterogeneity is a rather typical feature in living tissue and organs, continuous durotaxis may be generally induced in *in vivo* situations with a combination of different types of durotaxis under a certain stiffness pattern specific to each type of durotaxis. These different types of durotaxis would coincide in a mixotactic manner[45] depending on the features of stiffness heterogeneity that can induce continuous and sequential durotaxis. Frequent induction of durotaxis with cell-scale stiffness heterogeneity is a phenomenon in which the cells keep moving between regions of different stiffness. However, does such a phenomenon really occur *in vivo*? The answer is likely yes. Mesenchymal stem cells (MSCs) in bone marrow may be a good example. In bone marrow, MSCs are delocalized among regions of different stiffness of the osteoblastic (~GPa), vascular (~kPa), and stromal niches (~Pa)[46] and are considered to move around among them. How are these cells mechanically affected during this nomadic migration among regions of different stiffness? Such cellular behaviors of delocalization and nomadic migration among regions of different stiffness cannot be investigated unless we focus on the usual transient durotaxis. In general, systematic analysis of the relationship between the stiffness distribution of ECM and the responses of moving cells *in vivo* is extremely difficult due to a lack of methods for the nondestructive *in situ* measurement of both microscopic stiffness and cell motility *in vivo*. However, it is possible to approach and analyze this problem if we can freely manipulate continuous durotaxis by designing a matrix with cell-scale stiffness heterogeneity. To investigate the behavior of cells undergoing nomadic migration among regions of different stiffness, we have developed a continuous durotactic system that drives both competing usual and reverse durotaxis as a model platform.

A microelastically patterned gel matrix that simultaneously and repeatedly induces both usual and reverse durotaxis can be realized by designing an equilateral triangular stiff region[43] (Figure 12.5A). The apex of the stiff triangle is curved with a radius of curvature of less than 50 μm due to the resolution limitation of photolithography, but this apex instead becomes a stiffness boundary that induces reverse durotaxis. If the stiffness gradient in a single cell is set above a threshold intensity, forward durotaxis is induced on the side of the domain. Since these dual durotaxes, *i.e.* reverse and forward durotaxis, are induced separately at the apex and the side, respectively, the cells on this hexagonal superlattice pattern repeatedly move back and forth between stiff and soft regions.

The residence time of the cells in each stiff or soft region depends on the degree of shuttling of cells between the regions, which is regulated by the features of the patterning design such as the size of the stiff triangles and the distance between them. Larger triangle patterns make the cells concentrate inside, while smaller triangle patterns balance the residence probability between stiff and soft regions, inducing nomadic migration (Figure 12.5B–E). When the residence probability is balanced, forward and reverse durotaxes occur most frequently and continuously. Although the ratio of the stiff/soft region and the curvatures of the boundary should vary in a real *in vivo* situation, the small triangle pattern is useful as a conceptual model of systems that induce the highest degree of nomadic cell migration and is effective for examining the mechanical impact of the nomadic migration of cells between stiff and soft regions. The clear impact has been observed with long-term traction stress dynamics in an entire cell. During nomadic migration between stiff and soft regions, the time-dependent fluctuations of traction stress were revealed to be amplified because of the increased variation in cell morphology (Figure 12.5F, G). Since durotaxis itself is induced by the transient fluctuation in cell polarity resulting from

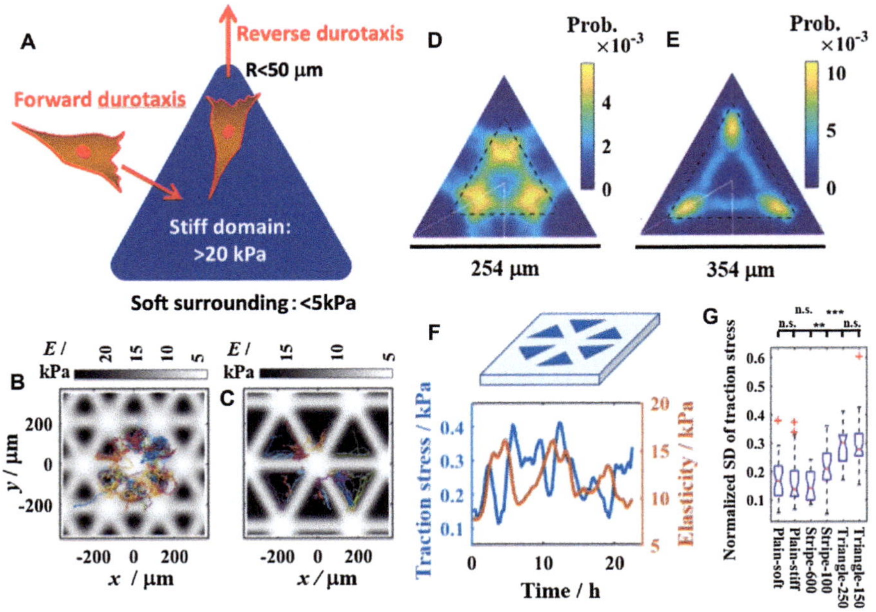

Figure 12.5 (A) Scheme of stiffness patterning to induce dual forward and reverse durotaxis. Forward durotaxis occurs at the linear stiffness boundary on the side of the domain, while reverse durotaxis is induced at the curved stiffness boundary at the corner of the domain with radius of curvature smaller than 50 μm. (B–E) Analysis of the migration of cells around a triangular stiff region. Trajectories on the triangular stiff region with a side length of 150 μm and unit size 254 μm (B, small triangle) and with a side length 250 μm and unit size 354 μm (C, large triangle). (Young's modulus of the stiff region, *ca.* 22 kPa; soft region, 5 kPa.) Probability distribution of the cell position on unit patterns of small (D) and large (E) triangles. On the small triangle, the probability of the cell location spreads around the corner of the domain, indicating nomadic migration between stiff and soft regions caused by dual durotaxis. (F) Example of the time course of the mean traction stress (blue curve) of a cell and mean stiffness (orange curve) under a cell on gels with small triangular stiff patterns. (G) Normalized standard deviation (SD) of the traction stress calculated from the time course of the mean traction stress of a cell. The standard deviation of the traction stress was divided by the time-averaged traction stress. The code number of each condition indicates the size of the unit pattern in μm. Fluctuation of the traction stress of a cell is strongly amplified through dual durotaxis-induced nomadic migration on the small triangular pattern. Reproduced from ref. 43 with permission from Elsevier, Copyright 2021.

changes in the FA distribution, CSK architecture, and internal stress distribution, it is natural that the amplification of the internal stress fluctuation occurs on a matrix with cell-scale stiffness heterogeneity where continuous durotaxis frequently occurs. Amplified and lasting fluctuation in intracellular stress dynamics means that tensional equilibrium in cells is avoided, while tensional homeostasis has been reported to be reached within the order of 1–2 hours for a single cell.[43] The cell-scale stiffness heterogeneity as seen *in vivo* may play an essential role for the cells migrating on it to keep them far from tensional equilibrium during nomadic migration through continuous durotaxis.

12.5 Conclusions and Perspectives

This review summarized stiffness gradient conditions and a technique for creating these conditions that are required to manipulate durotaxis. As an ideal platform for the manipulation of durotaxis, the application of photolithographic microstiffness patterning of photocrosslinked hydrogels was explained in detail. In particular, although durotaxis is only a transient phenomenon that is locally observed in the transition region of stiffness, it was clarified that cell-scale stiffness patterning can induce continuous and frequent durotaxis. To induce continuous durotaxis, the use of rectified durotaxis and reverse durotaxis through design of the elasticity distribution was introduced, and the role of cell-scale stiffness heterogeneity that can induce the combination of these durotaxes was highlighted. Cell-scale stiffness heterogeneity may play an important role in the coupling of durotactic migration and functional modulation of the cells *in vivo*, as it can strongly amplify internal stress fluctuations of moving cells during nomadic migration among regions of different stiffness. The induction and manipulation of continuous durotaxis, especially on a matrix with cell-scale stiffness heterogeneity, is a technique for placing cells far from tensional equilibrium and is expected to be useful for the development of novel cell manipulation technologies.

One of the critical requirements to manipulate continuous durotaxis during biomaterials design is to mechanically activate the cells in such a state that avoids tensional equilibrium, which can induce not only the nucleocytoplasmic shuttling of mechanosensitive nuclear translocating factors between locations of the cells in stiff and soft regions, but also the regulation of gene expression through amplified intracellular-stress-dynamics-dependent nuclear deformation. We will soon report, in another article, the functional modulations of the mesenchymal stem cells in the state of far-from-tensional-equilibrium including stemness maintenance and therapeutic activation. The material-based manipulation of mechanobiology in the durotaxing cells should be a sophisticated methodology to modulate the dynamic input of mechanical signals from extracellular matrix to the cells.

Abbreviations

ECM	extracellular matrix
FA	focal adhesion
IG	intrinsic stiffness gradient
MSC	mesenchymal stem cell

PAAm	polyacrylamide
PDMS	polydimethylsiloxane
TG	threshold stiffness gradient

References

1. C.-M. Lo, H.-B. Wang, M. Dembo and Y.-L. Wang, *Biophys. J.*, 2000, **79**, 144.
2. D. S. Gray, J. Tien and C. S. Chen, *J. Biomed. Mater. Res., Part A*, 2003, **66**, 605.
3. S. Kidoaki and T. Matsuda, *J. Biotechnol.*, 2008, **133**, 225.
4. S. Li, *et al.*, *Proc. Natl. Acad. Sci. U. S. A.*, 2002, **99**, 3546.
5. S. Li, N. F. Huang and S. Hsu, *J. Cell. Biochem.*, 2005, **96**, 1110.
6. M.-P. Valignat, *et al.*, *Nat. Commun.*, 2014, **5**, 5213.
7. J. Sikora, Z. Baranowski and M. Zajaczkowska, *Experientia*, 1992, **48**, 789.
8. S. B. Carter, *Nature*, 1967, **213**, 256.
9. J.-S. Park, D.-H. Kim and A. Levchenko, *Biophys. J.*, 2018, **114**, 1257.
10. D. T. Tambe, *et al.*, *Nat. Mater.*, 2011, **10**, 469.
11. D. Caballero, *et al.*, *Quant. Cell Biol.*, 2015, **25**, 815.
12. L. Pieuchot, *et al.*, *Nat. Commun.*, 2018, **9**, 3995.
13. E. Hadjipanayi, V. Mudera and R. A. Brown, *J. Tissue Eng. Regener. Med.*, 2009, **3**, 77.
14. J. Schrader, *et al.*, *Hepatology*, 2011, **53**, 1192.
15. Y.-T. Yeh, *et al.*, *PLoS One*, 2012, **7**, e46889.
16. A. Shkumatov, *et al.*, *Am. J. Physiol.*, 2015, **308**, 1.
17. A. J. Engler, S. Sen, H. L. Sweeney and D. E. Discher, *Cell*, 2006, **126**, 677.
18. A. D. Young, Y.-S. Choi, A. J. Engler and K. L. Christman, *Biomaterials*, 2013, **34**, 8581.
19. L. Wong, *et al.*, *Acta Biomater.*, 2019, **96**, 321.
20. H. Alexander, L. Qing, S. Guanbin and A. A. Dwi, *Stem Cell Rev. Rep.*, 2020, **16**, 661.
21. S. V. Plotnikov, A. M. Pasapera, B. Sabass and C. M. Waterman, *Cell*, 2012, **151**, 1513.
22. L. Smith, S. Cho and D. E. Discher, *Semin. Cell Dev. Biol.*, 2017, **71**, 84.
23. J. A. Burdick, A. Khademhosseini and R. Langer, *Langmuir*, 2004, **20**, 5153.
24. N. Zaari, P. Rajagopalan, S. K. Kim, A. J. Engler and J. Y. Wo, *Adv. Mater.*, 2004, **16**, 2133.
25. J. Hong, S. Lee, S. Park and J. Lee, *J. Micromech. Microeng.*, 2015, **25**, 045012.
26. S. Nemir, H. N. Hayenga and J. L. West, *Biotechnol. Bioeng.*, 2009, **105**, 636.
27. T. Kawano and S. Kidoaki, *Biomaterials*, 2011, **32**, 2725.
28. S. Kidoaki and H. Sakashita, *PLoS One*, 2013, **8**, e78067.
29. T. Kuboki, W. Chen and S. Kidoaki, *Langmuir*, 2014, **30**, 6187.

30. S. Wong, W.-H. Guo, I. Hoffecker and Y.-L. Wang, *Methods Cell Biol.*, 2014, **121**, 3.
31. P.-H. G. Chao, S.-C. Sheng and W.-R. Chang, *J. Mech. Behav. Biomed. Mater.*, 2014, **38**, 232.
32. B. Wang, *et al.*, *Microelectron. Eng.*, 2017, **175**, 50.
33. K. Moriyama and S. Kidoaki, *Langmuir*, 2019, **35**, 7478.
34. A. K. Yip, *et al.*, *Biophys. J.*, 2013, **104**, 19.
35. M. K. Gunnewiek, *et al.*, *Adv. Mater. Interfaces*, 2016, **3**, 1500456.
36. S. J. Attwood, *et al.*, *Sci. Rep.*, 2016, **6**, 34334.
37. C. D. Reyes and A. J. Garcia, *J. Biomed. Mater. Res., Part A*, 2003, **65**, 511.
38. P. Roca-Cusachs, R. Sunyer and X. Trepat, *Curr. Opin. Cell Biol.*, 2013, **25**, 543.
39. Y. S. Choi, *et al.*, *Biomaterials*, 2012, **33**, 6943.
40. C. L. Happe, *et al.*, *Mol. Biol. Cell*, 2017, **28**, 1950.
41. A. Ueki and S. Kidoaki, *Biomaterials*, 2015, **41**, 45.
42. A. P. Boresi, R. J. Schmidt and O. M. Sidebottom, *Advanced Mechanics of Materials*, Wiley, NJ, USA, 5th edn1993.
43. H. Ebata and S. Kidoaki, *Biomaterials*, 2021, DOI: 10.1101/2020.10.15.342006, in press.
44. T. Kuboki, *et al.*, *Cell Struct. Funct.*, 2020, **45**, 33.
45. E. H. Barriga and E. Theveneau, *Front. Physiol.*, 2020, **11**, 586432.
46. K. Maeda, *et al.*, *Sci. Rep.*, 2016, **6**, 22288.

CHAPTER 13

Engineered 3D Matrices with Spatiotemporally Tunable Properties

KOICHIRO UTO,[a] RIHO TANIMOTO[a,b] AND COLE A. DEFOREST*[c,d,e,f,g]

[a] Research Center for Functional Materials, National Institute for Materials Science (NIMS), 1-1 Namiki, Tsukuba, Ibaraki 3050044, Japan; [b] Graduate School of Pure and Applied Sciences, University of Tsukuba, 1-1-1 Tennodai, Tsukuba, Ibaraki 3058577, Japan; [c] Department of Bioengineering, University of Washington, Seattle, Washington 98105, USA; [d] Department of Chemical Engineering, University of Washington, Seattle, Washington 98195, USA; [e] Institute for Stem Cell & Regenerative Medicine, University of Washington, Seattle, Washington 98109, USA; [f] Molecular Engineering & Science Institute, University of Washington, Seattle, Washington 98195, USA; [g] Department of Chemistry, University of Washington, Seattle, Washington 98105, USA

*Email: profcole@uw.edu

13.1 Introduction

It has become widely appreciated that mechanical forces are ever present between cells and their surroundings and that these forces provide a crucial set of cellular signals that can control cell function and fate. In fact, cells sense their surrounding microenvironment mechanically through push/pull-type interactions and exhibit specific biological functions

Biomaterials Science Series No. 12
Material-based Mechanobiology
Edited by Jun Nakanishi and Koichiro Uto

Published by the Royal Society of Chemistry, www.rsc.org

(*e.g.* proliferation, differentiation) in response to these forces through a process well known as mechanotransduction.[1] Mechanobiology is the study of how cells perceive surrounding forces presented by the extracellular matrix (ECM) and how they are converted into intracellular biochemical signaling cascades to alter cellular function and fate. To date, our collective insights into cellular mechanobiology processes have been primarily obtained from experiments performed on flat culture substrates with non-physiological mechanics (*e.g.* polystyrene, glass). Since Engler *et al.* reported in 2006 that the differentiation of human mesenchymal stem cells (hMSCs) strongly depends on the elastic modulus of the substrate using polyacrylamide hydrogels with elastic modulus mimicking biological tissues,[2] significant efforts have sought to elucidate the relationship between matrix mechanics and cell behavior. However, *in vivo*, cells are embedded in a complex and information-overloaded environment consisting of multiple ECM components and cell types. Biochemical and biophysical signals displayed by and within the ECMs are also presented with spatiotemporal variations. Though static 2D culture platforms have proven useful for understanding cellular phenomena to some degree, these approaches are intrinsically limited in their ability to fully recapitulate the full 4D complexity of signaling cells experience *in vivo*. Thus there is a growing interest in the development of material systems enabling the 3D cell culture with dynamic properties, where changes in physical and biochemical properties of materials can be modulated in a user-defined and/or cell-dictated manner. In this chapter, we focus our attention on the development of 4D-programmable culture matrices that are actively catapulting mechanobiology into a new dimension. Specifically, we will discuss the importance of 3D biomaterials in studying cell function, what can be controlled in spatiotemporal fashion by proposed 4D culture systems, and what physiological phenomena can be recapitulated and explored from the perspective of materials science and mechanobiology.

13.2 The Importance of 3D Cell Culture

Cells are mechanically coupled with their extracellular environments; the cellular niche plays a crucial role in communicating between cell and ECM, as well as mediating cellular responses to various cues. Along with the biochemical and mechanical properties of the ECM, several studies have demonstrated the impact of the culture dimensionality on cell–ECM interactions for regulation of cellular behaviors. This comes as little surprise, given that cells are most often embedded within a 3D matrix *in vivo* that is markedly different from a 2D substrate with respect to cue presentation.[3] As a result of the difference in culture dimension (2D *versus* 3D, as shown in Figure 13.1A), cells behave quite differently in their adhesion and biological response. For example, when 3T3 fibroblasts are cultured on collagen-treated glass, the cells take on a spread morphology, develop actin stress fibers, and proliferate more rapidly.[4] Curiously, 3T3

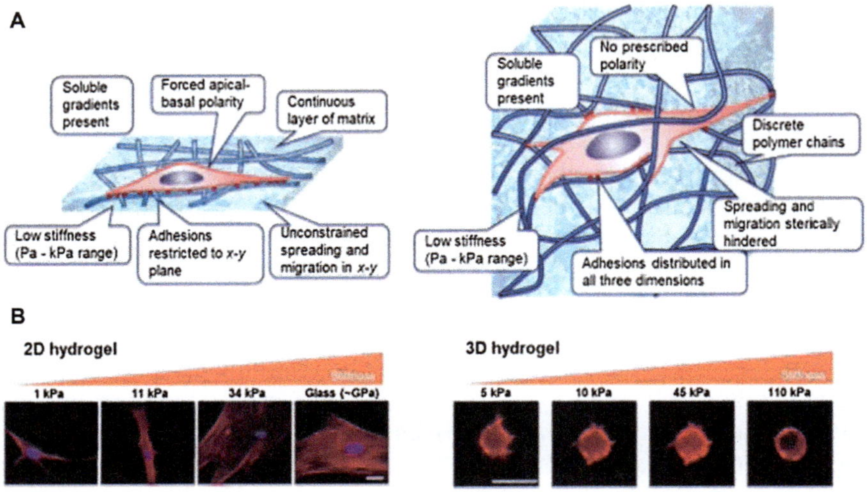

Figure 13.1 Microenvironmental effects on cell function in 2D and 3D culture. (A) Cells experience starkly different cues when cultured on and within materials. (B) Cell morphology depends on culture substrate in 2D but not in 3D. Mesenchymal stem cells cultured on increasingly stiff 2D hydrogel substrate increasing their spread area. Adapted from ref. 5 with permission from Springer Nature, Copyright 2016, while cells encapsulated in 3D hydrogel maintain a round morphology. Adapted from ref. 6 with permission from Springer Nature, Copyright 2010.

fibroblasts exhibit opposite behavior when cultured within a 3D collagen gel; though cell–substrate interactions are promoted through the same integrin–collagen binding, the dimension of culture ultimately defines how these interactions manifest. As previously noted, these cellular behaviors can be further influenced by matrix mechanics, though such mechanobiological responses also vary greatly with culture dimension. Similar seemingly paradoxical behavioral findings have been observed for mesenchymal stem cells (MSCs), where the cells cultured on increasingly stiff 2D hydrogel substrates show increasing spreading[5] that is unobserved in 3D culture (Figure 13.1B).[6] Furthermore, mechanosensitive Yes-associated protein (YAP) localization to the nucleus, morphological spreading, and level of stress fiber formation are good predictors of MSC lineage commitment, including both osteogenic and adipogenic differentiation, in 2D culture. In 3D culture, however, all of these variables do not have the same predictive power of MSC fate, highlighting that 3D mechanobiology studies remain largely underexplored.[7] The mechanism governing such cellular response regulation stems from many interrelated factors, including how the ECM transmits force to the cell and how the ECM/cell are mechanically/biochemically coupled. To better understand these complex mechanobiological phenomena, development of appropriately optimized 3D biomaterials for cell culture remains critical.

13.3 Design Strategy of Engineered Matrices for 3D Cell Culture

When designing engineered matrices for 3D cell culture, one must carefully consider several important aspects, including biomaterial mechanics, biochemistry, biodegradability, and overall biocompatibility.[8,9] Hydrogels are water-swollen 3D polymeric networks; mimicking various biochemical and tissue-relevant mechanical properties, they have proven to be excellent materials for supporting 3D cell growth. From a biocompatibility perspective, the synthetic polymers such as poly(ethylene glycol) (PEG) and naturally derived polymers (*e.g.* polysaccharides including hyaluronic acid and alginate, collagen, gelatin) have been widely employed. As such gels present average mesh sizes that are typically orders of magnitude smaller than the size scale of eukaryotic cells (~10–20 μm diameter), it is often challenging to postsynthetically introduce cells into preformed hydrogels. Hence, direct encapsulation strategies must be developed that enable hydrogels to be created from biocompatible polymers using cytocompatible chemistries in the presence of living cells. Though hydrogels have been synthesized using a variety of chemistries spanning a wide range of reaction conditions, biomaterial formation should avoid those that could cause excessive damage to the cells through excessive free-radical generation, high and large temperature/pH changes, or toxic biproduct formation. Careful consideration of the chemistries must be given to those employed in forming cell-laden hydrogels from biocompatible polymers. To date, several cyto- and biocompatible chemistries have been developed for cell encapsulation within the hydrogels in high viability or intact state (Figure 13.2).

Free-radical chain polymerization is initiated through thermal, photochemical, or oxidative/reductive initiator decomposition into active radicals that can form and crosslink functionalized polymers into 3D networks (Figure 13.2). These chemistries most typically utilize multifunctional acrylates or methacrylates (*e.g.* PEG-dimethacrylate). Though these chemistries are rapid and robust, consideration must be given to the effects of such free radicals, known to transfer nonspecifically onto bioactive substances including proteins and cells, on bioactivity and viability.[10] Under some conditions, such exothermic chemistries may induce localized heating that can damage cells during polymerization.[11] Despite these potential problems, several cytocompatible formulations have been devised that have enabled radical chain polymerization to be successfully utilized for cell encapsulation in gels. Bryant and Anseth have shown the encapsulation of chondrocytes into PEG hydrogels by photoinduced free radical polymerization using methacrylated PEG macromers (PEGDM) and water-soluble photoinitiator Irgacure 2959. The viability of the chondrocytes was maintained during the polymerization of hydrogel at lower macromer concentration (<30%), suggesting the biocompatibility of the crosslinking method.[12,13]

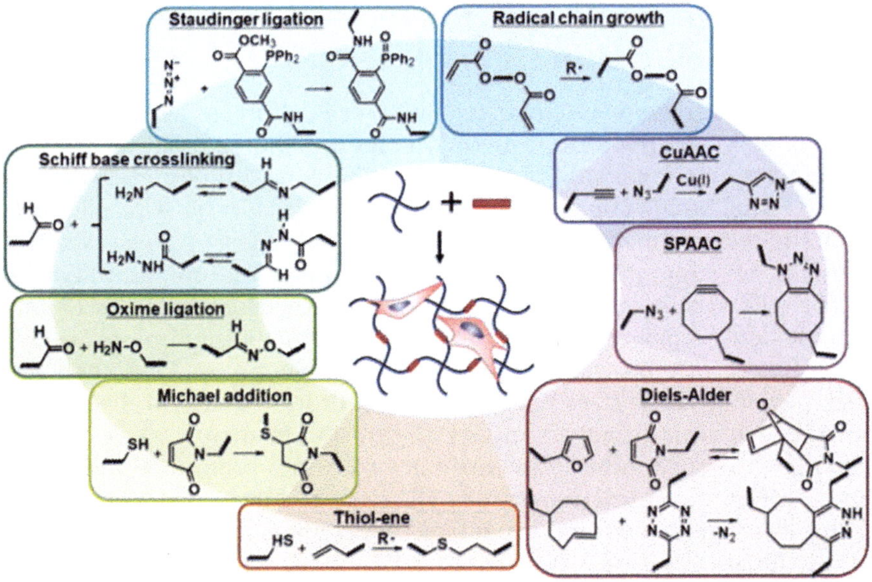

Figure 13.2 Chemistries used previously for efficient encapsulation of living cells within hydrogels. A variety of bioorthogonal and cell-friendly chemistries can be exploited to form and postsynthetically modify hydrogels for 3D cell culture.

Furthermore, recent studies have demonstrated that light conditions typically employed in free-radical polymerization of hydrogels are fully cytocompatible, causing no changes to the cellular proteome.[14] Though such chemistries are compatible with live cell encapsulation, free-radical polymerization results in a polydisperse molecular weight distribution of the polymer chains that manifests as a molecularly heterogeneous network with nonuniform mechanics.[15] Therefore, the development of methods to form hydrogels with more homogeneous or well-defined network structures, particularly in the context of mechanobiology, is of great interest for 3D cell culture.

13.3.1 Azide–Alkyne Cycloaddition for Engineered Hydrogel Crosslinking

The ideal mechanism used to crosslink engineered hydrogels would utilize chemistries that do not crossreact or interfere with the biology of the system. The field of bioorthogonal chemistry has emerged to address the need for highly specific and robust reactions in biological contexts, and the advantages of bioorthogonal crosslinking strategies can be exploited in hydrogel design to gently and efficiently encapsulate cells within a variety of gel systems.[16] From this perspective, researchers now regularly exploit bioorthogonal click chemistry for hydrogel formation because of its simplicity, extremely high reaction selectivity and efficiency, and its ability to be performed in water. The azide–alkyne cycloaddition reaction is the most common bioorthogonal chemistry used for cell encapsulation. The early study for hydrogel formation by azide–alkyne cycloaddition used copper(I) as a catalyst,[17] but the potential cytotoxicity of copper has limited the use of this reaction for cell encapsulation (Figure 13.2). However, several cell types, including human foreskin fibroblasts (HFF-1), human adipose-derived mesenchymal stem cells (hMSCs), and human umbilical vein endothelial cells (HUVECs), have been shown to tolerate copper exposure during the crosslinking process,[18] and it is possible that copper-binding ligands such as tris(hydroxypropyltriazolyl)methylamine (THPTA), which is known to reduce cytotoxicity, can be used in crosslinking for other cells.[19] To overcome this problem of copper cytotoxicity, DeForest *et al.* utilized a copper-free click chemistry based on strain-promoted azide–alkyne cycloaddition (SPAAC) reaction between azide and difluorinated cyclooctyne (DIFO) groups developed by Bertozzi *et al.*[20,21] to produce biocompatible PEG hydrogels for cell encapsulation (Figure 13.2).[22] In this early example, spontaneous hydrogel formation was proceeded by crosslinking azide-functionalized 4-arm PEG (4-arm PEG-N_3) with a bisDIFO-functionalized peptide that could be enzymatically cleaved by cell-secreted matrix metalloproteases (MMPs). Using this approach, NIH3T3 fibroblasts were encapsulated with PEG hydrogels through SPAAC, and they demonstrated high viability after 24 hours with the biomaterial. Subsequently, the group also demonstrated that the mechanical properties of such hydrogels can be tuned by changing the molecular weight of the 4-arm PEG-N_3 and the molar ratio of azide and DIFO groups.[23] Early cyclooctyne derivatives such as DIFO had comparatively slow reaction kinetics, requiring several tens of minutes to complete crosslinking by SPAAC reaction, and were challenging to synthesize.[24] To improve the kinetics for bioorthogonal crosslinking, dibenzylcyclooctyne (DBCO) derivatives were developed through fusion of two benzyl rings to and inserting an amide-linked nitrogen into the cyclooctyne ring. It has been reported that the reaction rate can be improved by up to one order of magnitude over other SPAAC reactions.[25–27] Hermann. *et al.* reported the rapid encapsulation of rmGremlin, a BMP inhibitor, through *in situ* crosslinking of hydrogels using azide-functionalized PEG and DBCO-functionalized PEG crosslinkers, which was completed within 2 min.[28] By using DBCO-based SPAAC, cells such as chondrocytes and MSCs were successively encapsulated within various matrices such as dextran and PEG hydrogels efficiently and with high viability.[29,30] Furthermore, the use of highly strained bicyclononyne (BCN), in which a cyclopropane ring is fused to cyclooctyne, has also led to SPAAC reactions with fast reaction kinetics comparable to that involving DBCO.[31] BCN has advantages over DBCO in terms of being more accessible synthesis, less hydrophobic, and more optically transparent. DeForest and Tirrell demonstrated SPAAC-crosslinked hydrogels using 4-arm PEG tetraBCN for encapsulation of hMSCs, a platform that was used to promote spatially controlled osteogenic differentiation in response to photopatterned protein immobilization.[32] Madl *et al.* reported the encapsulation of hMSCs, HUVECs, and murine neural progenitor cells (mNPCs) within elastin-like protein (ELP) hydrogels by SPAAC reaction between BCN- and azide-functionalized ELP.[33] By varying the amount of BCN groups per ELP, the mechanical property of SPAAC-crosslinked ELP hydrogels can be tuned; hMSC encapsulation was achieved with remarkably high cell viability (~100%). In terms of bioorthogonal reactions, the potential reactivity of cyclooctynes with free thiol, including those found on cysteine residues, presents some concern, and indeed some cyclooctynes are known to undergo such thiol-yne reactions under physiological conditions.[25,34] However, since few free thiols exist in the physiological environment and the second-order reaction rate of SPAAC is significantly higher than that of the thiol-yne reaction (10^{-1} M^{-1} s^{-1} and 10^{-4} M^{-1} s^{-1} for BCN-azide and BCN-thiol reaction, respectively), SPAAC is recognized practically as a true bioorthogonal reaction. In addition to SPAAC reactions, noncatalyzed azide–alkyne cycloaddition reactions in water using terminal alkynes with adjacent electron-withdrawing substituents have also been proposed.[35] Truong *et al.* succeeded in rapid gel formation at 37 °C by mixing azide-functionalized chitosan (CS-azide) with propiolic acid ester–functionalized PEG (PEG-alkyne) in the presence of hMSCs, establishing a promising alternative copper-free click chemistry to construct engineered 3D matrices.[36]

13.3.2 Diels–Alder Reaction for Engineered Hydrogel Crosslinking

The Diels–Alder (DA) reaction is also classified as a click reaction due to its simplicity, high efficiency, and selectivity; the reaction is widely used for hydrogel crosslinking. In general, when DA reaction is used for organic synthesis, high temperatures are required to achieve moderate reaction rates, though a hydrophobic effect accelerates the reaction dramatically when performed in water. Several diene–dienophile reactions have proven useful for bioorthogonal hydrogel crosslinking (Figure 13.2).

Wei *et al.* first reported the preparation of hydrogels by DA reaction using poly(*N*,*N*-dimethylacrylamide-*co*-furfuryl methacrylate) as the diene and *N*-[4-formyl polyethylene glycol ester] bismaleimide as the dienophile.[37] In this system, the DA reaction is accelerated in water, and the gelation time is about 1 h. Nimmo *et al.* also synthesized furan-modified hyaluronic (HA) acid and dimaleimide PEG and reported a facile method to prepare HA hydrogels *via* aqueous-based DA reaction.[38] Both the mechanical properties and degradability of HA hydrogels can be tuned by adjusting the molar ratio of furan to maleimide on the HA and PEG. The results of culturing human adenocarcinoma cells (MDA-MB-231) on HA hydrogels showed high viability

(>98%) even after 14 days of culture, indicating that the hydrogels have excellent cytocompatibility.

The furan/maleimide reaction is the most used DA chemistry for hydrogel formation, though it lacks complete bioorthogonality. In addition to their potential to crossreactivity with thiol-containing amino acids such as cysteine and the primary amine found on lysine side chains, maleimides have also been noted to be susceptible to hydrolysis.[39] Blackman *et al.* were the first to report that the inverse electron demand Diels–Alder (IED-DA) reaction between dipyridyl-tetrazine and *trans*-cyclooctene is useful for bioconjugation.[40] This reaction proceeded rapidly without a catalyst, proceeding up to three orders of magnitude faster than SPAAC. As an alternative, norbornene can be used as the dienophile in place of *trans*-cyclooctene. Zhou *et al.* developed the IED-DA reaction between norbornene and tetrazine for the preparation of PEG hydrogels.[41] Subsequently, Alge *et al.* successfully prepared cell-laden hydrogels using the IED-DA reaction.[42] Mixing of 4-arm PEG-tetrazine macromer and dinorbornene-modified peptide yielded stable gels within a few minutes by IED-DA and hMSC encapsulation with high viability. In addition to rapid kinetics, the IED-DA reaction is characterized by high bioorthogonality and yields irreversible crosslinking, which differs from the typical DA reaction.

13.3.3 Thiol–Ene Reaction for Engineered Hydrogel Crosslinking

Thiol–ene chemistry involves a reaction between a thiol and an unsaturated alkene (*e.g.* acrylate, maleimide, norbornene) through free-radical or catalyzed Michael addition (Figure 13.2). Unsaturated functional groups may react with thiol and amino groups on biomolecules, which yields incomplete bioorthogonality. Despite this, the reaction is commonly deemed to have "click" status because it occurs rapidly and with high selectivity/yield under mild conditions.[43] Rydholm *et al.* have successfully used poly(ethylene glycol)-*b*-poly(lactic acid) (PEG-PLA)-diacrylate and multivalent thiols to form hydrogels by thiol–ene click reaction *via* photoinitiation and mixed step-chain growth reactions.[44] Furthermore, a photopolymerization reaction using PEG-dimethacrylate and RGDS peptides containing four cysteine residues was successfully used to encapsulate hMSCs within hydrogels with high viability, demonstrating its usefulness for *in situ* crosslinking in the presence of cells.[45]

This thiol–ene click reaction is not limited to the reaction with thiol-acrylate or thiol-methacrylate. Lutolf *et al.* reported a base-catalyzed Michael-type addition reaction between thiol and vinyl sulfone using 4-arm PEG-vinyl sulfone and bis-cysteine-peptide to produce hydrogels under relatively mild conditions (pH of 8.0) that support cell encapsulation.[46–48] The Michael-type addition reaction does not involve the generation of cytotoxic free radicals or

UV irradiation but does require a nucleophilic buffer such as triethanolamine (TEA) to promote the reaction. Phelps *et al.* formed hydrogels using 4-arm PEG with terminal maleimides and bis-cysteine–containing peptides.[49] This Michael-type addition between maleimide-thiol showed higher reactivity than that of acrylate-thiol or vinyl sulfone-thiol, permitting rapid PEG hydrogel formation with less TEA. Although TEA is toxic to some cell types (*e.g.* endothelial cells[50]), the Michael-type addition of thiol-maleimide allows C2C12 murine myoblasts encapsulation within hydrogels with high survival. Hydrogels using step-growth polymerization, which occurs in this complementary reaction mechanism, have been shown to form a uniform network structure with superior mechanical properties compared to hydrogels with same crosslinking density obtained through chain-growth polymerization.[15,51] However, the alkaline conditions required to promote the kinetics of hydrogel formation may promote the formation of disulfide bonds, leading to reduced bioorthogonality and the formation of hydrogels with nonstoichiometric compositions. Fairbanks *et al.* reported PEG hydrogels formed by thiol–norbornene photopolymerization.[52] The photoinitiation of radicals enables rapid network formation by a step-growth mechanism, demonstrated to be cytocompatible with hMSC encapsulation. In addition, *in situ* crosslinking by the photopolymerized step-growth mechanism not only affords hydrogels with a uniform network structure but also enables temporal and spatial control over formation unattainable using spontaneous Michael-type chemistries.

13.3.4 Other Reactions for Engineered Hydrogel Crosslinking

Carbonyls such as aldehydes and ketones are absent from proteins and DNA, making them useful targets for bioorthogonal reactions. Ligation of carbonyls with primary amines through imine formation is difficult to perform in water or neutral pH conditions due to the poor thermodynamic stability of imines, reaction reversibility, and the need for harsh acidic dehydration condensation conditions. Conversely, oxime ligation occurring between an alkoxyamine/hydroxylamine and a carbonyl has proven useful for hydrogel formation (Figure 13.2). Grover *et al.* demonstrated the preparation of oxime-linked PEG hydrogels by mixing glutaraldehyde with 8-arm PEG containing alkoxyamine as an end group.[53] The formation of oxime bonds can be catalyzed by acid, but rapid gelation within 30 minutes was observed even at pH 7.2; murine MSCs were encapsulated within the hydrogel with high viability. Since glutaraldehyde, which was used as a crosslinking agent, can take on a variety of structures depending on pH and may react with primary amines in proteins and cells, optimization of conditions is necessary to achieve bioorthogonality and the click nature of the reaction. Furthermore, the hydrogels with excellent stability can be formed *via* oxime ligation between alkoxyamine and aldehyde due to its essentially irreversible nature at neutral pH, which is different from hydrazone and imine-based reactions (Figure 13.2). Notably, a photomediated oxime ligation exploiting a photocaged alkoxyamine has been reported that provides spatiotemporal control over gel formation/modification in a radical-free bioorthogonal manner.[32,54,55]

Staudinger ligation is a reaction that occurs between phosphine and azide, producing aza-ylide (Figure 13.2). In the presence of water, this intermediate undergoes spontaneous hydrolysis to yield a primary amine and the corresponding phosphine oxide. The reaction of phosphine and azide proceeds with high efficiency under aqueous solution conditions at room temperature. Saxon and Bertozzi found that electrophilic traps, such as methyl esters at appropriate positions in the phosphine structure, capture the nucleophilic aza-ylide by intramolecular cyclization, ultimately producing a stable amide bond rather than the hydrolyzed product.[56] Madl *et al.* designed elastin-like proteins (ELPs) containing the azide and triarylphosphine and prepared ELP hydrogels by Staudinger ligation.[33] Although the gelation rate was about 20 times slower than the SPAAC reaction between BCN and azide moieties, the gelation of ELPs can be achieved within 20 min by adjusting the molar ratio of azide to triarylphosphine. Gattás-Asfura *et al.* also reported the formation of hydrogels and encapsulation of mouse insulinoma (MIN6) cells by Staudinger ligation between azide-modified alginate and 4-arm PEG with triarylphosphines as the end group.[57] The encapsulated MIN6 cells showed high viability and proliferation in the hydrogel, indicating that the crosslinking of hydrogels using Staudinger ligation is cytocompatible and bioorthogonal.

13.4 Engineered Matrices for 4D Cell Culture

The extracellular microenvironment regulates cell behavior and function through a myriad of spatially and temporally coordinated cues, wherein mispresented signals can lead to dysfunction and disease. To better understand basic biology and engineer effective strategies for regenerative medicine, researchers must unravel the complex picture of the extracellular microenvironment and develop ways to mimic these functions *in vitro*. This section describes technologies for 4D biology – cell-laden matrices that recapitulate the heterogeneity of tissues and organs both in 3D space and time. In particular, cells *in vivo* are not only embedded within a 3D matrix but also actively remodel this ECM to enable spreading, migration, proliferation, and differentiation. For many applications, it is essential to outfit engineered biomaterials with the ability to undergo dynamic physiochemical alterations similar to those in the native ECM. In this section, we will discuss two distinct types of hydrogel materials that provide a dynamic environment for cells: programmable hydrogels and adaptable hydrogels.

13.4.1 Programmable Hydrogels Based on Degradability

The hydrogel systems introduced in Section 13.3 and most frequently utilized in the biomaterials community are generated through chemical crosslinking, resulting in permanent and stable bond formation. To support essential cell functions including spreading, migration, and differentiation, it is often necessary to introduce degradable units within these otherwise stable hydrogels. By incorporating appropriate labile moieties, chemically crosslinked hydrogels can be rendered degradable and can be controlled in cell-dictated, preprogrammed, or user-triggered manners (Figure 13.3A).

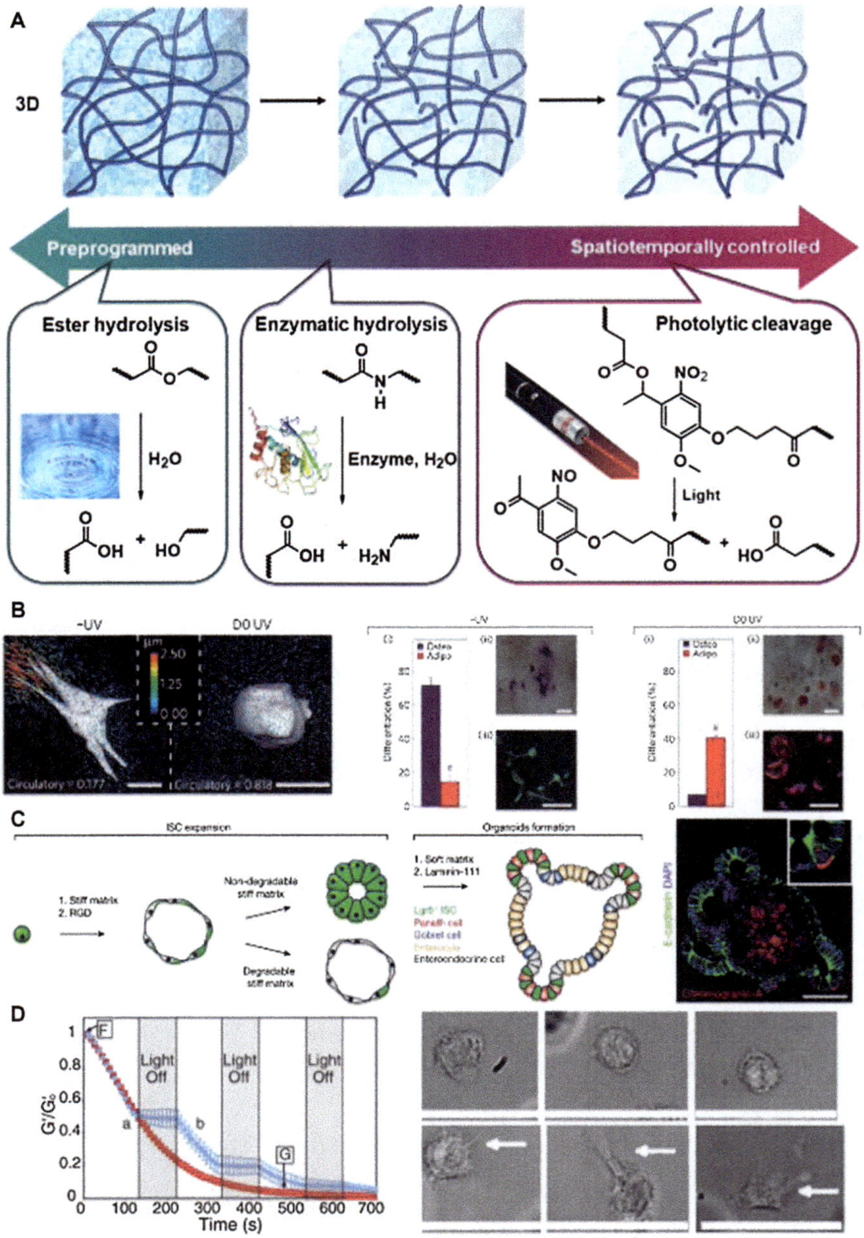

13.4.1.1 Proteolytically Degradable Hydrogels as Engineered Matrices

Proteolytically degradable hydrogels have been designed that respond to cell-secreted enzymes, allowing for the matrix softening and remodeling on a cell-dictated timescale. This has most frequently been achieved through inclusion of proteolytically sensitive degradable peptide crosslinkers, in which site-specific modifications to the substrate sequence can yield materials with very different degradative profiles.[46] To examine how the chemical crosslink density and gel affects encapsulated hMSCs' fate, the Burdick group employed RGD-modified methacrylated hyaluronic acid (MeHA) hydrogels.[58] Chain-growth photocrosslinking enabled hMSC encapsulation, with gels of varying elastic moduli attainable depending on the MeHA macromer concentration. When hMSC-laden hydrogels were cultured in bipotential osteogenic/adipogenic medium, hMSCs in MeHA gels of all elastic moduli (about 4–91 kPa) underwent almost exclusively adipogenic differentiation, implying that matrix mechanics and crosslink density have little effect on stem cell fate in nondegradable hydrogels. In addition, they synthesized methacrylate- and maleimide-modified HA (MeMaHA) and constructed a multistep crosslinkable hydrogel; here, the Michael-type reaction of maleimide-thiol yielded a primary crosslinking reaction for initial gel formation, while the photoinitiated chain-growth reaction of methacrylate groups could be independently triggered for secondary crosslinking. For hMSCs encapsulated within MMP-degradable HA hydrogels (~4 kPa) prepared by Michael-type reaction crosslinking, osteogenic outpaced adipogenic differentiation after 14 days in culture, though the opposite trend was observed when nondegradable secondary crosslinks were introduced immediately after cell encapsulation (Figure 13.3B). Furthermore, experiments using 3D traction force microscopy (TFM) demonstrate that differentiation of hMSCs encapsulated within covalently crosslinked HA hydrogels were induced by the generation of degradation-mediated cellular traction forces, independent of cell morphology and matrix modulus. Thus hMSCs encapsulated in HA hydrogels with comparable elastic moduli have higher (lower) degree of cell spreading and

Figure 13.3 Designing the degradability of 3D hydrogels and its application for manipulating cellular functions: (A) The degradability of chemically crosslinked hydrogels can be manipulated in preprogrammed, cell-dictated, or user-defined manner using different labile groups. Reproduced from ref. 8 with permission from the Royal Society of Chemistry. (B) examples of control of MSC differentiation using proteolytically degradable hydrogels: Adapted from ref. 58 with permission from Springer Nature, Copyright 2013. (C) Hydrolytically degradable hydrogels for organoid formation: Adapted from ref. 65 with permission from Springer Nature, Copyright 2016. (D) Spatiotemporal control of MSC morphology using photolytically degradable hydrogels: Adapted from ref. 71 with permission from AAAS, Copyright 2009.

traction force, which favors osteogenesis (adipogenesis). Interestingly, they revealed that switching the hydrogel from a permissive to a constrained state through staged secondary crosslinking suppressed hydrogel degradation and traction, triggering a shift from osteogenic to adipogenic differentiation while cells remained spread. These results clearly indicate the importance of matrix degradability on crosslinking and cell morphology, differing from 2D culture where elastic modulus dominantly affects hMSC differentiation.

Cell-degradable materials have also proven essential in maintaining organoid function *in vitro*. Organoids are formed through self-assembled stem cells and resemble their native counterparts in cellular composition, multicellular structure, and functional characteristics. Therefore, organoids have attracted much attention as a powerful tool for basic and translational research.[59,60] Current organoid culture methods are largely based on animal or naturally derived hydrogels (*e.g.* Matrigel, collagen); there is a growing need for the development of synthetic ECM analogs that offer better reproducibility and user-defined properties. Cruz-Acuña *et al.* established a robust and reproducible methodology for designing the chemically well-defined enzymatically degradable PEG hydrogels to promote the expansion and growth of human intestinal organoids (HIOs) *in vitro*.[61] These engineered hydrogels mainly consist of 4-arm PEG with maleimide group at each end, a cysteine-containing RGD peptide (GRGDSPC) as an adhesive ligand, and bis cysteine-protease-sensitive peptide (GCRDGPQG↓IWGQDRCG; ↓ denotes enzymatic cleavage site) as a degradable crosslinker. When components were mixed in HEPES buffer at pH 7.4 in the presence of cells, gel formation occurred spontaneously by Michael-type addition, and HIOs and human lung organoids (HLOs) could be encapsulated. The mechanical properties of the hydrogel can be tuned by changing the polymer concentration, with PEG hydrogels prepared at ~4.0 wt% displaying mechanics comparable to Matrigel ($G' = 78$ Pa, $G'' = 5.8$ Pa) were determined suitable for the long-term viability of encapsulated HIOs. In addition to the network's biochemical/mechanical properties, hydrogel degradability was demonstrated to be critical to the long-term viability of HIOs; HIOs cannot survive extended culture in non-degradable PEG hydrogels prepared with 1,4-dithiothreitol instead of MMP-degradable peptide crosslinkers. This fully synthetic hydrogel enabled the *in vitro* establishment of intestinal organoids from human pluripotent stem cell (hPSC)-derived spheroids and HLOs from human embryonic stem cells (hESCs). Furthermore, the dynamic and specific mechanical environment was reported as crucial in regulating epithelial cell cyst formation, polarization, and lumen formation, while cell–gel adhesivity regulates apicobasal polarity and lumenogenesis, leading to epithelial cell morphogenesis in a 3D culture system.[62] Finally, the authors demonstrated the improvement of HIO engraftment and colonic wound repair through gel injection within injured intestinal mucosa, highlighting the material's potential utility as an injectable therapy.

13.4.1.2 Hydrolytically Degradable Hydrogels as Engineered Matrices

While proteolytically sensitive materials can be degraded on timescales dictated by cell state, in many applications it is desirable to preprogram material response profiles into gels. Since hydrogels environments are inherently aqueous, this has been most commonly achieved through network hydrolysis. Several hydrolytically hydrogels have been developed in which hydrolyzable moieties are introduced into the backbone or crosslinking agent of the networks; ester units are most commonly included, decomposing into nontoxic carboxylic acids and alcohols. In such hydrolyzable hydrogels, the chemical properties of the network (*e.g.* crystallinity, hydrophobicity), crosslinking density, and local pH affect the degradation rate. Though material degradation can be tuned, hydrolyzable biomaterials have been most frequently designed to exhibit slower degradation kinetics than the enzymatic degradation just described, theoretically matching the rate of gel hydrolysis with that of encapsulated cell ECM production (Figure 13.3A). In an early study, Rice and Anseth utilized PEG-dimethacrylate (PEG-DM) and PLA-*b*-PEG-*b*-PLA-dimethacrylate (PEG-PLA-DM) as macromers to culture chondrocytes in 3D.[63] With PEG-based macromers, chondrocyte-laden PEG hydrogels were prepared by inducing chain-growth radical polymerization by UV irradiation at 365 nm. PEG-DM gives a nondegradable PEG hydrogel, whereas the hydrogel prepared from PEG-PLA-DM contains the hydrolytically degradable PLA moiety, enabling gradual increase in gel swelling, and finally solubilization over the period of approximately 10 days.[64] The mixture of these two PEG macromers can produce PEG copolymer hydrogels that exhibit bimodal degradation profiles in a narrow composition range (PEG-DM/PEG-PLA-DM = 19 : 81 ∼ 25 : 75 mol%) where the structural integrity of PEG hydrogels is minimally maintained after degradation, while allowing for a uniform spatial distribution of large ECM components. The stable PEG network provides mechanical support to encapsulated chondrocytes and promotes phenotype maintenance. In fact, the mechanical properties of formed tissues obtained by long-term culture were superior to those of PEG-PLA-DM hydrogels, which show fast and monomodal degradation, especially in the early stages of neocartilage development. Additionally, compared to the nondegradable PEG-DM hydrogel, the most rapidly degradable hydrogel yielded enhanced production and a more uniform spatial distribution of chondrocyte-secreted collagen type I. On the other hand, the slower degrading PEG copolymer hydrogels (PEG-DM/PEG-PLA-DM = 23 : 77 mol%) suppressed differentiation while maintaining moderate cell morphology and type II collagen distribution within the hydrogels. These results highlight the importance of preprogrammed hydrolytic degradation in 3D cell culture, specifically with chondrocytes for cartilage tissue formation.

Hydrolyzable hydrogels, whose degradation rate can be macroscopically tuned at the molecular level, can be applied to create intestinal organoids with a higher level of multicellular structure. Gjorevski *et al.* reported that intestinal stem cell (ISC) expansion, differentiation, and interstitial organoid formation were enhanced within hydrolytically degradable hydrogels (Figure 13.3C).[65] ISCs were encapsulated within PEG hydrogels enzymatically formed through thrombin-activated transglutaminase factor XIII (FXIIIa)-mediated ligation of 8-arm PEG modified with FXIIIa substrate peptides. Though stiffer hydrogels (shear modulus of about 1.3 kPa) supported more cell proliferation than softer ones (shear modulus of 0.2 kPa), stiff gel did not support cell differentiation or organoid formation. To achieve spheroid growth and morphogenesis, the hydrogels needed to be softened from 1.3 kPa to about 0.2 kPa after cell expansion. To do so, the Lutolf group introduced TG-FXIII substrate peptides into vinylsulfone-functionalized 8-arm PEG and acrylate-functionalized 8-arm PEG by Michael-type reaction and crosslinked them by adding thrombin-activated FXIIIa to form chemically stable hydrogels (sPEG) and hydrolyzable hydrogels (dPEG), respectively. Hybrid PEG hydrogels that gradually degrade and soften were prepared through copolymerization, with the profile of degradation and softening controlled by changing the ratio of sPEG to dPEG during the network formation. In addition, enzymatically crosslinked PEG hydrogels modified with RGD peptides derived from FN (fibronectin) and one from LMN (laminin) supported the growth but not differentiation of mouse intestinal organoids. With laminin-111 and enzymatically crosslinked hybrid PEG gels, they succeeded in creating a mechanically dynamic 3D matrix that has the high elastic modulus required for YAP activation and ISC expansion but then softens to relax the accumulation of compressive forces exerted by cells, allowing for *in vitro* organogenesis. Here, the dynamic changes in bulk mechanical property of the hydrogel would also be modulated by processes that simultaneously affect other environmental properties, such as ligand accessibility, clustering, and nanomechanical responses to cell-exerted forces after ligand binding.[66–69] The intestinal organoid, or "Mini-gut", produced using such designer materials is expected to have a wide range of applications not only in stem cell research but also in pathological models such as colorectal cancer and cystic fibrosis and regenerative medicine.[59] These results highlight the importance of hydrolytic degradability for 3D cell culture.

13.4.1.3 *Photolytically Degradable Hydrogels as Engineered Matrices*

As an alternative to enzymatic or hydrolytic degradable hydrogel systems, light can be used to induce the degradation of hydrogels and the associated changes in mechanical properties. Although degradation is induced in the presence of cells, as in enzymatic and hydrolytic degradation, light can be used to achieve user-defined temporal and spatial control in a highly controlled manner.[70] The Anseth group first demonstrated that 3D cell–material interactions could be dynamically and externally directed by photodegradable hydrogels.[71–73] Acrylated *ortho*-nitrobenzyl (*o*NB) ester-derived moieties (PDA) that can be radically polymerized but photolytically cleaved are appended onto the end of linear PEG molecules (yielding PEG-diPDA); the resulting gel mechanics can be tuned precisely with high spatiotemporal precision using light. Kloxin *et al.* prepared cell-laden photodegradable PEG hydrogels by mixing PEG-diPDA, PEG-monoacrylate (PEGA), fibronectin, hMSC, and free-radical redox initiators in phosphate-buffered saline.[71] Gel photolysis can be tuned by the wavelength (365 nm or 405 nm) and irradiation intensity. When light irradiation is ceased, photodegradation halts, and the mechanical properties are stably maintained; staged light exposures yield stepwise softening of the hydrogel. By changing gel mechanics through network photodegradation, the morphology of encapsulated hMSCs was dynamically regulated; photochemically softened gels yielded higher cell spreading than those maintained at high stiffness (Figure 13.3D). Functionalized PEG macromers modified with variably substituted *o*NB groups have also been utilized to tune PEG gel photodegradation rates and wavelength responsiveness; degradation rate was increased by decreasing the number of aryl ethers on the *o*NB group or changing the functionality from primary to secondary at the benzylic site.[74] Encapsulation of hMSCs with high viability (~90%) was achieved by redox-initiated free radical polymerization between PEG-acrylate (PEGA) and the developed PEG-diPDA, and wavelength-biased release of encapsulated cells was achieved using bi-layer PEG hydrogels crosslinked with two different degradable PEG-diPDA. These results showed that the operability of user-defined photolysis can be extended by material design.

Many of the photodegradable groups used in 3D culture systems, including those based on *o*NB, yield one-photon/one-event responsiveness. In this case, the photodegradable groups incorporated into the hydrogel absorb photons and undergo cleavage reactions, which require large light dosages or high quantum yields for rapid degradation. As many of the photodegradable groups are UV responsive and prolonged or intense irradiation may cause phototoxicity, there has been interest in constructing systems that exhibit more efficient degradation. To achieve this, Brown *et al.* reported on the amplification mechanism of photolysis *via* addition-fragmentation chain transfer (AFCT) reaction.[75] They focused on the use of allyl sulfides as efficient AFCT functionalities[76] and synthesized a new azide crosslinker containing allyl sulfide (PEG3-azide). A PEG hydrogel was prepared by SPAAC reaction of PEG3-azide with cyclooctyne-terminated 4-arm PEG (4-arm PEG-DBCO). The PEG hydrogel was cleaved by light irradiation in the presence of photoinitiator and monofunctional thiol, whereby the hydrogel reverted to a soluble branched macromer. In this process, photogenerated thiyl radicals propagate rapidly through thiol–ene addition reactions and chain transfer events. The thiyl–thiol chain transfer events allowed for the cleavage of multiple crosslinking points with a single absorbed photon, and the change of storage modulus (G') of the hydrogel upon photodegradation could be tuned based on the amount of free monofunctional thiol (PEG-SH) added.

This hydrogel degradation *via* the AFCT reaction was strongly influenced by the kinetics of the allyl sulfide exchange depending on the thiol's pK_a; the acetamide-type allyl sulfide (AS-AA, p$K_a = 10.4$) crosslinker, with a higher pK_a, exhibited faster degradation than propionamide types (AS-PA, p$K_a = 8.8$) and phenyl acetamide (As-PhAA, p$K_a = 6.6$). Based on this fact, they constructed a cell-friendly degradation system for PEG hydrogels using glutathione (GSH, p$K_a = 9.42$) as a free monofunctional thiol. In the presence of a low concentration of soluble thiol (GSH at 15×10^{-3} M) and photoinitiator [lithium phenyl-2,4,6-trimethylbenzoylphosphinate (LAP) at 1×10^{-3} M], the PEG hydrogel was completely degraded within 15 s by irradiation with 365 nm light [5 mW cm^{-2}]. Colony formation from single ISCs encapsulated within this PEG hydrogel was mechanosensitive, and the highest efficiency was achieved with RGD-functionalized hydrogels with moderate storage modulus of approximately 1.5 kPa. The rapid degradation of the PEG hydrogel also allowed the isolation of encapsulated ISC colonies, and colony formation was confirmed again when the colonies were reseeded within the allyl sulfide PEG hydrogel after isolation. This indicates that the allyl sulfide PEG hydrogel is useful for the growth and amplification of ISCs. This AFCT reaction-based degradation is potentially useful for particularly sensitive cell types since it significantly reduces the concentration of photoinitiating species and exposure to light and radicals and can be applied to the degradation of thicker hydrogels (scaffolds).

13.4.2 Engineered 3D Matrices with Spatiotemporally Programmable Biochemical Cues

In degradable hydrogels, the dynamic mechanical environment of surrounding cells can be created as the material degrades in preprogrammed, cell-dictated, and/or user-defined manners. Complementing these mechanical signals in regulating the function of cells and tissue *in vivo* are biochemical cues, whereby cues are presented and interact in complex and synergistic manners. In 2D culture systems, a dynamic interface (see Chapter 6) has been realized to enable switching of various biochemical factors. The ability to spatiotemporally modulate the presentation of proteins, peptides, and small molecules within 3D biomaterials represents an important step toward capturing the dynamic heterogeneity of native tissue.

The Anseth group demonstrated that 3D cell–material interactions could be dynamically and externally directed where peptides are photochemically released from gels with light.[71] They successfully demonstrated the ability to pattern the hydrogels in both 2D and 3D using masked flood irradiation and two-photon laser scanning microscopy, respectively. Using this photoreleasable strategy, Kloxin *et al.* established that the time-dependent presentation of the integrin-specific ligand RGDS significantly enhanced the differentiation of hMSCs into chondrocytes compared to hMSCs cultured with persistent RGDS signaling, suggesting the importance of temporal cue presentation. In their system, both biochemical and biophysical cues within the PEG hydrogel platform could be dynamically manipulated with photoirradiation, although these cues were only tuned independently of one another in this study. In a follow-up study, DeForest and Anseth successfully created a 3D system that allowed for independent spatiotemporal regulation of biochemical and biophysical cues that influence 3D cell migration.[22,73] They designed PEG hydrogels *via* SPAAC reaction between difluorinated cyclooctyne–functionalized 4-arm PEG (4-arm PEG-DIFO) and bisazide-functionalized peptides that allow for the photocleavage of crosslinks (biophysical cue) and photoconjugation of pendant functionalities (biochemical cue). With visible light irradiation with eosin Y as photoinitiator, thiol-containing biomolecules (*e.g.* RGD peptide) could be covalently bonded to vinyl functionalities in the hydrogel using thiol–ene reactions (Figure 13.4A). In the presence of UV light, however, *o*NB moieties embedded within the peptide crosslinker are cleaved to yield network photodegradation. Both photocleavage and photoconjugation reactions could be used orthogonally in a wavelength-dependent manner to independently regulate the biophysical and biochemical properties of the hydrogels. This system was exploited for real-time manipulation of cell function, including guiding 3T3 fibroblast migration and outgrowth behavior within peptide-functionalized microvascular channels (Figure 13.4B). Here, patterned control over RGD biochemical presentation and network biophysics proved essential for directing 3D fibroblast migration within the PEG hydrogel. Importantly, these studies also illustrate the ability of such wavelength-dependent chemistries to spatiotemporally define microenvironmental niches that could potentially recapitulate those found *in vivo*. Collectively, these studies are representative of the emerging field of 4D cellular biology, and other promising approaches have been utilized to control various cellular functions in a dynamic and spatially defined manner.[77–79]

Though the aforementioned photochemical strategies have enabled patterned immobilization of biochemical cues within synthetic gels, these systems lack the capability to reversibly present biochemical cues in a dynamic manner akin to that in native tissue. DeForest and Anseth first demonstrated that the combination of two orthogonal, biocompatible photoreactions based on the combination of the two shown in Figure 13.4C enables the reversible spatial presentation of biological peptide cues.[80] Peptides (*e.g.* RGD) were synthesized to contain both a thiol group for visible light-mediated photocoupling and an *o*NB moiety for UV-induced photocleavage; utilizing differently colored lights allowed peptides to be first attached and subsequently removed from PEG hydrogels on demand and with 4D control. They successfully demonstrated reversible spatiotemporal presentation of bioactive peptides, as well as patterned attachment and detachment of NIH 3T3 fibroblasts on the surface of these hydrogels. Building on this work, DeForest and Tirrell proposed a new strategy in which the presentation of full-length proteins could be reversibly patterned within a 3D space of PEG hydrogel by combining three bioorthogonal chemistries.[32] Specifically, they utilized SPAAC for gel formation, a photomediated oxime

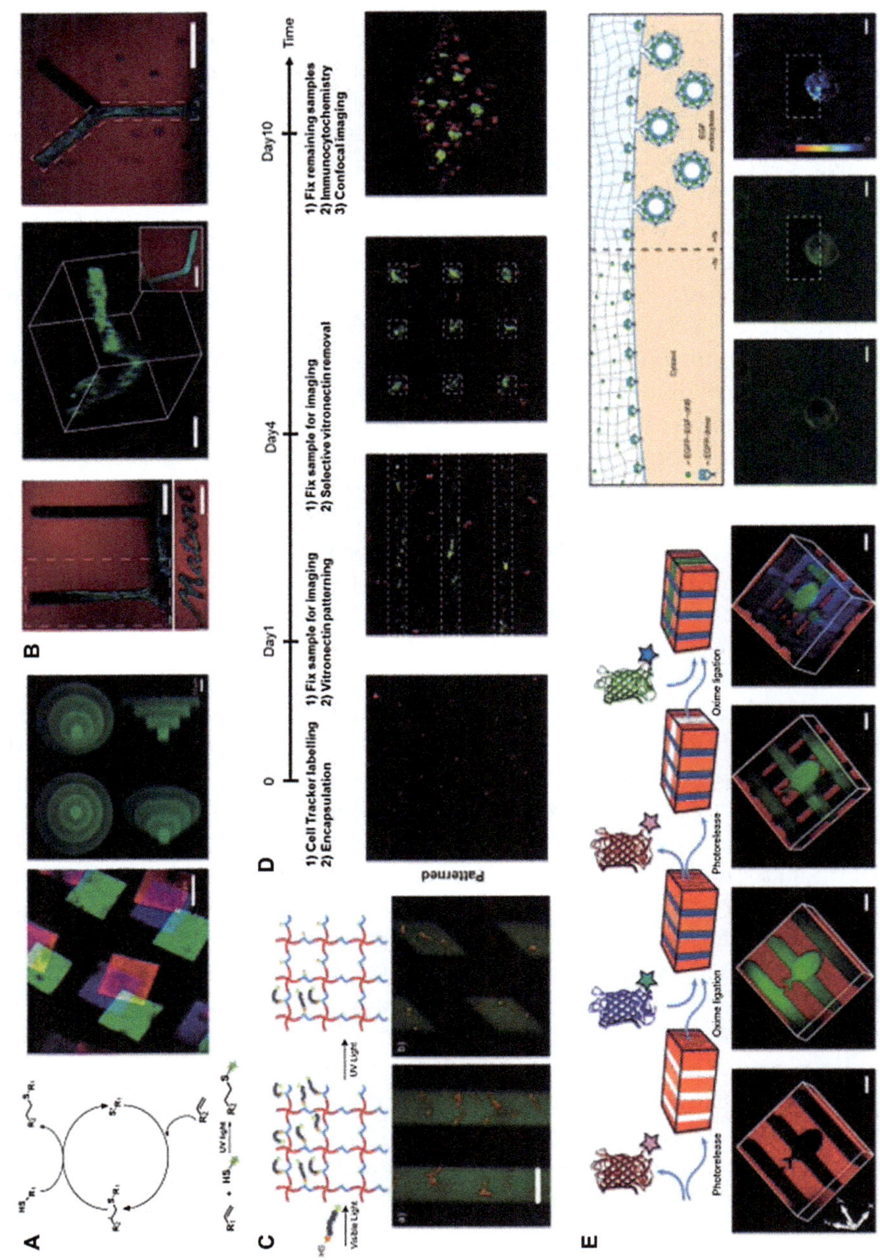
A
B
C
D
E
1) Cell Tracker labelling
2) Encapsulation
0
Day1
1) Fix sample for imaging
2) Vitronectin patterning
Day4
1) Fix sample for imaging
2) Selective vitronectin removal
Day10
1) Fix remaining samples
2) Immunocytochemistry
3) Confocal imaging
Time
Patterned
Visible Light
UV Light
Photorelease
Oxime ligation
Photorelease
Oxime ligation

ligation for protein immobilization, and *o*NB photoscission for protein removal. Introduction and subsequent removal of biochemical cues in a spatially defined manner were first demonstrated with aldehyde-functionalized proteins. Not only were specific regions of protein introduction achieved, but concentration gradients could also be produced based on a dose-dependent response to light irradiation. Furthermore, since the chemistries used for protein tethering and excision can be initiated using the same light source, intricate patterns comprised of multiple proteins can be created in 3D space. Interestingly, reversible presentation of vitronectin protein within hMSC-laden gels yielded dynamic presentation of osteogenic markers in a spatially controlled manner (Figure 13.4D).

Although these methods have enabled hydrogel modification using small peptides and stable proteins, it is not easy to modify fragile proteins with biological activity reversibly and in a user-defined manner, owing to the random nature in which proteins are conventionally modified (*e.g.* NHS-type chemistry). Toward this, the DeForest group recently demonstrated a modular and robust semisynthetic approach to reversibly pattern cell-laden hydrogels with site-specifically modified proteins.[81] The authors utilized a technique called sortase-tag enhanced protein ligation (STEPL) to design singly modified proteins with bioorthogonal reactive units for PEG hydrogel functionalization. STEPL was utilized to create photopatternable fluorescence proteins, model enzymes, and growth factors that each retain native bioactivity, far outperforming proteins modified using conventional NHS chemistry.[81,82] These proteins were patterned by mask-based and multiphoton-based laser-scanning lithographic strategies within PEG hydrogels (Figure 13.4E). Human epidermal A431 cells were encapsulated into the PEG hydrogel modified with a photoreleasable green fluorescent protein-epidermal growth factor fusion (EGFP-EGF-*o*NB-N_3). Since canonical EGF signaling requires protein internalization,[83] this behavior could be "turned on" with EGF's photochemical transition from a tethered growth factor to a locally solubilized one. Shadish *et al.* demonstrated the ability to

Figure 13.4 Engineered 3D hydrogel platforms with spatiotemporally programmable biochemical cues: (A) Cytocompatible and spatiotemporal biochemical patterning of 3D hydrogels. Adapted from ref. 22 with permission from Springer Nature, Copyright 2009. (B) Spatiotemporally directed 3D cell motility based on photolytic degradation and biochemical patterning. Adapted from ref. 73 with permission from Springer Nature, Copyright 2011. (C) Selective attachment and detachment of cells through photoreversible presentation of cell-adhesive peptides. Adapted from ref. 80 with permission from John Wiley and Sons, Copyright © 2012 WILEY-VCH Verlag GmbH & Co. KGaA, Weinheim. (D) Spatiotemporal control of hMSC differentiation by photoreversible biochemical patterning with full-length proteins. Adapted from ref. 32 with permission from Springer Nature, Copyright 2015. (E) Spatiotemporal modulation of epidermal intracellular signaling with a photoreleasable growth factors. Adapted from ref. 81 with permission from Springer Nature, Copyright 2019.

confine EGF receptor internalization and canonical activation to one side of a single encapsulate cell. This developed technology is generalizable and can be extended to hydrogels made of natural polymers such as collagen I and fibrin.[55] These results demonstrate that photochemical control over gel biochemistry can govern 3D biological fate with single-cell and/or subcellular resolution, offering exciting routes toward better understanding of how cells respond to transient extracellular biochemical cues.

13.4.3 Adaptable Engineered 3D Matrices for Mechanobiology

In general, some degree of biodegradability is required for cells to migrate through and proliferate within engineered hydrogels. The degradation strategies described in Section 13.4.1 result in an irreversible elimination of the polymer network and potentially undesired release of any immobilized biochemical cues over time. Addressing these limitations, while still allowing for cells to take on *in vivo*-like phenotypes through active matrix remodeling, has given rise to the concept of "adaptable hydrogels" – those crosslinked with reversible bonds that can be broken and reformed without changes in the external environment. Since 2D mechanobiology using adaptable hydrogels and interfaces have already been introduced in Chapters 6 and 11, we will highlight here the importance of an adaptable nature in hydrogel-based 3D culture environments.

In addition to stiffness, matrix viscoelasticity and stress-relaxation have proven important biophysical parameters in the mechanobiological regulation of cell fate (Figure 13.5A).[84] Formation of cell-laden hydrogels using equilibrium reactions of different strengths (*e.g.* guest–host chemistries,[85] hydrophobic interactions,[86] hydrogen bonding,[87] dynamic covalent linkage[88]) enables creation of hydrogels with tunable viscoelasticity. By comparing cell function in materials that are fully elastic with those that exhibit different degrees of stress relaxation, Chaudhuri and Mooney have demonstrated the substantial impact that stress relaxation has on 2D and 3D cell fate.[89,90] hMSCs cultured in ionically crosslinked alginate gels with different molecular weighted alginate and PEGylated alginate exhibited different levels of stress relaxation ($t_{1/2} = \sim 1$ h to ~ 1 min). Although the hMSCs encapsulated within nondegradable elastic hydrogel and alginate hydrogel with slower stress relaxation exhibited rounded cell morphologies, increasing the stress relaxation of alginate hydrogels enhanced spreading, proliferation, and osteogenic differentiation and maturation. The proposed mechanism is shown in Figure 13.5B. First, hMSCs within a 3D matrix exert strain on the matrix, resulting in forces/stresses resisting the strain, as determined by the initial elastic modulus of the matrix. In the case of an elastic (nondissipative) matrix, these forces are never relaxed, ultimately yielding no matrix remodeling. In viscoelastic matrixes, however, the forces exerted by cells in the matrix can be relaxed over time because of mechanical yielding and remodeling of the matrix, whereby the rate of matrix stress relaxation

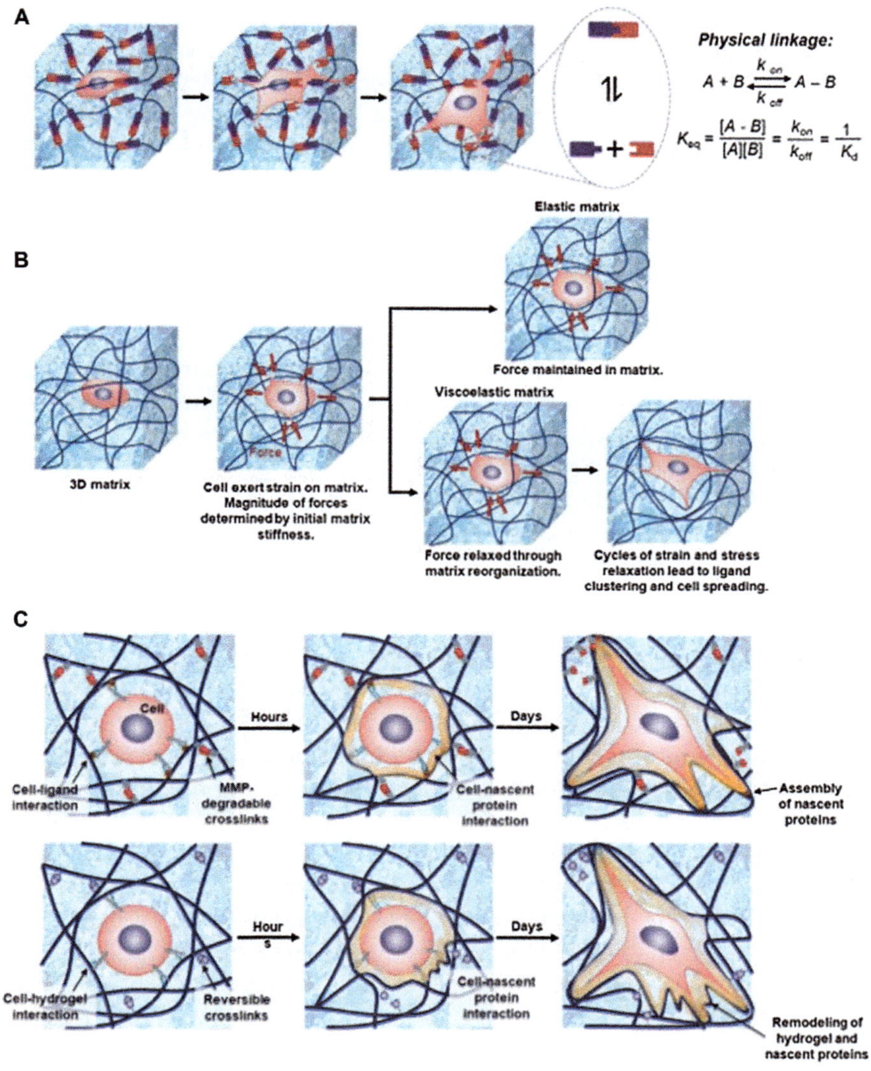

Figure 13.5 Engineered 3D reversibly crosslinked, adaptable hydrogel platforms: (A) Schematic of a conventional adaptable hydrogel incorporating reversible crosslinks. Reproduced from ref. 84 with permission from John Wiley & Sons, Copyright © 2015 WILEY-VCH Verlag GmbH & Co. KGaA, Weinheim. (B) Initial matrix stiffness and stress relaxability regulate MSC state in 3D hydrogels. Reproduced from ref. 89 with permission from Springer Nature, Copyright 2016. (C) Nascent protein adhesion and remodeling govern MSC spreading and fate in both degradable and adaptable hydrogels. Reproduced from ref. 92 with permission from Springer Nature, Copyright 2019.

determines the degree of mechanical remodeling. Similar effects have been observed in other adaptable hydrogels, including PEG-based materials covalently crosslinked through dynamic covalent hydrazone bonds.[91]

As discussed here, matrix degradation and remodeling play an important role in regulating the cell behaviors in 3D culture. Loebel *et al.* also demonstrated the impact of early protein deposition on cell behaviors within hydrogel matrices (Figure 13.5C).[92] Using bioorthogonal noncanonical amino acid tagging that replaces naturally occurring methionine residues in proteins with azidohomoalanine,[93] they visualized nascent proteins, which were secreted from hMSCs encapsulated in proteolytically degradable crosslinked hyaluronic acid (HA) hydrogel and viscoelastic physically crosslinked HA hydrogels based on host (cyclodextrin)–guest (adamantane) interactions, within a day of culture. Importantly, the thickness of protein deposit was influenced by hydrogel stiffness, suggesting the ability of the hydrogel to limit cell spreading influences on regulating deposition. They found that nascent protein deposition and remodeling were needed for mechanosensing (YAP/TAZ nuclear translocation) and osteogenic differentiation. Since the native ECM also exhibits some degree of stress relaxation, synthetic culture matrices of a viscoelastic or degradable nature that isolate these effects are of prime interest.

13.4.4 Conclusions and Future Perspective

In native tissue, a variety of biophysicals and biochemicals act synergistically in a complex and coordinated manner to govern cell function and fate. In this chapter, we discussed emerging trends in biomaterial design that afford user-defined programmability of network physiochemistry. At a minimum, biomaterials should enable direct cell encapsulation using methods that are bioorthogonal and cytocompatible. Desirable properties include the ability to mechanically and chemically tune bulk biomaterial properties at the time of encapsulation. Ideal material strategies further offer the ability to modulate biomaterial properties with spatiotemporal control, in an independent and reversible manner, and in the presence of live cells. Though some such systems have been developed and have been used for advanced cell and organoid culture, tremendous opportunities persist in the creation of biomaterial systems that further recapitulate *in vivo* complexity *in vitro*.

In addition to novel biomaterials, so too are new methods needed to better probe 4D cell behavior within these systems. From a mechanobiological perspective, high-throughput techniques that provide real-time information about the local gel microenvironment and how it changes through cell-based matrix remodeling remain in great need. Nondestructive methods to visualize and quantify traction forces exerted by cells on their surrounding matrices also remain of interest, particularly for those that could perform concurrently with user-triggered material alterations. As the ability to regulate cell function spatially continues to evolve, methods that provide single-cell information about cell state but with spatial control would also be tremendously powerful. When coupled with 4D-tunable biomaterials, these strategies are likely to yield many more landmark discoveries in mechanobiology and beyond.

Acknowledgements

This work was supported by a CAREER Award (DMR 1652141, C.A.D.) and a grant (DMR 1807398, C.A.D.) from the National Science Foundation, as well as a Maximizing Investigators' Research Award (R35GM138036, C.A.D.) from the National Institutes of Health. It is also supported in part by AMED-PRIME (JP19gm5810017, K.U.) from the Japan Agency for Medical Research and Development and Grant-in-Aid for Transformative Research Areas (A) (JP20329906, K.U.) from the Japan Society for Promotion of Science.

References

1. N. Wang, J. Butler and D. Ingber, *Science*, 1993, **260**(5111), 1124–1127.
2. A. J. Engler, S. Sen, H. L. Sweeney and D. E. Discher, *Cell*, 2006, **126**(4), 677–689.
3. B. M. Baker and C. S. Chen, *J. Cell Sci.*, 2012, **125**(13), 3015–3024.
4. J. A. Pedersen and M. A. Swartz, *Ann. Biomed. Eng.*, 2005, **33**(11), 1469–1490.
5. S. R. Caliari and J. A. Burdick, *Nat. Methods*, 2016, **13**(5), 405–414.
6. N. Huebsch, P. R. Arany, A. S. Mao, D. Shvartsman, O. A. Ali and S. A. Bencherif, *et al.*, *Nat. Mater.*, 2010, **9**(6), 518–526.
7. J. Zonderland and L. Moroni, *Biomaterials*, 2021, **268**, 120572.
8. P. M. Kharkar, K. L. Kiick and A. M. Kloxin, *Chem. Soc. Rev.*, 2013, **42**(17), 7335–7372.
9. R. C. Bretherton and C. A. DeForest, *ACS Biomater. Sci. Eng.*, 2021, **7**(9), 3997–4008.
10. A. D. Rouillard, C. M. Berglund, J. Y. Lee, W. J. Polacheck, Y. Tsui and L. J. Bonassar, *et al.*, *Tissue Eng., Part C*, 2010, **17**(2), 173–179.
11. J. A. Burdick, A. J. Peterson and K. S. Anseth, *Biomaterials*, 2001, **22**(13), 1779–1786.
12. S. J. Bryant and K. S. Anseth, *J. Biomed. Mater. Res.*, 2002, **59**(1), 63–72.
13. S. J. Bryant and K. S. Anseth, *J. Biomed. Mater. Res., Part A*, 2003, **64A**(1), 70–79.
14. E. R. Ruskowitz and C. A. DeForest, *ACS Biomater. Sci. Eng.*, 2019, **5**(5), 2111–2116.
15. A. M. Kloxin, C. J. Kloxin, C. N. Bowman and K. S. Anseth, *Adv. Mater.*, 2010, **22**(31), 3484–3494.
16. C. M. Madl and S. C. Heilshorn, *Adv. Funct. Mater.*, 2018, **28**(11), 1706046.
17. D. A. Ossipov and J. Hilborn, *Macromolecules*, 2006, **39**(5), 1709–1718.
18. A. M. Testera, A. Girotti, I. G. de Torre, L. Quintanilla, M. Santos and M. Alonso, *et al.*, *J. Mater. Sci.: Mater. Med.*, 2015, **26**(2), 105.
19. V. Hong, N. F. Steinmetz, M. Manchester and M. G. Finn, *Bioconjugate Chem.*, 2010, **21**(10), 1912–1916.
20. J. M. Baskin, J. A. Prescher, S. T. Laughlin, N. J. Agard, P. V. Chang and I. A. Miller, *et al.*, *Proc. Natl. Acad. Sci. U. S. A.*, 2007, **104**(43), 16793–16797.

21. J. A. Codelli, J. M. Baskin, N. J. Agard and C. R. Bertozzi, *J. Am. Chem. Soc.*, 2008, **130**(34), 11486–11493.
22. C. A. DeForest, B. D. Polizzotti and K. S. Anseth, *Nat. Mater.*, 2009, **8**(8), 659–664.
23. C. A. DeForest, E. A. Sims and K. S. Anseth, *Chem. Mater.*, 2010, **22**(16), 4783–4790.
24. E. A. Sims, C. A. DeForest and K. S. Anseth, *Tetrahedron Lett.*, 2011, **52**(16), 1871–1873.
25. J. C. Jewett, E. M. Sletten and C. R. Bertozzi, *J. Am. Chem. Soc.*, 2010, **132**(11), 3688–3690.
26. M. F. Debets, S. S. van Berkel, S. Schoffelen, F. P. J. T. Rutjes, J. C. M. van Hest and F. L. van Delft, *Chem. Commun.*, 2010, **46**(1), 97–99.
27. A. Kuzmin, A. Poloukhtine, M. A. Wolfert and V. V. Popik, *Bioconjugate Chem.*, 2010, **21**(11), 2076–2085.
28. C. D. Hermann, D. S. Wilson, K. A. Lawrence, X. Ning, R. Olivares-Navarrete and J. K. Williams, *et al.*, *Biomaterials*, 2014, **35**(36), 9698–9708.
29. J. Zheng, L. A. Smith Callahan, J. Hao, K. Guo, C. Wesdemiotis and R. A. Weiss, *et al.*, *ACS Macro Lett.*, 2012, **1**(8), 1071–1073.
30. X. Wang, Z. Li, T. Shi, P. Zhao, K. An and C. Lin, *et al.*, *Mater. Sci. Eng. C*, 2017, **73**, 21–30.
31. J. Dommerholt, S. Schmidt, R. Temming, L. J. A. Hendriks, F. P. J. T. Rutjes and J. C. M. van Hest, *et al.*, *Angew. Chem., Int. Ed.*, 2010, **49**(49), 9422–9425.
32. C. A. DeForest and D. A. Tirrell, *Nat. Mater.*, 2015, **14**(5), 523–531.
33. C. M. Madl, L. M. Katz and S. C. Heilshorn, *Adv. Funct. Mater.*, 2016, **26**(21), 3612–3620.
34. R. van Geel, G. J. M. Pruijn, F. L. van Delft and W. C. Boelens, *Bioconjugate Chem.*, 2012, **23**(3), 392–398.
35. Z. Li, T. S. Seo and J. Ju, *Tetrahedron Lett*, 2004, **45**(15), 3143–3146.
36. V. X. Truong, M. P. Ablett, H. T. J. Gilbert, J. Bowen, S. M. Richardson and J. A. Hoyland, *et al.*, *Biomater. Sci.*, 2014, **2**(2), 167–175.
37. H.-L. Wei, Z. Yang, L.-M. Zheng and Y.-M. Shen, *Polymer*, 2009, **50**(13), 2836–2840.
38. C. M. Nimmo, S. C. Owen and M. S. Shoichet, *Biomacromolecules*, 2011, **12**(3), 824–830.
39. S. Kirchhof, F. P. Brandl, N. Hammer and A. M. Goepferich, *J. Mater. Chem. B*, 2013, **1**(37), 4855–4864.
40. M. L. Blackman, M. Royzen and J. M. Fox, *J. Am. Chem. Soc.*, 2008, **130**(41), 13518–13519.
41. H. Zhou, J. Woo, A. M. Cok, M. Wang, B. D. Olsen and J. A. Johnson, *Proc. Natl. Acad. Sci.*, 2012, **109**(47), 19119–19124.
42. D. L. Alge, M. A. Azagarsamy, D. F. Donohue and K. S. Anseth, *Biomacromolecules*, 2013, **14**(4), 949–953.
43. A. Gress, A. Völkel and H. Schlaad, *Macromolecules*, 2007, **40**(22), 7928–7933.

44. A. E. Rydholm, C. N. Bowman and K. S. Anseth, *Biomaterials*, 2005, **26**(22), 4495–4506.
45. C. N. Salinas, B. B. Cole, A. M. Kasko and K. S. Anseth, *Tissue Eng.*, 2007, **13**(5), 1025–1034.
46. M. P. Lutolf, J. L. Lauer-Fields, H. G. Schmoekel, A. T. Metters, F. E. Weber and G. B. Fields, *et al.*, *Proc. Natl. Acad. Sci.*, 2003, **100**(9), 5413–5418.
47. M. P. Lutolf, N. Tirelli, S. Cerritelli, L. Cavalli and J. A. Hubbell, *Bioconjugate Chem.*, 2001, **12**(6), 1051–1056.
48. M. P. Lutolf, G. P. Raeber, A. H. Zisch, N. Tirelli and J. A. Hubbell, *Adv. Mater.*, 2003, **15**(11), 888–892.
49. E. A. Phelps, N. O. Enemchukwu, V. F. Fiore, J. C. Sy, N. Murthy and T. A. Sulchek, *et al.*, *Adv. Mater.*, 2012, **24**(1), 64–70.
50. A. Shikanov, R. M. Smith, M. Xu, T. K. Woodruff and L. D. Shea, *Biomaterials*, 2011, **32**(10), 2524–2531.
51. M. Malkoch, R. Vestberg, N. Gupta, L. Mespouille, P. Dubois and A. F. Mason, *et al.*, *Chem. Commun.*, 2006, **26**, 2774–2776.
52. B. D. Fairbanks, M. P. Schwartz, A. E. Halevi, C. R. Nuttelman, C. N. Bowman and K. S. Anseth, *Adv. Mater.*, 2009, **21**(48), 5005–5010.
53. G. N. Grover, J. Lam, T. H. Nguyen, T. Segura and H. D. Maynard, *Biomacromolecules*, 2012, **13**(10), 3013–3017.
54. P. E. Farahani, S. M. Adelmund, J. A. Shadish and C. A. DeForest, *J. Mater. Chem. B*, 2017, **5**(23), 4435–4442.
55. I. Batalov, K. R. Stevens and C. A. DeForest, *Proc. Natl. Acad. Sci.*, 2021, **118**(4), e2014194118.
56. E. Saxon and C. R. Bertozzi, *Science*, 2000, **287**(5460), 2007–2010.
57. K. M. Gattás-Asfura, C. A. Fraker and C. L. Stabler, *J. Biomed. Mater. Res., Part A*, 2011, **99A**(1), 47–57.
58. S. Khetan, M. Guvendiren, W. R. Legant, D. M. Cohen, C. S. Chen and J. A. Burdick, *Nat. Mater.*, 2013, **12**(5), 458–465.
59. T. Sato and H. Clevers, *Science*, 2013, **340**(6137), 1190–1194.
60. J. Kim, B.-K. Koo and J. A. Knoblich, *Nat. Rev. Mol. Cell Biol.*, 2020, **21**(10), 571–584.
61. R. Cruz-Acuña, M. Quirós, A. E. Farkas, P. H. Dedhia, S. Huang and D. Siuda, *et al.*, *Nat. Cell Biol.*, 2017, **19**, 1326.
62. N. O. Enemchukwu, R. Cruz-Acuña, T. Bongiorno, C. T. Johnson, J. R. García and T. Sulchek, *et al.*, *J. Cell Biol.*, 2015, **212**(1), 113–124.
63. M. A. Rice and K. S. Anseth, *J. Biomed. Mater. Res., Part A*, 2004, **70A**(4), 560–568.
64. S. J. Bryant, R. J. Bender, K. L. Durand and K. S. Anseth, *Biotechnol. Bioeng.*, 2004, **86**(7), 747–755.
65. N. Gjorevski, N. Sachs, A. Manfrin, S. Giger, M. E. Bragina and P. Ordóñez-Morán, *et al.*, *Nature*, 2016, **539**(7630), 560–564.
66. M. J. Wilson, S. J. Liliensiek, C. J. Murphy, W. L. Murphy and P. F. Nealey, *Soft Matter*, 2012, **8**(2), 390–398.
67. G. Maheshwari, G. Brown, D. A. Lauffenburger, A. Wells and L. G. Griffith, *J. Cell Sci.*, 2000, **113**(10), 1677–1686.

68. B. Trappmann, J. E. Gautrot, J. T. Connelly, D. G. T. Strange, Y. Li and M. L. Oyen, *et al.*, *Nat. Mater.*, 2012, **11**(7), 642–649.
69. V. Hernandez-Gordillo, T. Kassis, A. Lampejo, G. Choi, M. E. Gamboa and J. S. Gnecco, *et al.*, *bioRxiv*, 2019, 806919.
70. E. R. Ruskowitz and C. A. DeForest, *Nat. Rev. Mater.*, 2018, **3**, 17087.
71. A. M. Kloxin, A. M. Kasko, C. N. Salinas and K. S. Anseth, *Science*, 2009, **324**(5923), 59–63.
72. A. M. Kloxin, M. W. Tibbitt and K. S. Anseth, *Nat. Protoc.*, 2010, **5**(12), 1867–1887.
73. C. A. DeForest and K. S. Anseth, *Nat. Chem.*, 2011, **3**(12), 925–931.
74. D. R. Griffin and A. M. Kasko, *J. Am. Chem. Soc.*, 2012, **134**(31), 13103–13107.
75. T. E. Brown, I. A. Marozas and K. S. Anseth, *Adv. Mater.*, 2017, **29**(11), 1605001.
76. T. F. Scott, A. D. Schneider, W. D. Cook and C. N. Bowman, *Science*, 2005, **308**(5728), 1615–1617.
77. R. G. Wylie, S. Ahsan, Y. Aizawa, K. L. Maxwell, C. M. Morshead and M. S. Shoichet, *Nat. Mater.*, 2011, **10**(10), 799–806.
78. Y. Aizawa, R. Wylie and M. Shoichet, *Adv. Mater.*, 2010, **22**(43), 4831–4835.
79. N. Broguiere, I. Luchtefeld, L. Trachsel, D. Mazunin, R. Rizzo and J. W. Bode, *et al.*, *Adv. Mater.*, 2020, **32**(25), e1908299.
80. C. A. DeForest and K. S. Anseth, *Angew. Chem., Int. Ed.*, 2012, **51**(8), 1816–1819.
81. J. A. Shadish, G. M. Benuska and C. A. DeForest, *Nat. Mater.*, 2019, **18**(9), 1005–1014.
82. P. M. Gawade, J. A. Shadish, B. A. Badeau and C. A. DeForest, *Adv. Mater.*, 2019, **31**(33), 1902462.
83. P. R. Kuhl and L. G. Griffith-Cima, *Nat. Med.*, 1996, **2**(9), 1022–1027.
84. H. Wang and S. C. Heilshorn, *Adv. Mater.*, 2015, **27**(25), 3717–3736.
85. C. B. Rodell, A. L. Kaminski and J. A. Burdick, *Biomacromolecules*, 2013, **14**(11), 4125–4134.
86. Y. Liu, B. Liu, J. J. Riesberg and W. Shen, *Macromol. Biosci.*, 2011, **11**(10), 1325–1330.
87. H. Tan, C. Xiao, J. Sun, D. Xiong and X. Hu, *Chem. Commun.*, 2012, **48**(83), 10289–10291.
88. B. Yang, Y. Zhang, X. Zhang, L. Tao, S. Li and Y. Wei, *Polym. Chem.*, 2012, **3**(12), 3235–3238.
89. O. Chaudhuri, L. Gu, D. Klumpers, M. Darnell, S. A. Bencherif and J. C. Weaver, *et al.*, *Nat. Mater.*, 2016, **15**(3), 326–334.
90. O. Chaudhuri, L. Gu, M. Darnell, D. Klumpers, S. A. Bencherif and J. C. Weaver, *et al.*, *Nat. Commun.*, 2015, **6**, 6365.
91. D. D. McKinnon, D. W. Domaille, J. N. Cha and K. S. Anseth, *Adv. Mater.*, 2014, **26**(6), 865–872.
92. C. Loebel, R. L. Mauck and J. A. Burdick, *Nat. Mater.*, 2019, **18**(8), 883–891.
93. D. C. Dieterich, J. J. Lee, A. J. Link, J. Graumann, D. A. Tirrell and E. M. Schuman, *Nat. Protoc.*, 2007, **2**(3), 532–540.

CHAPTER 14

Combining Genetic and Mechanical Factors to Model Disease

JAIMIE MAYNER,[a] ELENA DEMEESTER[b] AND ADAM J. ENGLER*[a,c]

[a] Department of Bioengineering, University of California, San Diego, La Jolla, CA 92037, USA; [b] Division of Biological Sciences, University of California, San Diego, La Jolla, CA 92037, USA; [c] Sanford Consortium for Regenerative Medicine, La Jolla, CA 92037, USA
*Email: aengler@ucsd.edu

14.1 Introduction

The classic Mendelian approach to disease inheritance, in which hereditary, single-gene mutations result in disease by loss of function, does not explain most inherited diseases or genetic predisposition. The complexity of diseases, with patient-to-patient variability and heterogeneity within diseased tissue, remains poorly understood. Cell–cell variability in gene expression (and phenotype) within a population is ubiquitous in disease and caused by (1) genetics, *e.g.* heterogeneous mutations within a tumor, and (2) asymmetric application of nongenetic stimuli on a population. These symmetry-breaking events are at the core of developmental biology and create distinct phenotypes arising through both extrinsic and intrinsic factors. Extrinsic factors include the large variety of complex environmental cues from different tissue niches, which include both biochemical and mechanical stimuli and will be the focus

Biomaterials Science Series No. 12
Material-based Mechanobiology
Edited by Jun Nakanishi and Koichiro Uto

Published by the Royal Society of Chemistry, www.rsc.org

of this chapter. Intrinsic factors, although less understood, create the spontaneous diversification of clonal populations into distinct variants through some symmetry-breaking event propelled by a stochastic process that drives these discrete options, *e.g.* stochastic gene expression "noise".[1] Phenotype is driven by both genetic and nongenetic factors, and therefore studies in which these are independently investigated, while fruitful and tightly controlled, do not capture the entire picture of disease. Systems where both genetic and environmental factors and how they act together are modeled can elucidate novel pathways in disease development, prompting drug discoveries. This chapter will first focus on single-factor disease model systems in which genetics or mechanics alone are investigated. We will then conclude with a discussion of instances where both genetics and mechanics are jointly manipulated to model heart disease and cancer.

14.2 Modeling Disease with a Single Variable

14.2.1 Genetic Models

While disease development is a complex process that cannot be attributed to a single cause, significant progress can be made by developing models that alter a single variable and measure biological responses. Historically speaking, the medical field has long recognized the importance of genetics in disease risk, with physicians using family history to assess patient risk for a multitude of conditions. In addition to inherited genomic risk, acquired mutations, often attributed to environmental stressors,[2–4] have long been hypothesized to play a role in late onset diseases, such as cancer.[2,4,5] However, genomic regulation is multilayered and remains poorly understood. Next-generation sequencing (NGS) technologies such as Illumina Sequencing, Roche 453 Sequencing, and Ion Torrent: Proton/PGM sequencing, coupled with novel genetic editing tools such as CRISPR,[6] have ushered in a new era of disease modeling where the contribution of individual genes to disease pathogenesis can be modeled *in vitro*.

In more recent years, genetic reprogramming of primary cell lines into induced pluripotent stem cells (iPSCs) has become a valuable tool for studying a variety of cell types using recently developed differentiation protocols. Paired with the affordability and accessibility of modern next-generation sequencing, patient-specific genetics can be studied in a variety of diseases relevant cell types. However, when specific edits are required, gene editing has been made possible through nucleases, which create site-specific double-stranded breaks at the target location.[7,8] These breaks are repaired through homologous recombination (HR), which allows for a template sequence to be inserted, or nonhomologous end joining (NHEJ).[7,8] Early genetic engineering methods include zinc finger nucleases (ZFN), which cut DNA at defined triplet nucleotide locations[9] and were used extensively in iPSC reprogramming.[10] However, cost and off-target cleavages led to more modern gene-editing nucleases, *e.g.* transcription activator-like effectors attached to nucleases (TALENs)[11] and ultimately the clustered regulatory interspaced short palindromic repeats (CRISPR)-Cas9 nuclease system. Unlike TALENs and ZFNs, which rely on chimeric protein binding to DNA, CRISPR/Cas systems bind to target DNA sites using a complementary noncoding RNA (guide RNA). CRISPR-Cas9 was developed from a naturally occurring adaptive immune response in bacteria, by which DNA from invading viruses are captured to create DNA segments, or CRISPR arrays, which can be utilized by the bacteria to create RNA segments to target the virus if it invades again, as the guide RNAs also bind Cas9 nucleases.[6,8] CRISPR has proved to be more efficient and robust due to the simplicity of its construction and its flexibility for use in many gene loci.[12] These major advancements in genome editing have revolutionized both basic and translational research and are summarized in Figure 14.1; specific disease modeling examples are described next.

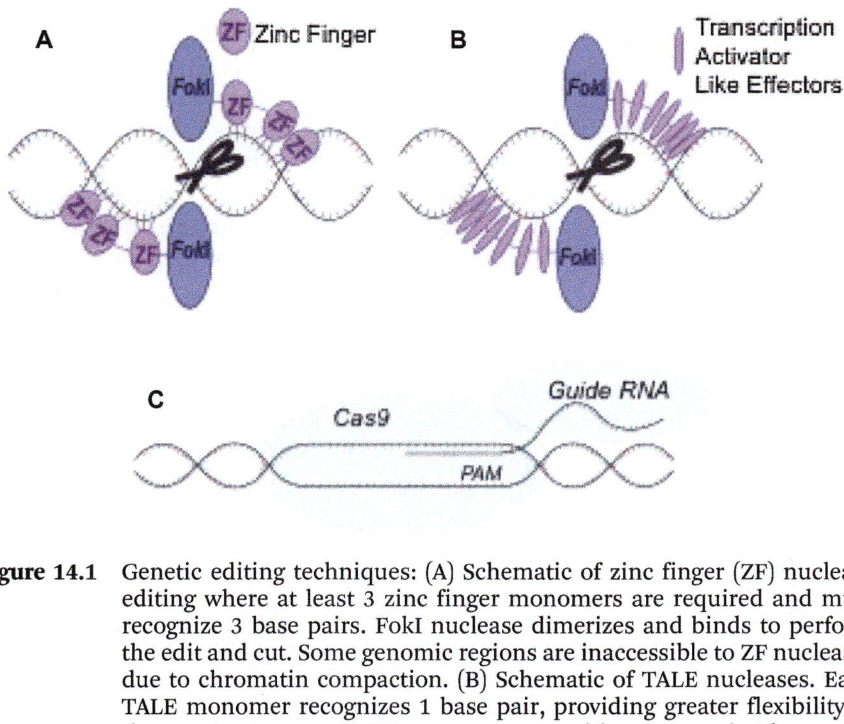

Figure 14.1 Genetic editing techniques: (A) Schematic of zinc finger (ZF) nuclease editing where at least 3 zinc finger monomers are required and must recognize 3 base pairs. FokI nuclease dimerizes and binds to perform the edit and cut. Some genomic regions are inaccessible to ZF nucleases due to chromatin compaction. (B) Schematic of TALE nucleases. Each TALE monomer recognizes 1 base pair, providing greater flexibility in the target sequence. TALE monomer recognition occurs *via* 2 hypervariable amino acid repeat variable di-residues. Again the FokI nuclease dimerizes and binds for editing, and, again, some genomic regions are inaccessible due to compact chromatin. (C) Schematic of the CRISPR/Cas9 system, which recognizes genomic DNA by a PAM sequence. A single guide RNA recognizes the target genomic sequence. CRISPR/Cas9 only needs a short guide RNA for DNA targeting, unlike ZF and TALE nucleases. Simultaneous genomic modifications at multiple independent sites are possible with CRISPR/Cas9.

14.2.1.1 Heart Disease

Cardiovascular disease (CVD) is the leading cause of death in the world,[13] and, of that, coronary artery disease (CAD) is responsible for one in every five deaths in the developed world.[14] Family history has long been used in medicine to predict CAD; however, the search for the associated genes has remained mainly fruitless until the advent of modern sequencing techniques. Genome-wide associate studies (GWAS) have not only confirmed that genetics account for 50% of susceptibility[15] but have also identified genetic variants that increase risk.[16–18] At least 10^6 disease-causing single-nucleotide polymorphisms (SNPs) have been identified in the genome, but most are located in noncoding regions, making the mechanism through which they enhance disease risk enigmatic.[19] Three separate GWAS studies in parallel identified variants at the 9p21 locus to be associated with CAD as having the greatest correlation with CAD, making it the most impactful genomic risk factor. Like many other associated SNPs, it does not contain any annotated genes and is in a noncoding region of DNA.[17,19–22]

To study how variants at 9p21 result in CAD, one group generated iPSCs from patients with and without homozygous risk variants (risk and nonrisk). By differentiating the iPSCs into vascular smooth muscle cells (VSMCs), the group found that phenotype of VSMCs was impacted by the risk variants, notably that risk patient's VSMCs had reduced contractility and adhesion strength. Because VMSCs function in blood vessels by contracting to regulate vascular tone and, during CAD progression and vascular remodeling, they become more proliferative and less contractile, the reduced contractility of the risk patient's VSMCs mirrored the disease status of patients. Isogenic comparison between VSMCs from the same risk patients with and without the SNPs was possible using TALENs to knock out (KO) the variant-containing region. Isogenic VSMCs containing SNPs also had reduced contractility and adhesion strength compared to their KO counterparts[23] (Figure 14.2A). This example demonstrates that editing methods can dramatically reduce background genetic variation to more easily highlight subtle mechanisms associated with enhanced disease risk.

Another common cause of heart failure and hypertension and the leading cause of heart transplants is dilated cardiomyopathy (DCM),[24] which is characterized by ventricular dilation and systolic dysfunction. Of all DCM cases, 20–35% are familial, although incomplete penetrance and the large set of associated genes (>20) make understanding its inheritance less clear.[25] A growing number of studies have similarly employed iPSCs to study the genetic causes of DCM, as iPSCs are an essentially infinite source of human cardiomyocytes and primary cardiomyocytes are both rare and difficult to isolate.[26,27] In one study, iPSC-derived cardiomyocytes were generated from DCM patients with a point mutation (R173W) in the *TNNT2* gene and from unaffected healthy family members without the mutation.[28] *TNNT2* encodes sarcomeric protein cardiac troponin T, one of the three subunits of the troponin complex, which regulates contraction and force

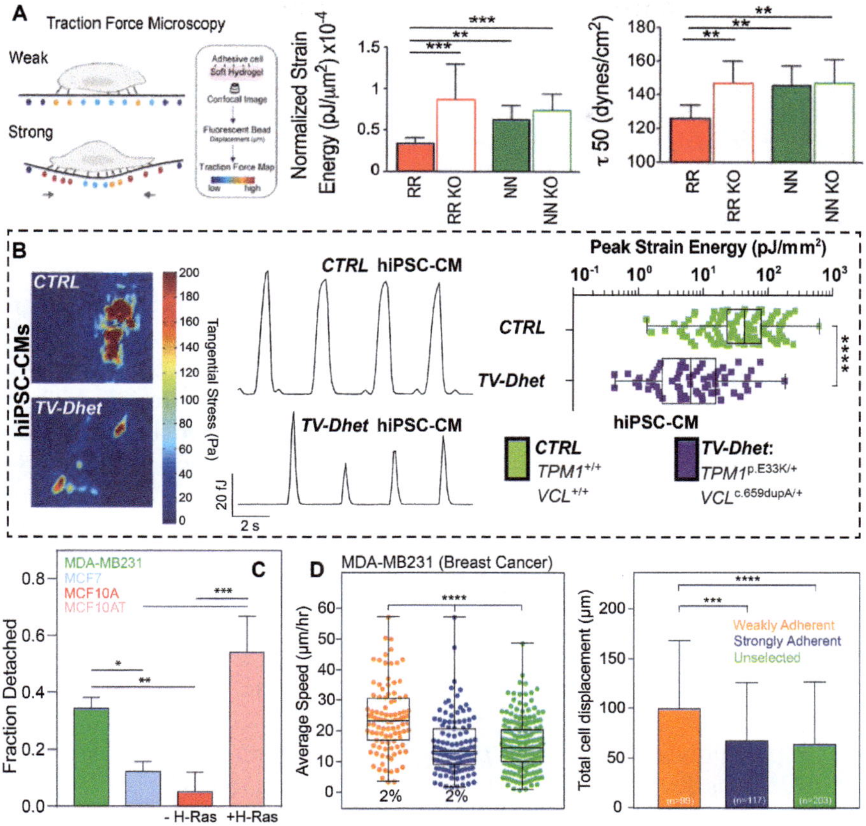

Figure 14.2 Genetic models to understand disease-in-a-dish. (A) Schematic of traction force microscopy (TFM) assay (left) applied to patients with (R/R) or without (N/N) variants in the 9p21 risk locus (center). Total strain energy was normalized to cell area and plotted ($n > 25$ in each assay; mean ± 95% CI). ***$p < 0.001$ one-way ANOVA-Bonferroni. Adhesion strength (right) assessed by spinning disk assay for SMCs of indicated genotypes ($n > 10$ in each assay; mean ± 95% CI.). **$p < 0.01$ one-way ANOVA-Bonferroni. Reproduced from ref. 23 with permission from Elsevier, Copyright 2018. (B) Single cardiomyocyte traction force microscopy studies of patient-derived iPSCs with double heterozygous variants in *TPM*1 and *VCL*, *i.e.* TV-Dhet, or control with normal copies of these genes. Tangential stress heat maps (left), strain energy time courses (center), and peak strain energy analyses (right) are shown. ****$P < 0.0001$; two-sided Student's *t*-test. Reproduced from ref. 32 with permission from Springer Nature, Copyright 2019. (C) Plot showing the fraction of detached cells from MDA-MB-231, MCF7, and MCF10A and their H-Ras–transformed counterparts MCF10AT after exposure to 250 dynes cm^{-2} of shear stress. *$P < 0.05$, **$P < 0.01$, and ***$P < 0.001$ for two-tailed unpaired *t*-test between lines. (D) Average speed (left) and total displacement (right) were plotted for MDA-MB-231 cells sorted by the indicated shear stress and allowed to migrate on collagen gels for 24 hours. Percentages reflect the portion of each population that detaches at a given stress. ***$P < 0.001$ and ****$P < 0.0001$ for two-tailed unpaired *t*-test between lines. Reproduced from ref. 37 with permission from AACR Publications, Copyright 2020.

production of cardiomyocytes through its regulation of sarcomere thin filament assembly and activity.[29] Cardiomyocytes from patients with the *TNNT2* mutations exhibited abnormal sarcomere structure, altered calcium ion signaling, and decreased contractility.[29] Other examples include but are not limited to using iPSC-derived cardiomyocytes from long QT patients with *CALM2* mutations to faithfully recapitulate reduced action potentials found in patients[30] and from patients with Danon disease and *LAMP* mutations who had impaired calcium handling.[31]

While this study serves as an example of how single-point mutations can model disease mechanisms and potentially outcomes, iPSC-derived progeny can also elucidate novel gene interactions. By sequencing a large DCM-affected family, another study found that heterozygous sequence variants in *TPM1*, the tropomyosin 1 gene, and *VCL*, the gene coding for vinculin, co-segregate in individuals with DCM diagnoses.[32] Tropomyosin binds troponin and is also an actin thin filament protein involved in contraction, whereas vinculin is an actin-binding protein involved in focal adhesion (cell–ECM) and adherens junctions (cell–cell).[33] Employing CRISPR to create a series of *TPM1* and *VCL* variant combinations in patient-derived iPSCs revealed that cardiomyocytes from patients with both *TPM1* and *VCL* variants had the greatest reduction in contractility, measured using TFM, and in disorganized sarcomeres[32] (Figure 14.2B). This study serves as an example of how combinatorial effects from genetic variants can be linked to disease using genetic models employing novel gene editing techniques like CRISPR and biomechanical assays, like TFM, when inheritance patterns are not as transparent as single-gene Mendelian inheritance patterns.

14.2.1.2 Cancer

In contrast to inherited heart disease, many genetic causes of cancer are acquired. Mutations are often acquired in tumor suppressor genes, *e.g.* DNA repair genes such as p53, and oncogenes, *e.g.* receptors and small GTPases such as HER2 and RAS, respectively. However, there are inherited germline variants that predispose one to developing cancer, *e.g.* tumor suppressor genes BRAC1 and BRAC2, which increase likelihood of acquiring melanoma and breast, ovarian, and prostate cancer.[34] To study oncogenes in disease models, such as the rat sarcoma (RAS) proto-oncogene family, including HRAS, KRAS, and NRAS, somatic cells can be transformed such that the exogenous oncogene of interest is incorporated and expressed. For example, to study breast cancer, nontransformed MCF10A breast epithelial cells can be transfected with constitutively active T24 H-Ras oncogene to create malignant MCF10CA1a cells that are malignant, or premalignant MCF10AT cells.[35] MCF10CA1 cells exert greater traction forces than MCF10AT and MCF10A,[36] helping them to metastasize; adhesion differences that result from these mutations also regulate metastasis (Figure 14.2C, D).[37]

This common strategy has been pervasive in the literature over past decades but has recently been enhanced by modern editing methods such that lines can easily carry multiple oncogene or tumor suppressor mutations as occurs *in vivo*. Once edited, the impact of these edits – such as on the mechanical properties of cells – can be more easily observed. Studies have shown that the viscoelastic properties of cells are modified in cancer cells, specifically that metastatic cancer cells become more deformable. That helps cells to pass through denser adjacent stroma than softer healthy tissue.[38,39] One study compared the elasticity of oral carcinoma cell lines (Cal-33, Cal-27, PC1-1, PC1-13) to that of healthy oral keratinocyte cell lines (OKF-4, OKF-6) using a microfluidic optical stretcher and found that tumor grade and thus mutation state scaled with deformability.[40] Similarly, the viscoelastic properties of nuclei isolated from different cancer cell lines (MCF-7, IMR-5, and HEK 293T) showed that chromatin organization is spatially inhomogeneous, specifically that the outer periphery of the nucleus had reduced stiffness and was more sensitive to mechanical stimuli.[41] These changes in nuclear structure and others associated with cancer help the cells to both squeeze through more narrow networks and in mechanosensing and recognition of these environmental changes.[42]

These findings on cancer cell deformability not only show promise for both novel diagnostics[43] but also as a research tool for identifying candidate genes (and mutations) in systematic genetic screens. Future applications of these methods could include CRISPR-based microfluidic screening system to identify novel tumor suppressor kinases. Loss of function aided by CRISPR/Cas9 and paired with a deformation-based microfluidic cell separation platform can sort and sequence flow-through cells, quickly identifying a library of kinases that enhance cell deformability and a potentially metastatic phenotype.[44] CRISPR/Cas9 system has also been used by other groups to screen for genes that regulate cell survival and drug resistance[45] or to investigate genetic networks through combinatorial perturbations,[46] hinting that these CRISPR screens are powerful strategies for systematic genetic analyses of cancer. Moreover, they can be utilized to identify not only cell cycle regulatory genes but also genes involved in cytoskeletal rearrangement and other relevant structures in cancer.

14.2.2 Environmental Effects on Cells

While genetics can impact cell phenotype, environmental cues can also impact cell fate in development and disease. Accurate disease models should mimic not only native tissue in the chemistry of soluble factors and scaffolding but also its mechanical properties. Gene expression and cellular phenotype more closely replicate healthy tissue by using biomimetic cell culture environments, and, importantly, the changes in the ECM environment during disease progression can also be modeled. The importance of replicating the mechanical signals from the environment, including from the ECM, neighboring cells, or surrounding fluid, is becoming a well appreciated feature of disease models (Figure 14.3).

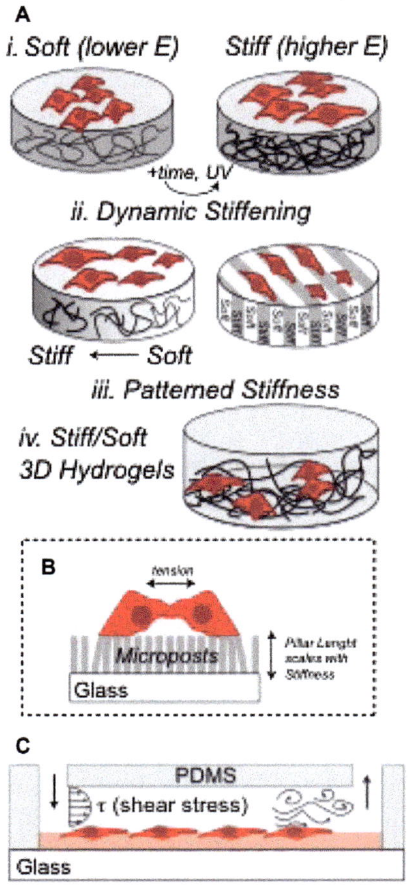

Figure 14.3 Schematic of mechanical environmental cues. (A) Cell–ECM interactions are commonly manipulated in culture by hydrogels (i) of different stiffness, (ii) of time-dependent stiffness, or (iii) that contain spatial patterns of stiffness. While these systems culture cells on top of the hydrogel generally, they can also be used (iv) in 3D systems where the material again can be soft or stiff depending on polymerization properties. In these systems, cells generate forces on the material. (B) Cell–cell interactions are also often probed in these systems. While panel A highlights continuous surfaces, discontinuous surfaces can also directly control and measure interactions, *e.g.* with microposts and cell–cell and cell–matrix tractions. (C) In contrast, forces can be directly applied to cells, such as with fluid shear stress. In microfluidic systems, a chamber is fabricated where cells reside and fluid is pumped through the chamber (see inflow and outflow arrows at the ends of the device made out of polydimethyl siloxane [PDMS]). Fluid motion obeys pipeflow and exerts a defined, laminar shear stress on cells, which can become turbulent in the presence of obstructions, *e.g.* stenosis.

The relationship between mechanics and disease status is becoming more widely studied. In some pathologies, cell phenotype is defined by reduction in mechanical output, *e.g.* reduced cardiomyocyte contractility in dilated cardiomyopathy (DCM).[28] In other diseases, *e.g.* cancer, the relationship is less understood. In both cases, the use of materials-based mechanobiological assessments adds valuable quantification to cell phenotype and is useful in genetic studies for uncovering molecular mechanisms linking genomic variants to disease predisposition. Mechanobiological measurements can include examining the extent of a cell's interaction with the extracellular matrix (ECM), such as the traction forces exerted by a cell onto its environment,[36] especially in muscle,[23,32] or the adhesion of the cell to ECM when exposed to fluidic shear stresses.[37] Cell–cell interactions can also be measured,[38,47–51] a notable property for diseases in which such bonds are disrupted, such as endothelial barrier function in cardiovascular diseases.[52,53] The elasticity or viscoelasticity of a cell and specific organelles[54] is also related to a cell's ability to squeeze through the surrounding dense ECM network, an especially relevant metric for cancer cell migration.[40,41,44] Continuing to move smaller in length scale, the tension on individual proteins can be measured, which can provide insight into gene expression, structure, and disease phenotype.[55,56]

14.2.2.1 *Modeling Cues from the ECM*

14.2.2.1.1 Hydrogels Mimic ECM Stiffness and Viscoelasticity. ECM is a three-dimensional fibrous network comprised of various amino acid and sugar-based macromolecules that provide functional and biochemical support for cells, control tissue structure, and diffusion of nutrients, growth factors, and metabolites. ECM is tissue specific with different protein compositions, resulting in unique integrin-binding profiles and different bulk mechanical properties, such as tissue stiffness. Synthetic hydrogels – which are covalently linked networks of polymers – are often used to model ECM in a reproducible manner because they can decouple chemical, physical, and biological properties in ways that native ECM cannot.[57] Natural hydrogels can also be made utilizing fibrous proteins, such as collagen, fibrin, glycosaminoglycans, and polysaccharides.[58] While biocompatible, these naturally derived polymers are less manipulatable because properties are coupled; as one increases protein concentration (a haptic cue), the matrix also stiffens (a mechanical cue). Therefore, synthetic polymers are often used in a reductionist manner to decouple properties.

Hydrogel mechanical properties depend on the rigidity of the original polymer chain, a product of the chemistry of its monomer, the density of the chains, the degree of the crosslinking, and the degree of swelling due to hydrophilic/hydrophobic balance. To mimic the elasticity of the specific tissue of interest, for a chosen hydrogel, the polymer density or the degree of crosslinking of the polymer can be adjusted.[59] Crosslinking can be either physical, using noncovalent hydrogen bonds to entangle polymers chains, or chemical, through covalent bonds between functional groups on the polymers. Adjusting the pH, temperature, or ion concentration of the surrounding solvent can all promote physical crosslinking, and the use of linker molecules, photo initiators and UV, and enzymes can all be utilized as crosslinking agents for chemical polymerization.[60] Cell responses to different stiffness, dimensionality, stiffness patterns, or temporal changes can all be used to study diseases, *e.g.* cancer and atherosclerosis,[61] which affect these parameters. For example, fibroblasts migrate faster on soft, 2D substrates.[62,63] Within the tumor environment, increased fibrillar collagen deposition and the resultant increased tensile strength both contribute to cell–EM interactions that promote cancer cell invasion,[64] as well as allow the tumor to grow in size and displace host tissue.[65] Similarly, atherosclerotic lesions are characterized by increased collagen composition and are stiffer than healthy vasculature.[66] While these examples present 2D surfaces to cells, stiffness-dependent behavior is also observed in 3D systems and depends on cross-linking type (covalent *versus* ionic) and pore size. Breast adenocarcinoma cells (MCF-7) seeded within 3D alginate hydrogels ranging between 150 and 4000 Pa show decreased viability with increasing stiffness.[67] MDA-MB 231 cells, another breast cancer carcinoma, are more invasive, traveling greater depths into 3D collagen gels of greater stiffness above a pore size threshold.[68] Stiffness can even promote cell plasticity where cells undergo epithelial-to-mesenchymal transition (EMT) and become more migratory and invasive in mammary,[69–72] lung,[73] hepatocellular,[74] and prostate[75] carcinomas, among others. Even squamous cell carcinomas, *e.g.* oral (OSCC), exhibit plasticity and EMT on stiffer substrates, making them more migratory and invasive.[76] These cells exhibited memory of the stiffness of their past environment,[76] an idea not unique to oral cancer, and has been noted in other tissues like mammary tumors[77] and colon cancer.[78]

The importance of elasticity is well appreciated, but the ECM is not strictly elastic. Many soft tissues are viscoelastic, exhibiting stress relaxation in response to deformation, creep in response to mechanical stress, or dissipating mechanical energy stored in the material.[79] Notably, viscoelastic properties, including the rate of stress relaxation, are altered in diseased tissues like breast cancer.[80] Hydrogels with tunable viscoelastic properties are increasingly being used to study how time-dependent relation and creep properties impact behavior, including stem cell differentiation,[81] cell spreading, proliferation,[79] and matrix formation. When the matrix is more plastic, cells actually switch between migration modes from protease-dependent to -independent mechanisms.[82] Cell spreading, traction forces, and focal adhesion size can indeed scale with stiffness but can also be independent of plasticity.[83] While many questions are still unanswered, viscoelasticity is being recognized for its role in regulating cell phenotype.[84]

14.2.2.1.2 Dynamically Stiffening Hydrogels. While materials can passively creep or relax over time, materials can also be actively driven to soften or stiffen as would occur during disease remodeling *in vivo*. Dynamic materials can mimic disease processes like fibrosis, including tumor stiffening, to interrogate the timescale over which cells probe their environment or even change shape to model developmental processes and disorders. For example, polyethylene glycol (PEG) hydrogels have been engineered to dynamically stiffen and mimic the stiffness transition from healthy to fibrotic heart muscle using anthracene photodimerization. On these gels, cardiac fibroblasts exhibit rapid, stiffness dependent localization of nuclear factor of activated T-cells (NFAT), a transcription factor downstream of intracellular calcium signaling that activates fibroblasts.[85] Similarly, cell response timescales can be studied by designing hydrogels with tunable viscoelastic spectra; reversible boronate ester crosslinking creates tunable viscoelasticity, and fibroblasts cultured on these gels had reduced cell and nuclear area, focal adhesion tension, and localization of YAP/TAZ compared to cells on elastic hydrogels with similar storage moduli;[86] frequency-dependent viscoelastic material properties impact fibroblast activation and YAP/TAZ translocation.[86] Overall, these highlight how materials with dynamic elasticity and viscoelasticity can elucidate the timescales at which cells sense their environment, specifically in the context of fibrosis.

Dynamic materials can also elucidate signaling pathways that are not observable in static stiffness systems in the context of cancer. Mammary epithelial cells (MCF10A) cultured on methacrylated HA hydrogels undergo EMT upon stiffening; both the degree of stiffening and time cultured prior to stiffening regulate this transformation, but importantly tumor spheroids do not exhibit memory of the mechanics of the prior niche given their multiple, overlapping mechanotransductive signaling pathways[70] (Figure 14.4A). While stiffness-driven EMT has been reported previously in static stiffness hydrogels,[72] this dynamic system was able to elucidate the cooperative effect of both paracrine signaling and stiffness-mediated changes leading to EMT.[70] Finally, dynamic materials can be used to study how shape changes that take place during development of direct cellular behavior and therefore also can be used in modeling developmental disorders. Dynamic shape changes occur during morphogenesis to give rise to more complex biological structures. To mimic such processes, such as formation of the heart tube or neurulation, one group designed a PEG diacrylate-based thin film incorporated with *ortho*-nitrobenzyl moieties, which through photodegradation induced swelling transforms from a flat 2D sheet to folded 3D structures *via* exposure to UV.[87] These and other photodeformable polymer gels[88] show potential for revealing the signaling pathways induced by the complex structural changes of morphogenesis and could shed light on associated disorders.

14.2.2.1.3 Mechanically Patterned Hydrogels. ECM changes with time, but it also changes spatially; cells move across stiffness gradients in a number of diseases, *e.g.* when cancer cells metastasize and migrate from the stiffer tumor to the softer surrounding stroma,[89] or when fibroblasts and mesenchymal cells enter injured tissue, or when smooth muscle cells migrate from the tunica media into the intima of the blood vessel.[90] Hydrogels with patterned stiffness can be used to study how cells are directed by matrix stiffness (durotaxis) for a variety of tissue niches and diseases. Patterned regions of different stiffnesses were initially created using uncontrolled droplet mixing.[63,91] Conventionally today, patterns

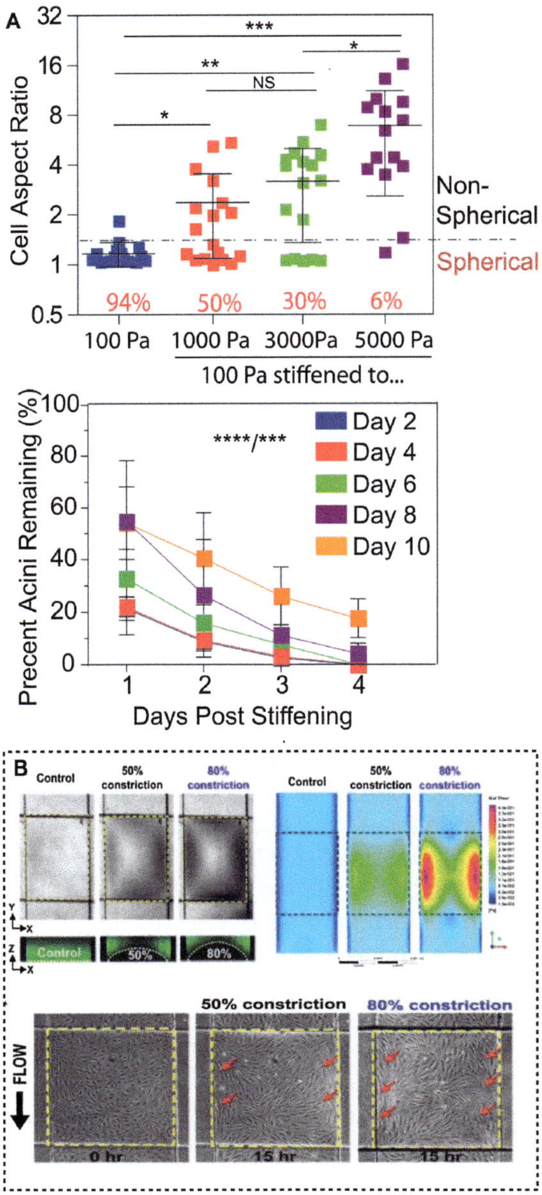

are typically created: (1) by controlled, radical-based photoinitiation using a photomask,[92] (2) by spatial composition of the precursor materials, or (3) by hydrogel thickness.[93] For light-based polymerization, photomasks can be designed with varying opacity or created by moving a mask during UV treatment.[94] Photoinitiators include those bound to polymers (*e.g.* *o*-nitrobenyzyl[95] or ketal-acetyl derivatives as photocleavable groups[96] and vinyl benzene, cinnamic acid, norbornene, or benzophenone as photocrosslinkable groups[93]) or those added exogenously, such as Irgacure.[97] For improved precision, the ratio of monomer to crosslinker can be controlled using hydrodynamic systems, like a combination of syringe pumps,[98] a microfluidic gradient generator,[99,100] or other unique devices.[101] Finally, stiffness patterns can be achieved through spatial control of hydrogel thickness. If the hydrogel is attached to a rigid substrate, the magnitude of the displacement field attenuates due to the no-slip boundary condition at the matrix substrate interface, and the amount of attenuation is inversely related to the matrix thickness.[102] Therefore a substrate-bound chemically homogeneous hydrogel can have a patterned stiffness locally varying the thickness of the hydrogel. This can be accomplished by using pattered photomasks and layering gels to create distinct soft and stiff striped regions[103] or by layering beads beneath compliant gels.[104]

Significant recent progress has been made in understanding cell behaviors on patterned substrates. For example, fibroblasts on collagen-coated hydrogels with 14 and 30 kPa regions exhibited increased migration speed when traveling "up" the stiffness gradient (soft to stiff) but turned away from

Figure 14.4 Dynamic Extracellular Cues. (A) Cell aspect ratio (top) is plotted with each color corresponding to its stiffening regiment indicated at bottom. The dashed line at 1.5 indicates an approximate transition point from spherical to nonspherical morphology. The percentage of data below the transition is shown, indicating the fraction of the population that remains spherical. $*P<0.05$, $**P<1\times10^{-2}$, and $***P<1\times10^{-3}$ from an unpaired Student t test ($n>15$ spheroids per condition). N.S., not significant. (Bottom) Cell clusters were also plated for the indicated days (colors in legend), stiffened, and tracked over 4 days post-stiffening. $***P<1\times10^{-3}$ and $****P<1\times10^{-4}$ for two-way ANOVA indicating significance when comparing the effects of time pre- and post-stiffening. Reproduced from ref. 70 with permission from National Academy of Science. (B; Top) At right are representative bright-field images showing different channel constrictions at the stenosis region (yellow dotted box) with corresponding confocal images of the cell channel loaded with FITC dye illustrating constrictions (white dotted lines). At left, fluid simulations highlighting the wall shear stress across the 3D constriction (dotted box). (B; Bottom) Phase contrast images of TNF-a treated HUVECs monolayer at different constrictions after 15 h of perfusion culture (1 dyn cm^{-2}). Active cell alignment parallel to the flow was observed for HUVECs along the high shear zones (red arrows) for 50% and 80% constrictions. Reproduced from ref. 129, https://doi.org/10.1063/1.4993762, under the terms of the CC BY 4.0 license https://creativecommons.org/licenses/by/4.0/.

the boundary when migrating on the stiff region;[63] this biased migration occurs in most adherent cell types and has been termed "durotaxis". Biased migration has been shown for fibroblasts and endothelial cells on fibronectin-coated hydrogels,[105] for various cancer cell lines on fibronectin-coated hydrogels,[89] and for smooth muscle on collagen-functionalized hydrogels.[106,107] Speed and persistence in these regions depend on gradient range and magnitude (slope of the gradient) but also on how the cells are interacting with the gradient, *i.e.* durotactic behavior is often ligand specific.[106] Finally, patterns need not be smooth gradients but rather can be stripes or other shapes with near step transitions. Such patterns have been used to position motor neurons and myoblasts to create neuromuscular junctions.[32] With each of these patterns, the goal is always to improve the functionality of disease-in-a-dish model.

14.2.2.2 3D Systems Structural Properties

While stiffness has been extensively studied, ECM structure, porosity,[108] fiber interconnectivity and tortuosity, fiber size,[109] and fiber surface topography and chemistry[110] are becoming increasingly studied given their influence on cell behavior in 3D. The following is a summary of the materials-based mechanobiological systems used to study these variables independently, as outlined in Figure 14.3.

14.2.2.2.1 Electrospun Fibrous Scaffolds. Hydrogels cannot recapitulate 3D fibrillar architecture as in collagen and fibrin.[111,112] To overcome this limitation, polymers are often electrospun into a fibrous morphology, which involves extruding a charged polymer solution through a blunt needle across a large potential difference. A variety of polymers with different properties have been electrospun, including poly(glycolic acid) (PGA), poly(lactic acid) (PLA), poly(caprolactone) (PCL), polydioxanone (PDO), and even natural polymers like collagen (types I, II, III, and IV), gelatin, elastin, and fibrinogen; mixing synthetic and natural polymers is also possible.[112] Electrospinning manipulates a variety of fibrous scaffold features ranging from fiber topography to mesh or network size of the spun fibers.[113] For example, by adjusting the viscosity or concentration of the polymer solution, the fiber diameter can be adjusted (increasing concentration increases diameter).[114] To adjust the scaffold's thickness, the volume spun can be adjusted to the size and shape of the scaffold, and the ground targets geometry can be varied.[115] By rotating the ground target, fibers can be aligned to produce anisotropy of both the fiber orientation and bulk physical properties.[116] These properties have been investigated for their influence on cellular behavior. For example, PLA fiber alignment can guide migration of fibroblasts,[117] filament curvature is sufficient to influence the direction of nerve cell outgrowth,[109] and the expression of collagen genes and proliferation markers in fibroblasts increases with the decreasing diameters of silk fibers.[118]

14.2.2.2.2 3D Printing. Three-dimensional (3D) printing, in contrast to traditional manufacturing processes, enables the fabrication of complex and diverse architectures through computer-aided design and layer-by-layer application of materials. 3D printing allows for spatial patterning of cells and biomaterials to recapitulate the complexity of tissues and organs and has the potential for patient-specific disease modeling using computed tomography (CT) and magnetic resonance imaging (MRI) data.[119] To identify the minimum design requirements that allow for optimal tissue growth and function, such as size, choice of biomaterial ink, vascularization, and cell type composition, significant research efforts are still required. Furthermore, the variety of materials suitable for 3D printing, which must be mechanically suitable for extrusion yet stiff enough to maintain the desired architecture while layers are deposited are limited.[120] However, the spatial precision allowed by 3D printing methods has been useful for disease modeling of high-order tissues, in which cell–cell interactions are critical.

3D printing has been used to study a number of tissues and organs, such as bone,[121] blood vessels,[122] and skin.[123] For example, extrusion-based 3D-printed gelatin methacryloyl (GelMa) scaffolds have been shown to trigger mineral deposition of osteoblasts in the absence of exogenous osteogenic factors, but osteoblasts seeded onto the same formulation in film failed to deposit minerals, suggesting that the 3D architecture of the scaffold promoted differentiation rather than its chemical cues.[124] In addition to recapitulation of architecture, 3D printing can also add precision to the placement of cells by using cell-laden viscous polymers. Manipulation of cell placement can be used to promote cell–cell interactions mimetic of native tissue, *e.g.* chondrocytes and MSCs embedded in alginate and printed with high spatial viability.[125]

Conversely, modeling multicellular organ systems is complicated by tissue size, with larger organs requiring vascularization for their cells to maintain metabolic requirements.[122] Therefore, there has been a growing demand for functional vasculature for perfusion of biofabricated organs, with the ability to perform gas and nutrient exchange as well as waste product removal.[126] The precise placement of multiple cell types *via* 3D printing highlights its promise for creating more complex organs by simultaneously constructing an integrated vascular system. For example, multiple bioprinted and perfusable networks of capillaries and arteries have been achieved and are a significant step in overcoming this long-standing disease modeling limitation.[127] Vascular systems have also been fabricated by 3D printing and shown to support biopsied tumors, a useful attribute, as traditional tumor models are not able to support the highly vascularized nature of tumors *in vivo*. This was accomplished by 3D printing sacrificial PVA scaffolds embedded with different combinations of ECM (such as collagen and fibrin) and different geometries that supported endothelial cell infiltration and migration along the channels, resulting in functional microvasculature that was also able to support fluid flow.[128]

14.2.2.3 Active Forces Applied to Cells

Environmental cues are largely passive and sensed by cells pulling or pushing on their niche. Conversely, external application of forces on cells, by matrix, other cells, or fluid, can also significantly impact cell function. The most common forces experienced are hemodynamic forces, and understanding how they affect vasculature has become increasingly important to model diseases such as CAD, strokes, and metastatic cancer. The use of 3D systems has been particularly useful to better understand the interaction of hemodynamic forces and cell tissue. Typically, blood flow is laminar and pulsatile. However, changes in vessel construct, such as bifurcations, turns, stenosis (*i.e.* stiffening and narrowing of vessel), and other disease phenotypes can cause disruptions to flow and change the shear stresses experienced by cells and surrounding tissue. For example, diseases such as atherosclerosis and thrombosis will characteristically show disturbed flow instead of laminar flow due to buildup of plaque, excessive clotting and stiffening of the arterial walls. The hemodynamic response is complex and requires the interaction of different fluid forces and multiple levels of tissue. As such, studying diseases that are impacted by hemodynamics requires reductionist systems that can emulate both the relevant fluid and the tissue-induced mechanical forces (Figure 14.4B).[129] Such 3D systems make it possible to study the complexity of native-like tissues in response to hemodynamic forces *in vitro*. For example, vascular endothelial cells, which are in direct contact with blood flow and are the first to detect and implement a response, detect and respond to changes in shear stress and flow.[130,131] Disturbed flow induces their inflammatory response, reorients actin stress fibers, and changes nuclear morphology within endothelial cells. These behaviors are mediated by the YAP/TAZ pathway responding to disrupted flow, as shown in atheroprone models. While these examples use continuous flow, oscillatory shear is more representative and can sensitize cells even further. For example, atheroprone oscillatory shear decreases in inhibitory phosphorylation of YAP/TAZ, increasing activation of its downstream targets, whereas laminar shear stresses do not.[132] A similar study, also using HUVECs, showed that disturbed flow activated YAP activity and decreased its cytoplasmic retention, whilst unidirectional shear stresses (USS) suppressed YAP activity thus inhibiting atherogenesis.[133] While shear stresses classically effect endothelial cells, it also impacts many important blood components, *e.g.* platelets, which reduce their adhesion and clotting ability with increasing blood shear stress.[134] Despite our vascular focus, active forces and stresses are extremely prevalent environmental cues; this larger topic, the material-based mechanobiological tools used to measure them and their larger influence on cell function, is more thoroughly reviewed elsewhere.[135–137]

One final point to emphasize is that, regardless of tissue, multiple active forces or multiple cell types can be present. While the systems just discussed isolate single forces, it is important to note that disease mechanisms are often dependent on multiple mechanical forces interacting with one another, *e.g.* shear stress and ECM stiffness, or multiple cell types, *e.g.* endothelial cells

(ECs) and smooth muscle cells (SMCs). To examine these more complex systems, ECs and SMCs have been cocultured in microfluidic devices to study cell–cell interaction upon exposure to fluid flow. This EC-SMC-signaling-on-a-chip was exposed to 1–1.5 Pa shear stress and strain of 5–8%, which resulted in the endothelial layer adopting a "cobblestone" morphology, while the smooth muscle cells aligned themselves perpendicularly to direction of flow. This study represents one of the first co-cultured systems able to mimic the construction of the vasculature and respond to its mechanical environment *in vitro*.[138] Subsequently, several other multiforce/cell systems have been fabricated. For example, atheroprone vasculature develops inflammatory plaques in part from disturbed flow and recruited immune cells. Bifurcated flow-through slides seeded with HUVECs have been used to model this system. Nonuniform shear stress and inflammatory cytokines greatly enhanced monocyte cell adhesion and changed endothelial gene expression; telmisartan was able to inhibit this response *via* downregulation in VCAM-1, a protein involved in leukocyte adhesion.[139] These studies ultimately show the importance of using multivariable systems to elucidate cell–cell responses to their mechanical microenvironment.

14.3 Combining Genetic and Environmental Factors to Model Disease

While the previous section focused individually on genetic or environmental factors, their combinatorial effects are becoming well recognized for their importance in initiating and regulating disease pathologies.[140] Their continued study can improve our ability to estimate patient risk attributes,[141] improve understanding of biological pathways and dose–response relationships, provide clarity on heterogeneity across studies,[142] and identify novel candidate genes.[143] Environmental risks that can aggravate genetic causes of disease are not restricted to traditional concepts, such as smoking or UV exposure, increasing the risk for cancer but also include endogenous molecules' more subtle changes within a tissue that develop over the natural progression of disease and aging.[141] Often environmental factors will "reactivate" and upregulate expression of noncoding DNA, which itself can contain single-nucleotide polymorphisms (SNPs) that exert control over signaling and result in a higher likelihood of disease.[17,144] The vast majority of SNPs are present in noncoding regions leading to incomplete penetrance[145] and highlights the need for disease models to capture conditions that lead to the disease phenotype in order to study mechanism. Next we review major organ systems and diseases where combinatorial effects have been identified thus far.

14.3.1 Heart Disease

Significant risk for coronary artery disease is inherited, although 23 of the 33 genetic risk variants for CAD identified by GWAS act through unknown mechanisms.[17] Clearly, approaches that model *in vivo* mechanical signals along with patient-derived iPSC present potentially fruitful methodology for resolving signaling mechanisms. For example, iPSC-derived endothelial cells cultured with microfluidic devices respond to proangiogenic growth factors and shear stresses to form functional and perfusable capillaries;[146] forces also enhance EC differentiation[147] and make their behavior more representative and patient specific for drug toxicity studies.[148] Overall, iPSC-ECs have been shown to be a reliable and reproducible source of ECs for disease modeling using 3D vascular networks, providing opportunities for genetic studies in which microphysiological conditions are mimicked. In this section, we summarize several recent case studies of gene–environment interactions, organized by their system and specific cardiovascular disease.

As a first representative example, patient-derived iPSCs are differentiated (in this case into SMCs) and fashioned into a 3D tissue (in this case self-assembled rings).[149] Here, iPSCs were obtained from a patient with a loss-of-function, premature termination in exon 10 of the elastin gene (*ELN*), resulting in supravalvular aortic stenosis syndrome (SVAS). SVAS is characterized by SMC hyperproliferation that narrows the ascending aorta.[150] Patient iPSC-VSMCs had reduced contractile strength and were also more proliferative[149] in these rings when subjected to force. This example, nearly six years old now, represents an early, albeit superficial analysis of disease mechanisms induced by a combination of gene–environment stimuli. More complex bioreactors with a fiber mesh and pulsatile flow conditions[151] or even decellularized blood vessels[152] under flow are technological improvements on the environment that should be applied to the *ELN* truncation to further probe it.

A second, more complex and recent example employs tissue-engineered blood vessels made from collagen, lined with endothelial cells, and perfused with media using a peristaltic pump. iPSC-SMCs from patients with Hutchison–Gilford progeria syndrome (HGPS) were used in this system to study how their accelerated aging phenotype was affected by perfusion and the presence of collagen. HGPS is a rare genetic disorder caused by an altered form of the Lamin A gene, *LMNA,* resulting in nuclear accumulation of progerin, a truncated lamin A protein.[153] The main cause of death of HGPS patients is cardiovascular disease, but our understanding of its cardiac-specific mechanisms is limited. With this study, authors investigated the effect of progerin and aging in general, using iPSC-derived SMCs from a HGPS patient in this tissue-engineered blood vessel system. HGPS iPSC-SMCs had reduced vessel wall diameter, increased medial wall thickness, and increased calcification and apoptosis compared to TEBVs fabricated from healthy patient iPSC-VSMCs.[154] HPGS patient-derived iPSCs but not healthy patient-derived cells had an exacerbated inflammatory and DNA damage responses to the strain in this system caused by pulsatile flow. This phenotype could be reversed by lovastatin, a drug that is used in combination with other statins to extend HGPS patient life.[155] To further mimic blood vessels, this same group differentiated patient-derived iPSCs into both

ECs and SMCs rather than just the SMCs. When patient matched, HGPS ECs aligned with flow but exhibited reduced gene expression responses to flow. HGPS vessels also exhibited a reduction in both vasodilation and vasoconstriction, while mixed vessels only exhibited reduction in vasodilation.[156] Overall, this study served as an example in which coculture of iPSC-derived ECs and VMSCs can elucidate how genetics influence responses to mechanical forces, such as those induced by fluid flow. It also points out the importance of isogeneity.

While the previous examples focus on vascular models, iPSC-derived cardiomyocytes have also been used extensively to investigate the role of biomechanical cues in genetic heart diseases. For example, cardiac microtissues (CMT) engineered from iPSCs from patients with DCM have been used to investigate the pathogenicity of genetic variants. Mutations causing truncations of the massive sarcomere protein titin (TTNtv) are the most common genetic cause of DCM. 3D CMTs suspended between PDMS pillars, as discussed in previous sections, more faithfully recapitulate native cardiomyocyte architecture and mechanics, promote maturity, and improve sarcomere alignment. Contractile function of iPSC-derived CMTs from patients with mutant titin TTNtv was reduced compared to those from wild type patients. By combing this sort of functional analysis with RNA-seq analysis, the group also found that certain missense mutations were more tolerated than others. Namely truncations in the A-band domain of the TTN impair contraction more than truncations in the I-band domain, which are less likely to result in DCM. A-band truncations also resulted in impaired response to mechanical stresses, accomplished by altering the PDMS post-stiffness. To ensure that observed functional differences were not due to differences in background genetics, TTNtv was also introduced into independent iPSC lines using CRISPR/Cas9, to create isogenic comparisons with only the truncation of titin varied. Overall, this study provided genomic insights into DCM, finding that while both A- and I- band titin truncations result in contractile deficits, alternative exon splicing mitigates pathogenicity of I-band TTNtv.[157]

Finally, while these genetic studies using biomimetic culture conditions have alluded to the importance of physiologically relevant mechanical stimuli, another study demonstrated that genetically linked disease phenotype may only be detectable under the appropriate environmental conditions. While GWAS identified numerous common polymorphisms associated with CAD, the majority are in noncoding genomic regions, making how they regulate cell function and contribute to disease largely unknown. To study how SNPs at the 9p21 locus, one such noncoding region with the greatest disease association leading to CAD, iPSC technology and dynamic PA hydrogels with controllable stiffness were used. In a physiological environment mimetic of healthy tissue, cultured cardiomyocytes from iPSCs from patients who are homozygous for cardiovascular-risk alleles (R/R) and cardiomyocytes from iPSCs from healthy, homozygous for nonrisk alleles (N/N) both contracted synchronously, independently of genotype.

However, once the hydrogels were stiffened, to mimic fibrosis, only the R/R iPSC-cardiomyocytes contracted asynchronously, while the N/N iPSC-cardiomyocytes were able to remain synchronous (Figure 14.5A). These mechanically induced phenotypic changes were associated with increased expression of a short isoform of ANRIL, the long noncoding RNA located in the 9p21 risk locus. ANRIL induced a c-Jun *N*-terminal kinase (JNK) phosphorylation-based mechanism that impaired gap junctions, specifically resulting in loss of connexin-43 expression. TALEN-mediated deletion of the

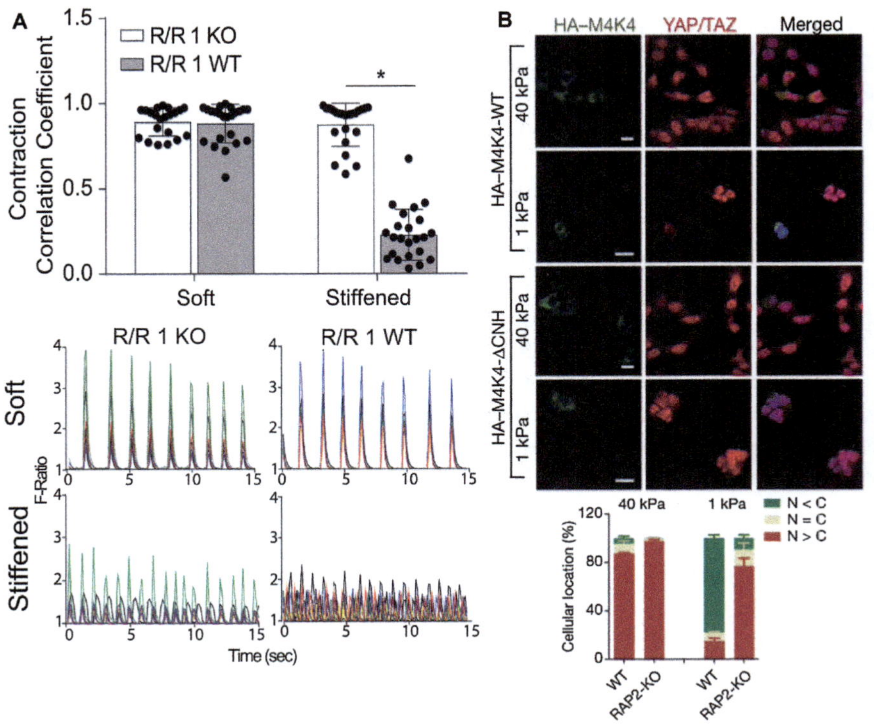

Figure 14.5 Combining gene-environment studies to understand their interplay. (A) At top is plotted the contraction correlation coefficient for R/R cardiomyocytes, with or without the 9p21 locus knocked out. $^*P < 0.01$ for all comparisons; one-way analysis of variance (ANOVA) followed by Tukey's multiple comparison test. At bottom are representative spontaneous Ca2 + transients plotted as the fluorescence intensity F-ratio over a 10-s time interval for the corresponding genotypes. The different colors represent transients from different cells. Reproduced from ref. 158 with permission from Springer Nature, Copyright 2019. (B) Expression of wild-type MAP4K4 (M4K4-WT), but not the citron domain-deleted mutant (M4K4-ΔCNH), rescued localization of YAP and TAZ in MM-8KO (MST1/2-MAP4K1/2/3/4/6/7-8KO) cells at low stiffness. HA, hemagglutinin. Combined signals from HA–MAP4K4 (green), YAP and TAZ (red), and DAPI (blue) are shown. Below is a plot that quantifies YAP and TAZ localization. Reproduced from ref. 165 with permission from Springer Nature, Copyright 2018.

risk locus was sufficient to prevent the stiffness-induced asynchronous contractions, in addition to a JNK antagonist.[158] Overall, this study revealed the potential for unveiling previously unknown functions of noncoding loci by employing microenvironmental models with relevant mechanical stimuli and iPSC technology. As the relevance of noncoding loci in disease becomes increasingly evident, the utility of strategies that employ both mechanical and genetic factors to model disease becomes increasingly important.

14.3.2 Cancer

While the proportion of cancer phenotype attributed to germline genotype, or inherited cases of cancer, is not as great as inheritance of heart disease, it is still a considerable proportion of heritable cases (for example 30% for breast cancer in Nordic countries[159,160]) and acquired genetic mutations with combined effects from gene–environment account for even greater proportions. Studies investigating how genetic factors, both acquired and hereditary, interact with the environmental factor have the potential to unveil novel mechanistic information and even identify new genetic risk factors. Explorative gene–environment interaction studies aim to identify such pathways.[161] In this section, we summarize several recent case studies of gene–environment interactions.

Exploration of gene–environment interactions date back several decades, with one early example to understand how H-RAS transformation leads to substrate adhesion–independent growth. While substrate adhesion plays a role in regulating cell growth of normal cells, it has long been recognized that oncogene-transformed cells grow and proliferate independently of substrate anchorage regulation. To investigate this, H-RAS-transformed fibroblasts and nontransformed controls were cultured on different-stiffness collagen-coated PA substrates. Nontransformed cells had decreased rates of DNA synthesis, increased rates of apoptosis, decreased cell spreading area, and decreased traction forces on flexibility compared to stiffer substrates. However, transformed cells lost this stiffness response, maintaining proliferation and apoptosis rates as well as spreading and traction forces. Similarly, the tumor-initiating glioblastoma (GBM) cells are insensitive to stiffness cues of laminin-functionalized hydrogels, evading inhibition of spreading, proliferation, and migration typically imposed by softer ECM. Constitutive activation of RhoA restricted 3D invasion of GBM spheroids and restored stiffness-dependent spreading and motility.[162] Overall, this suggests that oncogenic mutations may contribute to a loss in mechanosensing, explaining their unregulated growth.[163]

Epithelial carcinomas are the most common type of cancer and the focus of recent research, including that in which both mechanical and genetic cues are modulated. For example, Hippo pathway effectors YAP and TAZ, which are known for their role in mechanosensing ECM composition and mechanics, were investigated for their role in regulating focal adhesion (FA) formation. First, the authors found that, by controlling FA formation and cell spreading area through micropatterning fibronectin onto different stiffness surfaces, FA formation and cell size correlated with YAP nuclearization. Cells transiently overexpressing YAP and TAZ mutants had either cytoplasmic or nuclear localization through their inability to either be phosphorylated – trapping them in the nucleus – or, lacking the PDZ domain required for nuclear shuttling, sequestering them in the cytoplasm. Moreover, transcriptionally active YAP but not TAZ increased FA signaling. This study, by finely tuning both the mechanical environment the cells experience as well as their gene expression, was able to provide new insights into the mechanism of YAP mechanosensing.[164] Another study investigating the Hippo pathway in breast cancer cells found that by deleting different GTPases and monitoring YAP/TAZ, the group identified RAP2 as a key intracellular signal transducer; deletion of RAP2 blocked regulation of YAP and TAZ by stiffness signals and promoted aberrant growth (Figure 14.5B). By deleting or overexpressing targets and measuring nuclear translocation of YAP/TAZ on different stiffnesses, the study determined that matrix stiffness activates PDZGEF1 and PDZGEF2 to then activate RAP2. In low-stiffness ECM, RAP2 binds MAP4K4/5/7, which inhibits YAP and TAZ through LATS1 and LATS2.[165] Overall, these studies validated the importance of YAP and TAZ in mechanical regulation of transcription and that YAP has a more context-dependent role in mechanotransduction.[164,165]

Finally, the cooperative effects of oncogenic transformation and matrix stiffness have also been investigated in the context of EMT using isogenic comparisons of nontransformed and H-RAS-transformed MECs (MCF10A *versus* MCF10AT) cells on PA gels of different stiffness. While both cell lines exhibited spreading, indicative of EMT, on stiff hydrogels, only transformed 10ATs exhibited heterogeneous behavior on soft gels. Within this heterogeneous population, those with stretched/mesenchymal morphology exhibited nuclear localization of EMT transcription factor TWIST1. While traditional mechanosensitive pathways, like YAP and TGF-β, did not mediate this spreading, the study found ERK activation induced the spreading on soft hydrogels *via* microtubule dynamics.[166] This is another example of tuning genetics (*via* transformation) and mechanics (ECM stiffness) in order to elucidate novel disease pathways.

14.4 Conclusions and Perspectives

Investigating the effect of a single variable, such as inherited genetic risk, by using, for example, iPSC technology and/or novel gene editing techniques, has improved our understanding of the pathology of several diseases. This chapter highlighted genetic studies in heart disease and cancer. For example, the role of SNPs in a noncoding locus, 9p21, in enhancing risk for CAD was investigated using iPSCs and TALENs to study smooth muscle cell phenotype.[23] Genetic risk for cardiomyopathies, such as DCM, has also more fruitfully been studied using iPSCs differentiated into cardiomyocytes. For example, utilizing CRISPR, the combinatorial effects of heterozygous sequence variants in TPM1 and VCL genes, which cosegregate in patients with DCM, was demonstrated to disrupt sarcomere structure and function.[32] In addition to inherited genetic risk, acquired mutations have also been investigated. Oncogenes, such as RAS, have been investigated for their effect on cellular phenotype in cancer progression, which remains less defined. Comparing cell lines with different oncogenes and tumor suppressor gene mutations and correspondingly with different metastatic potentials, different biomechanical properties have been observed, *e.g.* differences in strength of adhesion to the ECM,[37] traction forces,[36] and cellular deformability,[40] to name a few.

In addition to genetics, this chapter highlighted the use of material-based mechanobiology in which mechanical stimuli, whether from the ECM, surrounding fluid, or neighboring cells, were modulated to study their role in disease. For example, hydrogels have been widely used to investigate the effects of matrix elasticity on several processes, as stiffness changes have been implicated in diseases such as cancer and atherosclerosis, which are both characterized by fibrosis.[61,64,66] In addition to comparing cellular responses to hydrogels of different stiffnesses, stiffness patterns have been investigated, with durotactic behavior of many cell types being noted, and temporal changes in stiffness have also been investigated. For example, mammary epithelial cells cultured on hydrogels that stiffen over time demonstrate EMT, important in cancer metastasis.[70] Stiffness-dependent behavior is also observed in 3D systems. In addition to stiffness, porosity, fiber interconnectivity, and topography are important in their influence on cell behavior in 3D. Both 3D printing and electrospinning are valuable methodologies for studying such variables systematically. 3D systems have also been important in their role in studying the impact of hemodynamics on phenotype, which is important in the study of diseases in which flow is disturbed, such as during atherosclerosis in which plaque buildup leads to stenosis.[129] Microfabricated systems that model these changes in arterial structure have been successfully utilized to demonstrate morphological changes in endothelial cells in response to fluidic shear stress,[130,131] as well as activation of important mechanotransduction pathways, such as the Hippo pathway.[132]

While focusing individually on genetics or mechanics is important, their combinatorial effects are increasingly becoming recognized for their importance in disease pathology. Most genetic risk variants identified through GWAS for CAD act through unknown mechanisms[20] and therefore present a perfect opportunity for novel approaches in which patient-derived iPSCs, gene editing, and biomimetic microfluidic devices or tissue-engineered constructs may be used in conjunction. Both iPSC-derived endothelial cells and SMCs have successfully been used in such systems to unveil novel signaling mechanisms activated through hemodynamic forces such as shear stress or flow-induced vessel strain.[146–149,151] Similarly, iPSC-derived cardiomyocytes have been utilized in more biomimetic cardiac tissue constructs.[157] Dynamically stiffened hydrogels have been used with

TALEN-edited iPSC-derived cardiomyocytes to elucidate mechanisms through which SNPs at a noncoding locus result in disease in response to fibrosis.[158] In the context of cancer, oncogenic mutations have been investigated in different stiffness conditions, with studies highlighting the resultant loss in mechanosensing.[163] Simultaneously controlling gene expression and the focal adhesion area through microcontact printing has also been used to study Hippo pathway effectors YAP and TAZ.[164] Deletion of various GTPases and culturing gene-edited breast cancer cells on dynamically stiffened hydrogels have also revealed novel regulators of the Hippo pathway.[157]

Overall, these studies show evidence for the power of more cross-disciplinary studies in which both genetics and mechanical stimuli are manipulated. Combinatorial effects from the environment and genetics have long been appreciated in disease, and therefore systems that systematically manipulate these parameters in conjunction can uncover novel mechanisms that would otherwise be missed if modeled separately. Studying genetics or biomechanics in an isolated system, though useful, can run the risk of oversimplifying the effect of each factor. As we further develop the field of materials-based mechanobiology, combined systems will become crucial to further elucidate the interplay between the multiple factors in a biological system.

Acknowledgements

A.J.E. would like to acknowledge funding from the National Institutes of Health (R01AG045428, R01NS116802, R01CA206880 and R21CA217735) and National Science Foundation (1763139) for financial support.

References

1. R. Allada, *Principles and Practice of Sleep Medicine*, 2011, vol. 5, p. 151.
2. A.-M. Martin, *J. Natl. Cancer Inst.*, 2000, **92**, 1126–1135.
3. W. D. Foulkes, *N. Engl. J. Med.*, 2008, **359**, 2143–2153.
4. C. Li and S. M. Williams, *Curr. Genet. Med. Rep*, 2013, **1**, 212–218.
5. P. Rizzolo, V. Silvestri, M. Falchetti and L. Ottini, *Appl. Clin. Genet.*, 2011, **4**, 145–158.
6. R. Barrangou and L. A. Marraffini, *Mol. Cell*, 2014, **54**, 234–244.
7. T. Gaj, C. A. Gersbach and C. F. Barbas, *Trends Biotechnol.*, 2013, **31**, 397–405.
8. A. A. Nemudryi, K. R. Valetdinova, S. P. Medvedev and S. M. Zakian, *Acta Nat.*, 2014, **6**, 19–40.
9. Y.-G. Kim, J. Cha and S. Chandrasegaran, *Hybrid Restriction Enzymes: Zinc Finger Fusions to Fok I Cleavage Domain (Flavobacterium okeanokoites/chimeric restriction endonuclease/protein engineering/recognition and cleavage domains)*, 1996, vol. 93.
10. N. Ma, *et al.*, *J. Biol. Chem.*, 2015, **290**, 12079–12089.

11. Method of the Year 2011, *Nat. Methods*, 2012, **9**, 1.
12. M. Adli, *Nat. Commun.*, 2018, **9**, 1–13.
13. The top 10 causes of death. https://www.who.int/news-room/fact-sheets/detail/the-top-10-causes-of-death.
14. A. Cassar, D. R. Holmes, C. S. Rihal and B. J. Gersh, *Mayo Clin. Proc.*, 2009, **84**, 1130–1146. Elsevier Ltd.
15. T. Kessler, J. Erdmann and H. Schunkert, *Cardiovasc. Genomics*, 2013, **15**, 8.
16. E. J. Topol, J. Smith, E. F. Plow and Q. K. Wang, *Hum. Mol. Genet.*, 2006, **15**, R117–R123.
17. R. Roberts and A. F. R. Stewart, *J. Am. Coll. Cardiol.*, 2012, **60**, 1715–1721.
18. F. Cambien and L. Tiret, *Circulation*, 2007, **116**, 1714–1724.
19. M. A. Schaub, A. P. Boyle, A. Kundaje, S. Batzoglou and M. Snyder, *Genome Res.*, 2012, **22**, 1748–1759.
20. R. Roberts and A. F. R. Stewart, *Clin. Chem.*, 2012, **58**, 104–112.
21. N. J. Samani and H. Schunkert, *Circ.: Cardiovasc. Genet.*, 2008, **1**, 81–84.
22. R. McPherson, *J. Am. Coll. Cardiol.*, 2013, **62**, 1382–1383.
23. V. Lo Sardo, *et al.*, *Cell*, 2018, **175**, 1796–1810, e20.
24. S. Roura and A. Bayes-Genis, *Nat. Rev. Cardiol.*, 2009, **6**, 590–598.
25. B. J. Maron, *et al.*, *Circulation*, 2006, **113**, 1807–1816.
26. C. Yang, *et al.*, *Stem Cells*, 2015, **33**, 2643–2651.
27. G. Guo, *et al.*, *J. Transl. Med.*, 2018, **16**, 288.
28. N. Sun, *et al.*, *Sci. Transl. Med.*, 2012, **4**, 130ra47.
29. A. J. Sehnert, *et al.*, *Nat. Genet.*, 2002, **31**, 106–110.
30. Y. Yamamoto, *et al.*, *Hum. Mol. Genet.*, 2017, **26**, 1670–1677.
31. S. I. Hashem, *et al.*, *Stem Cells*, 2015, **33**, 2343–2350.
32. D. C. Deacon, *et al.*, *Nat. Biomed. Eng.*, 2019, **3**, 147–157.
33. X. Peng, E. S. Nelson, J. L. Maiers and K. A. DeMali, *International Review of Cell and Molecular Biology*, Elsevier Inc., 2011, vol. 287, pp. 191–231.
34. E. H. Baugh, H. Ke, A. J. Levine, R. A. Bonneau and C. S. Chan, *Cell Death Differ.*, 2018, **25**, 154–160.
35. P. J. Dawson, S. R. Wolman, L. Tait, G. H. Heppner and F. R. Miller, *Am. J. Pathol.*, 1996, **148**, 313–319.
36. C. M. Kraning-Rush, J. P. Califano and C. A. Reinhart-King, *PLoS One*, 2012, **7**, e32572.
37. P. Beri, *et al.*, 2020, 10.1158/0008-5472.CAN-19-1794.
38. Z. Liu, *et al.*, *Proc. Natl. Acad. Sci. U. S. A.*, 2010, **107**, 9944–9949.
39. V. Gkretsi and T. Stylianopoulos, *Front. Oncol.*, 2018, **8**, 145.
40. J. Runge, *et al.*, *Oral Dis*, 2014, **20**, e120–e127.
41. M. Lherbette *et al.*, 2017, 10.1038/s41598-017-08517-6.
42. D. Zink, A. H. Fischer and J. A. Nickerson, *Nat. Rev. Cancer*, 2004, **4**, 677–687.
43. N. K. Gill, *et al.*, *Lab Chip*, 2019, **19**, 343–357.
44. X. Han, *et al.*, *Angew. Chem., Int. Ed.*, 2016, **55**, 8561–8565.

45. T. Zhan, N. Rindtorff, J. Betge, M. P. Ebert and M. Boutros, *Semin. Cancer Biol.*, 2019, **55**, 106–119.
46. J. P. Shen, *et al.*, *Nat. Methods*, 2017, **14**, 573–576.
47. P. Panorchan, *et al.*, *J. Cell Sci.*, 2006, **119**, 66–74.
48. A. Sancho, I. Vandersmissen, S. Craps, A. Luttun and J. Groll, *Sci. Rep.*, 2017, **7**, 1–14.
49. P. H. Puech, K. Poole, D. Knebel and D. J. Muller, *Ultramicroscopy*, 2006, **106**, 637–644.
50. W. Baumgartner, *et al.*, *Cadherin Interaction Probed by Atomic Force Microscopy*, www.pnas.org.
51. Y.-S. Chu, *et al.*, *J. Cell Biol.*, 2004, **167**, 1183–1194.
52. C. F. Buchanan, S. S. Verbridge, P. P. Vlachos and M. N. Rylander, *Cell Adh. Migr*, 2014, **8**, 517–524.
53. W. J. Polacheck, *et al.*, *Nature*, 2017, **552**, 258–262.
54. I. Ivanovska, J. Swift, T. Harada, J. D. Pajerowski and D. E. Discher, *Methods Cell Biol.*, 2010, **98**, 207–220, (NIH Public Access).
55. W. Baumgartner, G. J. Schütz, J. Wiegand, N. Golenhofen and D. Drenckhahn, *J. Cell Sci.*, 2003, **116**, 1001–1011.
56. C. Grashoff, *et al.*, *Nature*, 2010, **466**, 263–266.
57. M. S. Jhon and J. D. Andrade, *J. Biomed. Mater. Res.*, 1973, **7**, 509–522.
58. J. Zhu and R. E. Marchant, *Expert Rev. Med. Devices*, 2011, **8**, 607–626.
59. K. Y. Lee, *et al.*, *Macromolecules*, 2000, **33**, 4291–4294.
60. Z. Jiang, A. Bhaskaran, H. M. Aitken, I. C. G. Shackleford and L. A. Connal, *Macromol. Rapid Commun.*, 2019, **40**, 1900038.
61. F. B. Kai, H. Laklai and V. M. Weaver, *Trends Cell Biol.*, 2016, **26**, 486–497.
62. R. J. Pelham and Y. L. Wang, *Proc. Natl. Acad. Sci. U. S. A.*, 1997, **94**, 13661–13665.
63. C. M. Lo, H. B. Wang, M. Dembo and Y. L. Wang, *Biophys. J.*, 2000, **79**, 144–152.
64. I. Acerbi, *et al.*, *Integr. Biol.*, 2015, **7**, 1120–1134.
65. C. Voutouri, *et al.*, *Role of Constitutive Behavior and Tumor-Host Mechanical Interactions in the State of Stress and Growth of Solid Tumors*, 2014, 10.1371/journal.pone.0104717.
66. J. C. Kohn, *et al.*, *Biophys. J.*, 2015, **108**, 471–478.
67. M. Cavo, *et al.*, *Sci. Rep.*, 2016, **6**, 1–13.
68. N. R. Lang, *et al.*, *Acta Biomater.*, 2015, **13**, 61–67.
69. S. C. Wei, *et al.*, *Nat. Cell Biol.*, 2015, **17**, 678–688.
70. M. G. Ondeck, *et al.*, *Proc. Natl. Acad. Sci. U. S. A.*, 2019, **116**, 3502–3507.
71. H. Y. Jung, *et al.*, *Nat. Cell Biol.*, 2019, **21**, 359–371.
72. S. C. Wei, *et al.*, *Nat. Cell Biol.*, 2015, **17**, 678–688.
73. R. W. Tilghman, *et al.*, *PLoS One*, 2010, **5**, e12905.
74. Y. Yangben, *et al.*, *J. Biomater. Sci. Polym. Ed.*, 2013, **24**, 148–157.
75. M. Moazzem Hossain, X. Wang, R. C. Bergan and J. P. Jin, *FEBS Open Bio*, 2014, **4**, 627–636.
76. B. F. Matte, *et al.*, *J. Cell Sci.*, 2019, **132**, jcs224360.

77. S. Nasrollahi, *et al.*, *Biomaterials*, 2017, **146**, 146–155.
78. X. Tang, *et al.*, *Mol. Cancer*, 2014, **13**, 131.
79. E. E. Charrier, K. Pogoda, R. G. Wells and P. A. Janmey, *Nat. Commun.*, 2018, **9**, 449.
80. I. Levental, K. R. Levental, E. A. Klein, R. Assoian, R. T. Miller, R. G. Wells and P. A. Janmey, *J. Phys.: Condens. Matter*, 2010, **22**, 194120.
81. O. Chaudhuri, *et al.*, *Nat. Mater.*, 2016, **15**, 326–334.
82. K. M. Wisdom, *et al.*, *Nat. Commun.*, 2018, **9**, 1–13.
83. L. A. Lautscham, *et al.*, *Biomaterials*, 2014, **35**, 3198–3207.
84. N. F. Huang, *et al.*, *APL Bioeng*, 2020, **4**, 10906.
85. K. A. Günay, *et al.*, *Angew. Chem., Int. Ed.*, 2019, **58**, 9912–9916.
86. I. A. Marozas, K. S. Anseth and J. J. Cooper-White, *Biomaterials*, 2019, **223**, 119430.
87. E. Kä, S. M. Delgado and A. M. Kasko, *Shape-Changing Photodegradable Hydrogels for Dynamic 3D Cell Culture*, 2016, 10.1021/acsami.6b05527.
88. J. Wei and Y. Yu, *Soft Matter*, 2012, **8**, 8050–8059.
89. B. J. DuChez, A. D. Doyle, E. K. Dimitriadis and K. M. Yamada, *Biophys. J.*, 2019, **116**, 670–683.
90. S. F. Louis and P. Zahradka, *Exp. Clin. Cardiol.*, 2010, **15**, e75–e85.
91. M. Raab, *et al.*, *J. Cell Biol.*, 2012, **199**, 669–683.
92. M. T. Frey and Y. L. Wang, *Soft Matter*, 2009, **5**, 1918–1924.
93. J. Kim and M. Whang, *Tissue Eng. Regener. Med*, 2016, **13**, 126–139.
94. R. Sunyer, A. J. Jin, R. Nossal and D. L. Sackett, *PLoS One*, 2012, **7**, e46107.
95. A. M. Kloxin, A. M. Kasko, C. N. Salinas and K. S. Anseth, *Science*, 2009, **324**, 59–63.
96. S. Kaneko, *et al.*, *Phys. Chem. Chem. Phys.*, 2011, **13**, 4051–4059.
97. R. A. Marklein and J. A. Burdick, *Adv. Mater.*, 2010, **22**, 175–189.
98. V. E. G. Diederich, *et al.*, *Biotechnol. Bioeng.*, 2013, **110**, 1508–1519.
99. J. A. Burdick, A. Khademhosseini and R. Langer, *Langmuir*, 2004, **20**, 5153–5156.
100. N. Zaari, P. Rajagopalan, S. K. Kim, A. J. Engler and J. Y. Wong, *Adv. Mater.*, 2004, **16**, 2133–2137.
101. N. Rao, *et al.*, *Integr. Biol.*, 2013, **5**, 1344–1354.
102. J. M. Maloney, E. B. Walton, C. M. Bruce and K. J. Van Vliet, 2008, 10.1103/PhysRevE.78.041923.
103. Y. S. Choi, *et al.*, *Biomaterials*, 2012, **33**, 6943–6951.
104. C.-H. R. Kuo, J. Xian, J. D. Brenton, K. Franze and E. Sivaniah, *Adv. Mater.*, 2012, **24**, 6059–6064.
105. D. S. Gray, J. Tien and C. S. Chen, *J. Biomed. Mater. Res.*, 2003, **66A**, 605–614.
106. B. C. Isenberg, P. A. DiMilla, M. Walker, S. Kim and J. Y. Wong, *Biophys. J.*, 2009, **97**, 1313–1322.
107. J. Y. Wong, A. Velasco, P. Rajagopalan and Q. Pham, *Directed Movement of Vascular Smooth Muscle Cells on Gradient-Compliant Hydrogels*, 2003 10.1021/la026403p.

108. D. K. Cullen, M. C. Lessing and M. C. Laplaca, *Ann. Biomed. Eng.*, 2007, **35**, 835–846.
109. R. M. Smeal, R. Rabbitt, R. Biran and P. A. Tresco, *Ann. Biomed. Eng.*, 2005, **33**, 376–382.
110. J. Qiu, *et al.*, *J. R. Soc., Interface*, 2014, **11**, 20130852.
111. P. A. Janmey, J. P. Winer and J. W. Weisel, *J. R. Soc., Interface*, 2009, **6**, 1–10.
112. C. P. Barnes, S. A. Sell, E. D. Boland, D. G. Simpson and G. L. Bowlin, *Adv. Drug Delivery Rev.*, 2007, **59**, 1413–1433.
113. J. Lannutti, D. Reneker, T. Ma, D. Tomasko and D. Farson, *Mater. Sci. Eng., C*, 2007, **27**, 504–509.
114. Z. M. Huang, Y. Z. Zhang, M. Kotaki and S. Ramakrishna, *Compos. Sci. Technol.*, 2003, **63**, 2223–2253.
115. Q. P. Pham, U. Sharma and A. G. Mikos, Electrospinning of Polymeric Nanofibers for Tissue Engineering Applications: A Review, Tissue Engineering, www.liebertpub.com, 2006.
116. A. R. Tan, *et al.*, *J. Biomed. Mater. Res., Part A*, 2008, **87**, 1034–1043.
117. J. Xue, T. Wu and Y. Xia, *APL Mater.*, 2018, **6**, 120902.
118. T. Hodgkinson, X.-F. Yuan and A. Bayat, *J. Tissue Eng.*, 2014, **5**, 2041731414551661.
119. M. Heller, *et al.*, *Int. J. Comput. Dent.*, 2016, **19**, 323–339.
120. A. E. Jakus, A. L. Rutz and R. N. Shah, *Biomed. Mater.*, 2016, **11**, 014102.
121. S. Bose, S. Vahabzadeh and A. Bandyopadhyay, *Materials Today*, 2013, **16**, 496–504.
122. C. Lee Ventola, *Medical Applications for 3D Printing: Current and Projected Uses*, 2014, vol. P T 39, pp. 704–711.
123. S. Vijayavenkataraman, W. F. Lu and J. Y. H. Fuh, *3D Bioprinting of Functional Human Skin: Production and In Vivo Analysis Related Content 3D Bioprinting of Skin: A State-of-the-art Review on Modelling, Materials, and Processes*, 2016, 10.1088/1758-5090/9/1/015006.
124. C. McBeth, J. Lauer, M. Ottersbach, J. Campbell, A. Sharon and A. F. Sauer-Budge, *Biofabrication*, 2017, **9**, 015009.
125. N. E. Fedorovich, *et al.*, *Tissue Eng. - Part C Methods*, 2012, **18**, 33–44.
126. H. Bae, *et al.*, *Sci. Transl. Med.*, 2012, **4**, 160ps23.
127. L. E. Bertassoni, *et al.*, *Lab Chip*, 2014, **14**, 2202–2211.
128. M. Hu, *et al.*, *Facile Engineering of Long-Term Culturable Ex Vivo Vascularized Tissues Using Biologically Derived Matrices*, 2018, DOI: 10.1002/adhm.201800845.
129. N. Venugopal Menon, *et al.*, *APL Bioeng*, 2018, **2**, 016103.
130. A. K. Wong, P. Llanos, N. Boroda, S. R. Rosenberg and S. Y. Rabbany, *Cell. Mol. Bioeng.*, 2016, **9**, 127–138.
131. N. DePaola, M. A. Gimbrone, P. F. Davies and C. F. Dewey, *Arterioscler. Thromb.*, 1992, **12**, 1254–1257.
132. K. C. Wang, *et al.*, *Proc. Natl. Acad. Sci. U. S. A.*, 2016, **113**, 11525–11530.
133. L. Wang, *et al.*, *Nat. Publ. Gr*, 2016, **540**, 579–582.
134. E. Yeom, J. H. Park, Y. J. Kang and S. J. Lee, *Sci. Rep.*, 2016, **6**, 1–11.

135. K. Sugimura, P. F. Lenne and F. Graner, *Dev*, 2016, **143**, 186–196.
136. N. C. Heer and A. C. Martin, *Development*, 2017, **144**, 4249–4260.
137. D. Needleman and Z. Dogic, *Nat. Rev. Mater.*, 2017, **2**, 1–14.
138. N. C. A. Van Engeland, *et al.*, *Lab Chip*, 2018, **18**, 1607–1620.
139. I. Cicha, K. Urschel, W. G. Daniel and C. D. Garlichs, *Clin. Hemorheol. Microcirc.*, 2011, **48**, 65–73.
140. N. I. Simonds, *et al.*, *Genet. Epidemiol.*, 2016, **40**, 356–365.
141. P. Boffetta, *et al.*, *Int. J. Epidemiol.*, 2012, **41**, 686–704.
142. D. Thomas, *Nat. Rev. Genet.*, 2010, **11**, 259–272.
143. D. J. Hunter, *Nat. Rev. Genet.*, 2005, **6**, 287–298.
144. H. Giral, U. Landmesser and A. Kratzer, *Front. Cardiovasc. Med.*, 2018, **5**, 181.
145. T. M. Frayling, *Clin. Med.*, 2014, **14**, 428–431.
146. D. G. Belair, *et al.*, *Stem Cell Rev. Rep.*, 2015, **11**, 511–525.
147. Y. K. Kurokawa, *et al.*, *Tissue Eng. - Part C Methods*, 2017, **23**, 474–484.
148. H. Vazão, *et al.*, *Proc. Natl. Acad. Sci. U. S. A.*, 2017, **114**(15), E3022–E3031.
149. B. C. Dash, *et al.*, *Stem Cell Rep.*, 2016, **7**, 19–28.
150. X. Ge, *et al.*, *Circulation*, 2012, **126**, 1695–1704.
151. S. Sundaram, *et al.*, *Stem Cells Transl. Med*, 2014, **3**, 1535–1543.
152. E. Karamariti, *et al.*, *Circ. Res.*, 2013, **112**, 1433–1443.
153. P. Scaffidi and T. Misteli, *Nat. Med.*, 2005, **11**, 440–445.
154. L. Atchison, H. Zhang, K. Cao and G. A. Truskey, *Sci. Rep.*, 2017, **7**, 1–12.
155. J. Ribas, *et al.*, *Small*, 2017, **13**, 1603737.
156. L. Atchison, *et al.*, *Stem Cell Reports*, 2020, **14**, 325–337.
157. J. T. Hinson, *et al.*, *Science*, 2015, **349**, 982–986.
158. A. Kumar, *et al.*, *Nat. Biomed. Eng.*, 2019, **3**, 137–146.
159. L. A. Mucci, *et al.*, *JAMA - J. Am. Med. Assoc*, 2016, **315**, 68–76.
160. P. Aul *et al.*, The New England Journal of Medicine ENVIRONMENTAL AND HERITABLE FACTORS IN THE CAUSATION OF CANCER Analyses of Cohorts of Twins from Sweden, Denmark, and Finland ABSTRACT Background The contribution of hereditary factors. vol. 343, 2000.
161. A. Rudolph, J. Chang-Claude and M. K. Schmidt, *Br. J. Cancer*, 2016, **114**, 125–133.
162. S. Y. Wong, *et al.*, *Tumor and Stem Cell Biology Constitutive Activation of Myosin-Dependent Contractility Sensitizes Glioma Tumor-Initiating Cells to Mechanical Inputs and Reduces Tissue Invasion*, 2015, 10.1158/0008-5472.CAN-13-3426.
163. H. B. Wang, M. Dembo and Y. L. Wang, *Am. J. Physiol. – Cell Physiol*, 2000, **279**, C1345–C1350.
164. G. Nardone, *et al.*, *Nat. Commun.*, 2017, **8**, 15321.
165. Z. Meng, *et al.*, *Nature*, 2018, **560**, 655–660.
166. C. Plunkett, *et al.*, *Adv. Healthcare. Mater.*, 2020, **9**, 1901366.

Subject Index

ABPs. *See* actin-binding proteins (ABPs)
actin-binding proteins (ABPs), 6
addition-fragmentation chain transfer (AFCT) reaction, 297
AFCT reaction. *See* addition-fragmentation chain transfer (AFCT) reaction
altered peptide ligands (APL), 59
3-aminopropyl triethoxysilane (APTES), 92
antigen presenting cell (APC), 48
APTES. *See* 3-aminopropyl triethoxysilane (APTES)
AuNRs. *See* gold nanorods (AuNRs)
azide–alkyne cycloaddition, 286–288
    bioorthogonal chemistry, 286
    DBCO, 287
    NIH3T3 fibroblasts, 287
    SPAAC reaction, 287–288

β-CD. *See* β-cyclodextrin (β-CD)
BCMN. *See* block copolymer micellar nanolithography (BCMN)
β-cyclodextrin (β-CD), 251, 253
block copolymer micellar nanolithography (BCMN), 139
BMP-2. *See* bone morphogenetic protein-2 (BMP-2)
bone morphogenetic protein-2 (BMP-2), 139
bovine serum albumin (BSA), 126
BSA. *See* bovine serum albumin (BSA)

CAD. *See* coronary artery disease (CAD)
calmodulin (CaM), 251
CaM. *See* calmodulin (CaM)
cancer cultures, controlling shape and topology in
    cancer cells micropatterning, 100
    3D microenvironments, 102
    HA and cancer cells, 100
    malignant transformation, 102
    microengineered hydrogels, 100
    replica moulding method, 101
carbonyls, 290
cardiac microtissues (CMT), 327
cardiovascular disease (CVD), 319
CC. *See* corpus callosum (CC)
CD molecules. *See* cyclodextrin (CD) molecules
cell–cell adhesion, 69–70
cell culture, fabricating substrate topographies for use in
    template-assisted fabrication
        adhesive biomaterials, 159
        photolithography, 158–159
    template-free fabrication
        biocompatible thermo-responsive materials, 161
        chondrocytes, 160
        e-beam lithography, 159–160
        electrospinning, 160
        electrospun structures, 160
        homogeneous features, 161
        magnet-responsive materials, 161
        nanofibre microspheres, 160

cell–ECM adhesion, force
measurement
micropillar arrays
major geometrical parameters, 68–69
micron-scaled diameters, 69
microposts, 67
radius and height, 68–69
traction force measurement, 67–68
TFM
cell removal, 67
polyacrylamide gels, 65–66
silicone film, 66
cell–extracellular matrix interface
adhesion
direct cell shape and architecture, 137–138
matrix ligands, nanopatterning of
BCMN, 139
BMP-2, 139–141
cell membrane receptor clustering, 139
RGD, 138
cell–extracellular matrix interface, chemomechanical coupling, 135–136
cell–matrix adhesions, mechanobiology of, 147
cell–extracellular matrix adhesion
direct cell shape and architecture, 137–138
matrix ligands, nanopatterning of, 138–141
cell–extracellular matrix interface, 135–136
cell-generated force measurements
cell–matrix adhesion sites, 145–147
probe cell rigidity sensing and measure cellular traction forces, 142–144
cell–matrix interface, mechanosensing
cellular membrane, 24
FAs, 26–28
mechanical signals, origin of
cellular mechanosome, 23
ECM, 21–22
mechanosensing–mechanotransduction apparatus, 23
mechanosome, 23
mechanosensitive ion channels, 24–26
cellular curvature sensing
cellular-scale curvature, 221–222
current perspectives, 227
curvature recognition *via* molecular affinity
curvature-sensitive proteins, 223–224
MS channels, 224–225
intercellular connections, role of, 227
intracellular components, role of
FAs, 225
nuclei, 226–227
SFs, 225–226
cellular force nanoscopy (CFN) method, 59
cellular mechanobiology, molecular basis of, 36
cell–matrix interface, mechanosensing
cellular membrane, 24
FAs, 26–28
mechanical signals, origin of, 21–24
mechanosensitive ion channels, 24–26
mechanotransduction at cytoskeleton
cytoskeleton network, 28–29
Rho/ROCK pathway, 29–31

cellular mechanobiology, molecular basis of (*continued*)
transferring mechanical cues to nucleus
cellular adaptation, nucleo-cytoskeleton coupling in, 33
linker of nucleoskeleton and cytoskeleton (LINC) complex, 31–32
nucleoskeleton and transcriptional activity control, 32–33
nucleus as mechanosensor, 31
shuttling mechanosensors and mechanosensitive transcription factors, 33–36
chemomechanical coupling, 135–136
clustered regulatory interspaced short palindromic repeats (CRISPR)-Cas9, 318
CMT. *See* cardiac microtissues (CMT)
compliant interfaces
dissipative interfaces, 112–113
CD molecule, 122
fibrous interfaces, 125
lipid bilayers, bioinert feature of, 123
nanofibres, use of, 125
SLBs, 123
force-actuating materials, 128–129
liquid–liquid interfaces, 125–128
coronary artery disease (CAD), 319
corpus callosum (CC), 185
CRISPR-Cas9. *See* clustered regulatory interspaced short palindromic repeats (CRISPR)-Cas9
culture, enhance muscle cells in
applications, 170
assaying electrophysiological function, 172
MEAs, 172
patterned cardiomyocyte case studies
anisotropic topographic substrates, 165–166
DMD, 163
dystrophin, loss of, 164
engineered topographic surfaces, 165
human iPSC-derived cardiomyocytes, 163
mutant cells, 164
normal and mutant cardiomyocytes, 163
skeletal muscle structure
confining cells, 167
electrospinning, 169
larger linewidth patterns, 167
mature and aligned myotubes, 167–168
skeletal muscle fibres, 166–167
stem cell technology, 169
stem cell-derived cardiomyocytes and skeletal muscle, 156
curvature mechanobiology, 232–233
cellular curvature sensing
cellular-scale curvature, 221–222
current perspectives, 227
curvature recognition *via* molecular affinity, 223–225
intercellular connections, role of, 227
intracellular components, role of, 225–227
engineering curved surface
basic moulding techniques, 216–219
controlling interfacial geometries, 219–220
lithography, 220–221
machining and etching, 220
technical perspectives, 221

fundamental geometrical parameter, 214–216
investigate cellular behaviours, 213–214
modelling cellular behaviours
continuum modelling approaches, 228–231
discrete modelling approaches, 232
CVD. *See* cardiovascular disease (CVD)
cyclodextrin (CD) molecules, 122

DBCO. *See* dibenzylcyclooctyne (DBCO)
4D cell culture, engineered matrices for
hydrolytically degradable hydrogels, 295–296
mechanobiology, adaptable engineered 3D matrices for, 302–304
photolytically degradable hydrogels, 296–298
proteolytically degradable hydrogels, 293–294
spatiotemporally programmable biochemical cues
biochemical and biophysical cues, 298–299
degradable hydrogels, 298
STEPL, 301
visible light irradiation, 299
DCM. *See* dilated cardiomyopathy (DCM)
DGC. *See* dystrophin-glycoprotein complex (DGC)
dibenzylcyclooctyne (DBCO), 287
Diels–Alder reaction, 288–289
difluorinated cyclooctyne (DIFO), 286
DIFO. *See* difluorinated cyclooctyne (DIFO)
dilated cardiomyopathy (DCM), 324
dioleoylphosphorylcholine (DOPC), 123
dipalmitoylphosphorylcholine (DPPC), 123
directing cell fate, geometric cues for, 103
biomedical field, micropatterning application in, 86–87
biophysical aspects, 87
cell actin filaments remodeling, 86
cell adhesion, 86
controlling tissue geometry in 3D, 94–96
deconstructing biological behavior *via* micropatterning
cancer cultures, controlling shape and topology in, 100–103
geometric confinement, 97
NPC, 99
PLGA, 99
single epidermal stem cells, 99
stem cell studies, patterning for, 97–98
microengineering techniques to shape of cells and tissue
advantages and disadvantages, 88–89
basic steps, 88
dielectrophoresis, 93
electron beam lithography, 93
electron microscopy grids, 93
inkjet cell printing, 92
laser-based cell patterning, 92–93
masked ion beam lithography, 93
optical tweezers and optoelectronic tweezers, 92
photolithography, 88
soft lithography, 88, 90–91
microtubule networks, 86

directional photofluidization lithography (DPL), 195
distinct strain types, 73–75
3DMA. *See* 3D matrix adhesion (3DMA)
3D matrix adhesion (3DMA), 99
DMD. *See* Duchenne muscular dystrophy (DMD)
DNA-based molecular force sensing
APLs, 59
irreversible DNA tension probes, 57–58
MFM, 59
reversible DNA force probes, 57
SLB, 58
TGT, 59
DNA mechanics
DNA duplex peeling under force, 55
DNA duplex rupture under force, 53–55
DNA hairpin unfolding under force, 53, 58
DOPC. *See* dioleoylphosphorylcholine (DOPC)
DPL. *See* directional photofluidization lithography (DPL)
DPPC. *See* dipalmitoylphosphorylcholine (DPPC)
3D systems structural properties
3D printing, 323
electrospun fibrous scaffolds, 322
Duchenne muscular dystrophy (DMD), 163
durotaxis, manipulation of, 279
combined durotaxis-induced nomadic cell migration
bone marrow, 277
cell-scale stiffness heterogeneity, 276, 279
FAs, 276
larger triangle patterns, 2277
MSCs, 276–277
small triangle pattern, 277
definition, 265
rectified durotaxis, 269–272
reverse durotaxis
fibroblasts, 273
living tissues and organs, 272
stress distribution, 275
stiffness gradient substrate, fabrication of
adherent cells, 266
collagen triple helix, 268
determination of, 268–269
gelatin, 268
hydrogels, 266
molding/layering patterning, 267
monomer solutions, 267
PAAm, 267–268
dynamically tunable topography, engineered substrate with, 207–208
acute liver injury, 185
anatomical features, 187
basement membrane, 184–185
material systems, 188
micro- and nanofabrication, silicone elastomer system for, 188–194
organs, 185
physicochemical properties, 185
shape-memory polymer system
guiding cell functions, 201–207
programmed topography change, design of, 200–201
surface relief grating system
guiding cell functions, 196–199
photoinduced surface relief, 195–196
TACS-1, 185
dynamic hydrogel, 262–263
hydrogel classification, 239–240
reversible stiffness changes, hydrogels capable of
chemical stimuli, 248–253
physical stimuli, 253–257

stress relaxation, cell scaffolds capable of
ABA-type triblock copolymer, 259, 261
adaptive hydrogels, 258
hMSCs, 257
pNIPAAm, 259
time-dependent mechanical parameters, 257
switchable hydrogels
decreasable elastic modulus, 240–245
design principles, 239–240
increasable elastic modulus, 245–248
dystrophin-glycoprotein complex (DGC), 163

ECM. *See* extracellular matrix (ECM)
ECs. *See* endothelial cells (ECs)
elastin-like protein (ELP), 288
ELP. *See* elastin-like protein (ELP)
embryonic stem cells (ESCs), 94–95, 155
endothelial cells (ECs), 324–325
engineered 3D matrices, 304
biocompatibility perspective, 285
4D cell culture, engineered matrices for
hydrolytically degradable hydrogels, 295–296
mechanobiology, 302–304
photolytically degradable hydrogels, 296–298
proteolytically degradable hydrogels, 293–294
spatiotemporally programmable biochemical cues, 298–302
2D *vs.* 3D, 283–284
ECM, 283
engineered hydrogel crosslinking
azide–alkyne cycloaddition for, 286–288
carbonyls, 290
Diels–Alder reaction, 288–289
phosphine and azide, reaction of, 291
thiol–ene reaction, 289–290
engineered hydrogel cross-linking, azide-alkyne cycloaddition for
DBCO, 287
NIH3T3 fibroblasts, 287
SPAAC reaction, 287–288
free-radical chain polymerization, 285–286
engineered hydrogel crosslinking
azide–alkyne cycloaddition for bioorthogonal chemistry, 286
DBCO, 287
NIH3T3 fibroblasts, 287
SPAAC reaction, 287–288
carbonyls, 290
Diels–Alder reaction, 288–289
phosphine and azide, reaction of, 291
thiol–ene reaction, 289–290
engineering curved surface
basic moulding techniques, 216–219
controlling interfacial geometries, 219–220
lithography, 220–221
machining and etching, 220
technical perspectives, 221
ESCs. *See* embryonic stem cells (ESCs)
extracellular matrix (ECM), 5, 21, 46, 65, 135, 239

FAK. *See* focal adhesion-associated kinase (FAK)
FAs. *See* focal adhesions (FAs)
FBA. *See* fibrillar adhesion (FBA)
fibrillar adhesion (FBA), 99
fluorescence resonance energy transfer (FRET), 14

focal adhesion-associated kinase (FAK), 34
focal adhesions (FAs), 26–28, 46, 225
FRET. *See* fluorescence resonance energy transfer (FRET)

genome-wide associate studies (GWAS), 319
gold nanorods (AuNRs), 120
GWAS. *See* genome-wide associate studies (GWAS)

HA. *See* hyaluronic acid (HA)
HAIRS. *See* hydrogel-actuated integrated responsive system (HAIRS)
hASC. *See* human adipose-derived stem cells (hASC)
HGPS. *See* Hutchison–Gilford progeria syndrome (HGPS)
hMSCs. *See* human mesenchymal stem cells (hMSCs)
human adipose-derived stem cells (hASC), 206
human mesenchymal stem cells (hMSCs), 192, 241
Hutchison–Gilford progeria syndrome (HGPS), 326
hyaluronic acid (HA), 100, 304
hydrogel-actuated integrated responsive system (HAIRS), 120

induced pluripotent stem cells (iPSCs), 97, 122, 155, 317
INM. *See* inner nuclear membrane (INM)
inner nuclear membrane (INM), 32
iPSCs. *See* induced pluripotent stem cells (iPSCs)

large latent complex (LLC), 128–129
LAT. *See* linker for activation of T cells (LAT)
LCST. *See* lower critical solution temperature (LCST)
LFA-1. *See* lymphocyte function-associated antigen 1 (LFA-1)
LINC complex. *See* linker of nucleoskeleton and cytoskeleton (LINC) complex
linker for activation of T cells (LAT), 48
linker of nucleoskeleton and cytoskeleton (LINC) complex, 31–32
LLC. *See* large latent complex (LLC)
lower critical solution temperature (LCST), 255
lymphocyte function-associated antigen 1 (LFA-1), 48

MA. *See* methacrylic anhydride (MA)
materials-based mechanobiology, 15–16
- actin-based intracellular architectures, 5–6
- actin cytoskeleton, 5
- N- and C-terminals, 5
- cell–ECM interface, 4–5
- force detection, 14–15
- force transmission mechanisms, 5
- form
  - cell migration, 8
  - cell shape, 8
  - cross-linked material, 7
  - 3D environments, 8
  - ECM, 9–10
  - microcontact printing technology, 7
  - PDMS, 7–8
- historical background
  - mechanosensory molecules, 3–6
  - molecular assemblies, 3–6
  - molecular mechanobiology, 2–3
- matrix mechanics
  - cells sense, 11
  - chemotaxis, 13
  - 2D and 3D culture systems, 13–14
  - durotaxis, 13

mechanical memory, 11
mechanical micro-environment, 11
mechanical properties, diversity of, 11
molecular clutch model, 11–12
mechanosensitive channel (MS channel), 4–5
overview of, 1–2
patch clamp method, 3–4
stretch-activated channel, 5
tension-induced changes, 6
matrix metalloproteases (MMPs), 287
matrix metalloproteinase- (MMP-) cleavable sites
cleavable sites, 240
MEA. *See* multielectrode arrays (MEA)
mechanical measurement using silicone elastomers, 79–80
cell–cell adhesion, 69–70
cell–ECM adhesion
micropillar arrays, 67–69
traction force microscopy, 65–67
cell stiffness measurement using microfluidics, 70–72
mechanical forces, 64–65
mechanical manipulation
combination of, 77–79
distinct strain types, 73–75
sustained or cyclic, 75–77
mechanical stress, 33
mechanobiological studies, dynamic interfaces for, 129
biointerfacial design, 112
cell–substrate interface, 112
compliant interfaces
dissipative interfaces, 122–125
force-actuating materials, 128–129
liquid–liquid interfaces, 125–128
ECMs, 110–111
inside-out regulations, 112
outside-in regulation, 110
switchable interfaces
cell adhesion, spatio-temporal control of, 113–117
exploring dynamic mechanobiology, 119–122
subcellular and molecular resolutions, 117–119
mechanosensitive (MS) channels, 4–5, 10, 224–225
mechanotransduction at cell surface, 60–61
DNA-based molecular force sensing
APLs, 59
irreversible DNA tension probes, 57–58
MFM, 59
reversible DNA force probes, 57
SLB, 58
TGT, 59
DNA mechanics
DNA duplex peeling under force, 55
DNA duplex rupture under force, 53–55
DNA hairpin unfolding under force, 53, 58
integrins, 46
and mechanosensing, 44
T cell receptor
general overview of, 46–48
initial TCR–pMHC recognition, 48–49
TCR triggering, mechanical forces in, 49–51
mechanotransduction at cytoskeleton
cytoskeleton network, 28–29
Rho/ROCK pathway, 29–31
mesenchymal stem cells (MSCs), 116, 276–277, 284

methacrylic anhydride (MA), 249
MFM. *See* molecular force microscopy (MFM)
micron- and submicron-scale physical ECM microenvironment, 175–176
cell culture
template-assisted fabrication, 157–159
template-free fabrication, 159–162
culture, enhance muscle cells in applications, 170
assaying electrophysiological function, 172
MEAs, 172
patterned cardiomyocyte case studies, 163–166
skeletal muscle structure, 166–170
stem cell–derived cardiomyocytes and skeletal muscle, 155–156
historical background, 153–154
maturation puzzle, 170–171
muscle biology, structure-function relationship in, 154–155
muscle tissue engineering and applications
cardiovascular disease, 155
ECM fibres, scale of, 156–157
three-dimensional applications, 173–175
MLC. *See* myosin regulatory light chain (MLC)
MMP-cleavable sites. *See* matrix metalloproteinase- (MMP-) cleavable sites
MMPs. *See* matrix metalloproteases (MMPs)
modeling cues, ECM
dynamically stiffening hydrogels, 318–319
hydrogels mimic ECM stiffness and viscoelasticity, 317–318
mechanically patterned hydrogels, 319–322
modeling disease, 330–331
combining genetic and environment
cancer, 329–330
heart disease, 325–329
environmental effects on cells
active forces applied to cells, 324–325
3D systems structural properties, 322–323
modeling cues, ECM, 317–322
genetic models
cancer, 321
heart disease, 319–321
modelling cellular behaviours
continuum modelling approaches
internal stress, cellular alignment analysed based on, 229–230
multidisciplinary tissue growth models, 230–231
discrete modelling approaches, 232
molecular force microscopy (MFM), 59
molecular tension fluorescence microscopy (MTMF), 145
molecular tension sensors (MTS), 145
MPA. *See* multiphoton absorption (MPA)
MS channels. *See* mechanosensitive (MS) channels
MSCs. *See* mesenchymal stem cells (MSCs)
multielectrode arrays (MEA), 172
multiphoton absorption (MPA), 94
myosin regulatory light chain (MLC), 29

neural precursor cells (NPC), 99
next-generation sequencing (NGS), 317
NGS. *See* next-generation sequencing (NGS)
N-methylpyrrolidone (NMP), 193
NMP. *See* *N*-methylpyrrolidone (NMP)
NOA-63. *See* Norland Optical Adhesive 63 (NOA-63)
Norland Optical Adhesive 63 (NOA-63), 201
NPC. *See* neural precursor cells (NPC)
NPCs. *See* nuclear pore complexes (NPCs)
nuclear pore complexes (NPCs), 31, 32

ONM. *See* outer nuclear membrane (ONM)
outer nuclear membrane (ONM), 31

patch clamp method, 3–4
PBS. *See* phosphate-buffered saline (PBS)
PDGF. *See* platelet-derived growth factor (PDGF)
PDMS. *See* polydimethylsiloxane (PDMS)
PDPA. *See* poly(2-(diisopropylamino) ethyl methacrylate) (PDPA)
pDR1m. *See* poly-Disperse Red 1-methacrylate (pDR1m)
PEG. *See* polyethylene glycol (PEG)
perfluorocarbon (PFC), 15
PFC. *See* perfluorocarbon (PFC)
phosphate-buffered saline (PBS), 251
photoinduced surface relief (PSR), 195
photolithography, 88
piconewtons (pN), 46, 145
platelet-derived growth factor (PDGF), 129
PLGA. *See* poly(lactic-*co*-glycolic acid) (PLGA)
pMHCs. *See* presented by a major histocompatibility complex (pMHCs)
PMPC. *See* poly(2-(methyacryloyloxy)-ethyl phosphorylcholine) (PMPC)
pN. *See* piconewtons (pN)
pNIPAAm. *See* poly(*N*-isopropylacrylamide) (pNIPAAm)
polydimethylsiloxane- (PDMS-) based elastic materials
  lithography-free reconfigurable wrinkles, 188–190
  reconfigurable topography
    micro- and nanofabrication techniques, 191
    parallel wrinkling pattern, 193
    perpendicular wrinkling pattern, 193
    reversible topographical features, 192
    SiO_2-PDMS, 193
    strain-responsive wrinkling pattern, 192
    wavy surface features, 191
polydimethylsiloxane (PDMS), 3, 65, 137, 154, 267
poly-Disperse Red 1-methacrylate (pDR1m), 198
polyethylene glycol (PEG), 52, 94, 137
polyhedral oligomeric silsesquioxane (POSS), 206
poly(2-(diisopropylamino) ethyl methacrylate) (PDPA), 248
poly(lactic-*co*-glycolic acid) (PLGA), 99
poly(2-(methyacryloyloxy)ethyl phosphorylcholine) (PMPC), 248
poly(*N*-isopropylacrylamide) (pNIPAAm), 113, 259
POSS. *See* polyhedral oligomeric silsesquioxane (POSS)
presented by a major histocompatibility complex (pMHCs), 47
PSR. *See* photoinduced surface relief (PSR)

rat bone marrow mesenchymal stem cells (rBMSCs), 205
rBMSCs. *See* rat bone marrow mesenchymal stem cells (rBMSCs)

reversible stiffness changes, hydrogels capable of
chemical stimuli
ABA-type triblock copolymer, 248
CaM structure, 251
β-CD, 251, 253
hMSCs, 249
PDPA, 248–249
PMPC, 248
SrtA, 251
stimuli-responsive hydrogels, 249
physical stimuli, 253–257
Rho-associated coiled-coil kinases (ROCK) pathway, 29–31
ROCK pathway. *See* Rho-associated coiled-coil kinases (ROCK) pathway

SAMs. *See* self-assembled monolayers (SAMs)
self-assembled monolayers (SAMs), 113
SFs. *See* stress fibers (SFs)
shape-memory polymer system
guiding cell functions
crosslinked PCL, 203, 205
hASCs, 206
NOA-63, 201–202
semicrystalline PCLs, 202
tert-butyl acrylate and butyl acrylate, 206
YAP/TAZ, 203
programmed topography change, design of, 200–201
single-nucleotide polymorphisms (SNPs), 319
SLA. *See* stereolithography (SLA)
SLBs. *See* supported lipid bilayers (SLBs)
SMCs. *See* smooth muscle cells (SMCs)
smooth muscle cells (SMCs), 170, 325
SNPs. *See* single-nucleotide polymorphisms (SNPs)
soft lithography, 88, 90–91
sortase A (SrtA), 251
sortase-tag enhanced protein ligation (STEPL), 301
SPAAC. *See* strain-promoted azide–alkyne cycloaddition (SPAAC)
SrtA. *See* sortase A (SrtA)
STEPL. *See* sortase-tag enhanced protein ligation (STEPL)
stereolithography (SLA), 102, 221
strain-promoted azide–alkyne cycloaddition (SPAAC), 287
stress fibers (SFs), 28, 77–78, 225–226
stress relaxation, cell scaffolds capable of
ABA-type triblock copolymer, 259, 261
adaptive hydrogels, 258
hMSCs, 257
pNIPAAm, 259
time-dependent mechanical parameters, 257
supported lipid bilayers (SLBs), 58, 123
supravalvular aortic stenosis syndrome (SVAS), 326
surface relief grating system
PSR, reconfigurable topography, 195–196
reconfigurable topography, PSR
azobenzene, 197
hMSCs, 198
MDA-MB-231 cells, 199
pDR1m, 198
photoresponsive systems, 196–197
U2OS cells, 199
sustained or cyclic, 75–77
SVAS. *See* supravalvular aortic stenosis syndrome (SVAS)
switchable hydrogels
decreasable elastic modulus
external stimuli, 241–243
hMSCs, 241
photodegradable gels, 241
polymer gel, 241
RGD motif, 241
design principles, 239–240

increasable elastic modulus, 243–248
chemical structures, 250
hMSCs, 249
low-phototoxicity visible light, 247
matrix stiffening, 249
switchable interfaces
cell adhesion, spatiotemporal control of
cell migration studies, photoactivatable interfaces for, 114–115
dynamic interfaces, 116–117
ECM remodelling, 116
hydroquinone/quinone chemistry, 115
noncell-adhesive to cell-adhesive, 113
RGD, 116
SAMs, 113
exploring dynamic mechanobiology, 119–122
subcellular and molecular resolutions, 117–119

TACS. *See* tumor associated collagen signatures (TACS)
TALENs. *See* transcription activator-like effectors attached to nucleases (TALENs)
TAN line. *See* transmembrane actin-associated nuclear (TAN) line
T cell receptor, mechanotransduction
general overview of
adhesion ligands, 48
CD4 or CD8, 47–48
LAT, 48
TCR–CD3 complex, 47–48
TCR–pMHC binding, 47
initial TCR–pMHC recognition, 48–49
TCR triggering, mechanical forces in, 49–51
TCP. *See* tissue culture plate (TCP)
TEM. *See* transmission electron microscopy (TEM)
tension gauge tethers (TGT), 57–58
tethered model, 26
TFM. *See* traction force microscopy (TFM)
TG. *See* threshold gradient (TG)
TGT. *See* tension gauge tethers (TGT)
thiol–ene reaction, 289–290
threshold gradient (TG), 269
TIRF. *See* total internal reflection fluorescence (TIRF)
tissue culture plate (TCP), 100
tissue-engineered skeletal muscle, 154
total internal reflection fluorescence (TIRF), 67
TP-LSL. *See* two-photon laser scanning lithography (TP-LSL)
traction force microscopy (TFM), 14, 65–67, 143, 293
transcription activator-like effectors attached to nucleases (TALENs), 318
transferring mechanical cues to nucleus
cellular adaptation, nucleo-cytoskeleton coupling in, 33
linker of nucleoskeleton and cytoskeleton (LINC) complex, 31–32
nucleoskeleton and transcriptional activity control, 32–33
nucleus as mechanosensor, 31
shuttling mechanosensors and mechanosensitive transcription factors, 33–36
actin-associated proteins, 34
Nesprin-1 depletion, 35
YAP/TAZ activity, 34–36
transient receptor potential (TRP), 25
transmembrane actin-associated nuclear (TAN) line, 32
transmission electron microscopy (TEM), 137

TRP. *See* transient receptor potential (TRP)
tumor associated collagen signatures (TACS), 185
two-photon laser scanning lithography (TP-LSL), 94

unidirectional shear stresses (USS), 324
USS. *See* unidirectional shear stresses (USS)

vascular smooth muscle cells (VSMCs), 319
vasodilator-stimulated phosphoprotein (VASP), 27
VASP. *See* vasodilator-stimulated phosphoprotein (VASP)
VSMCs. *See* vascular smooth muscle cells (VSMCs)

WLC. *See* worm-like chain model (WLC)
worm-like chain model (WLC), 53

YAP. *See* Yes-associated protein (YAP)
YAP1. *See* Yes-associated protein (YAP1)
Yes-associated protein (YAP), 203, 284
Yes-associated protein (YAP1), 34